Ramanujan Rediscovered

Ramanujan Mathematical Society Lectures Notes Series

Vol. 1: Number theory

Vol. 2: The Riemann zeta function and related themes

Vol. 3: Formal language aspects of natural computing

Vol. 4: Commutative algebra and combinatorics

Vol. 5: Convexity in discrete structures

Vol. 6: Number theory and discrete geometry

Vol. 7: Discrete mathematics

Vol. 8: Lectures on operator theory

Vol. 9: Essays on geometric group theory

Vol. 10: Techmüller theory and moduli problems

Vol. 11: Perspectives in geometry and topology

Vol. 12: The Iwasawa theory of totally real fields

Vol. 13: Advances in discrete mathematics and applications

Vol. 14: Ramanujan Rediscovered

Vol. 15: Number Theory

Vol. 16: Symmetry: A Multi-Disciplinary Perspective

Ramanujan Mathematical Society

Lecture Notes Series

Volume 14

Ramanujan Rediscovered

Proceedings of a Conference on Elliptic Functions, Partitions, and q-Series in Memory of K. Venkatachaliengar, Bangalore, June 2009

Volume editors

Bruce C. Berndt (University of Illinois at Urbana)
Shaun Cooper (Massey University, Albany, Auckland, New Zealand)
Nayandeep Deka (Baruah, Tezpur University, Assam, India)
Tim Huber (University of Texas Pan-American)
Michael J. Schlosser (University of Vienna, Austria)

International Press
www.intlpress.com

Ramanujan Mathematical Society
Lecture Notes Series, Volume 14
Ramanujan Rediscovered

Volume editors:
Bruce C. Berndt (University of Illinois at Urbana)
Shaun Cooper (Massey University, Albany, Auckland, New Zealand)
Nayandeep Deka (Baruah, Tezpur University, Assam, India)
Tim Huber (University of Texas Pan-American)
Michael J. Schlosser (University of Vienna, Austria)

Published in 2012 by International Press, Somerville, Massachusetts, U.S.A. under license from the Ramanujan Mathematical Society.

This work is also published in the Republic of India, exclusively by the Ramanujan Mathematical Society.

ISBN: 978-1-57146-245-9

Printed in the United States of America.

16 15 14 13 12 1 2 3 4 5 6 7 8 9

Preface

On 1–5 June, 2009, a conference in number theory was held on the beautiful campus of Infosys in Bangalore. The impetus for organizing this meeting was to recognize and commemorate K. Venkatachaliengar, an outstanding, well-known mathematician, who taught primarily at universities in Bangalore and Mysore for most of his career. He was born on 8 December, 1908, and so the meeting marked the centenary of Venkatachaliengar's birth. In the last several decades of his long life of 95 years, KV, as he was affectionately known to most of his friends, had become keenly interested in the life and work of India's greatest mathematician, Srinivasa Ramanujan, and so it was natural for Ramanujan's first loves of theta functions, partitions, and q-series to be the focus of the conference. Accordingly, over 50 mathematicians gathered for the presentation of 32 lectures in memory of both Ramanujan and KV. This volume comprises 13 papers by mathematicians who lectured at the meeting. In addition, three papers on the life and work of KV, along with a complete list of his publications, are offered.

All the participants, especially the organizers, are extremely grateful to Infosys Corporation for making available its exceptional facilities, staff, and financial support for the conference. The International Institute of Information Technology, Bangalore, and especially its Director, Professor Sadagopan, deserve our warm thanks for their generous support. Lastly, it is a pleasure to acknowledge the munificent financial support of the Indian Mathematical Society.

Foreword

To commemorate the birth centenary of Prof. K. Venkatachaliengar, an International Conference "Ramanujan Rediscovered" was held at Infosys, Bangalore on June 1–5, 2009. It was jointly organized by the International Institute of Information Technology, Bangalore and the Indian Mathematical Society. The Conference was inaugurated with the lighting of the traditional lamp by Professors S. Sadagopan (Director, IIIT, Bangalore), B. C. Berndt (co-chair, University of Illinois at Urbana-Champaign, USA), A. K. Agarwal (co-chair, Panjab University, Chandigarh), Ravichandran (Indian Institute of Management, Indore), Mr. Dinesh (Infosys) and Mr. Srikantan Moorthy (VP, Infosys). In their inaugural speeches, the speakers highlighted the mathematical contributions of S. Ramanujan and K. Venkatachaliengar. The inauguration function was followed by three keynote addresses delivered by Professors B. C. Berndt, A. K. Agarwal and Ravichandran. After the keynote addresses Professor A. K. Agarwal gave a talk on the history, objectives and activities of the Indian Mathematical Society.

In all there were about 250 participants. There was a large number of overseas participants. Prof. G. N. Srinivasa Prasanna, IIIT, B was the convener of the Conference. There were two parallel sessions-one for mathematics and the other for Information Technology (IT). In mathematics, the main topics discussed were: elliptic functions, q-series, partitions and related number theory and in IT session the main topics covered were: discrete mathematics, game theory, bioinformatives, optimization techniques and formal methods in software engineering. There were 18 invited speakers in mathematics session and 7 in IT session. There were about 10 contributed talks and 4 poster presentations in mathematics session.

On the first day of the Conference a film "God, Zero and Infinity" on Ramanujan's story was screened in the evening. The film directed by Santosh Dhavala and narrated by Tom Alter was liked by all participants. The other highlights of the Conference were: (1) a skype session on June 3 in which Prof. Zhi-Guo Liu presented his paper as he could not attend the Conference physically, (2) a cultural program and a banquet at Manipal County in the evening of June 3, and (3) on June 4, Prof. S. Sadagopan met all the participants over a cup of tea and presented them momentos. The Conference was concluded on June 5 with a popular talk on Ramanujan by Prof. B. C. Berndt.

In my opinion the Conference was very stimulating and indeed achieved its goals. The proceedings of the conference will be published by the Ramanujan Mathematical Society in its Lecture Notes Series.

Prof. A. K. Agarwal
Centre for Advanced Study in Mathematics
Panjab University, Chandigarh 160 014, India
E-mail: aka@pu.ac.in

Members of the Advisory Board

K. Venkatachaliengar

G. N. Srinivasa Prasanna[1] and K. V. Venkataramu[2]

[1]International Institute of Information Technology, Bangalore
e-mail: gnsprasanna@iiitb.ac.in
[2]Retd. Scientific Officer, Department of Atomic Energy, Govt. of India
e-mail: venkataramu@yahoo.com

K. Venkatachaliengar (KVI) was born on 8 December 1908 in a small village called Kadaba about sixty kilometers northwest of Bangalore. He was the last of four brothers and two sisters all born to K. Venkatarama Iyengar (father) and Venkatalaxamma (mother). He lost his mother when he was still a toddler and thus missed motherly love. His father was a farmer by profession and was well read.

From early childhood, KVI showed great aptitude at studies, and his only limitation was lack of books and learned teachers in the small village. When he was around ten years his father also passed away and the burden of raising the family was thrust on the young shoulders of his elder brother, K. Anantha Char. After completing primary schooling in Kadaba, he continued his higher schooling in Tumkur and Bangalore. He was always at the top of his class and, apart from brilliance at mathematics, had great liking for the classics and Kannada language.

After the school final, he entered the Central College, and all his student colleagues and teachers alike immediately took note of his extraordinary talents. At this time, it is pertinent to note that generally the standards of mathematics among the Indian universities were very low, and teaching was like school teaching and research unknown.

Luckily for KVI, he had as teachers two very bright mathematicians, Prof. K. S. K. Iyengar and Prof. B. S. Madhava Rao, who were his seniors by a few years. Prof. KSK probably had studied in Cambridge and was very proficient in pure mathematics. Prof. B. S. Madhava Rao had completed his Masters from the Calcutta University and was more inclined towards applied mathematics and theoretical physics. Along with these teachers, he also had a senior fellow student, also very talented in mathematics, V. R. Tiruvenkatachar. It was very lucky for university mathematics in India that a very talented quartet was in the Central College as contemporaries. Abundant talents they had in mathematics but soon they learnt that the books in modern pure mathematics were not in English but mostly in French and German. India, being a British colony, usually had books only from England and in English. This quartet immediately went about repairing this sad state of affairs and started getting books to learn French and German from the continent. They self-acquired reading ability in these languages and subsequently started getting from the continent works of the great mathematicians of the French and German schools.

Combined with their innate brilliance and studying the great French and German works, they entered eagerly to pursue research and teaching of mathematics. They were even conducting courses in French and German for mathematics students!

After completing his B.Sc. (Hons) from the Central college, KVI went to Calcutta University for his Masters. Here his mathematical brilliance was immediately noticed by his teachers. During his time in Calcutta, an incident is worth recording. A British gentleman, who was good at mathematics, proposed a hard problem regarding some property of a triangle (the famous Morley Trisector Theorem) and remarked that it was beyond the ability of Indian students to solve. When KVI heard of this in the evening, he solved the same before going to bed and gave two different solutions the next morning!

KVI completed his Masters and as usual stood first. Some years later, he also obtained his D.Sc. from the Calcutta University. The thesis examiners for his D.Sc. thesis were Hermann Weyl, Garrett Birkhoff and F. W. Levi. Weyl went out of his way to praise the thesis and predicted a bright future for the candidate.

When he returned to Bangalore after his Masters, the immediate problem was getting a job. Since in those days the number of positions was rigidly fixed, one had to wait for the retirement of a senior member before one could get a job! For some time, KVI worked as a lecturer in a private college in Belgaum.

KVI had known Sir C. V. Raman probably even from Calcutta days and when Sir C. V. Raman was the director of the Indian Institute of Science, he finally offered some position to KVI at the Indian Institute of Science, with a salary around Rs. 80/- per month, a princely sum in those days. The first task Sir C. V. Raman assigned KVI was to conduct a course on integral equations at the Institute. During this time, KVI also worked out the mathematical theory of some problem connected with a vibrating string which had attracted the interest of Sir C. V. Raman, and a detailed paper resulted from this. Most of this was during World War II, during which time Max Born was also at the Indian Institute of Science.

KVI finally got a proper teaching job at the Central College after a vacancy arose. Here he had the excellent company of the other three mathematicians (KSK, BSMR, and VRT), and they started courses of high standard in modern mathematics and built up an excellent library. This was also the golden age of their research work. Since, fortunately, four of them were there, they collaborated on many problems and a number of papers of the highest standard followed. KSK passed away early around the 1940's, but the other three carried on their work vigorously and went on publishing results. One such was on particle spins and was published in the Proceedings of the Royal Society in 1946 [1].

Unfortunately, due to a variety of reasons, this collaborative work was put to an end, due to KVI being transferred out of Bangalore to a nearby small town called Tumkur, where only junior classes were held. In spite of lacking good students of senior classes and a library, KVI went about his research work ceaselessly. In the Tumkur College where he had been posted, he was actively looking for some talented students to teach some advanced mathematics. Luckily for him he found a fellow physics lecturer, K. N. Srinivasa Rao, who showed some mathematical ability. KVI immediately caught on to him and started a course for his colleague. Finding that he had not learnt group theory, he first taught him group theory and then complex variables, etc. After finding that Srinivasa Rao was very sincere and learnt all his teaching well, around 1948–50, he managed to get him a fellowship from the Council of Scientific and Industrial Research. This allowed KNS to devote all his time for mathematical studies under KVI without the burden of teaching students elementary physics at the college. This happened when both were posted to a college in Mysore.

In 1953, KVI was transferred back to Bangalore and this time to the Engineering college. Since there were no postgraduate courses here, to satisfy his intellectual prowess, he introduced advanced mathematics for those students who showed an above average interest in mathematics. This was besides the normal curriculum.

Around 1962, the University of Mysore created a postgraduate centre in Mysore called Manasa Gangothri, and KVI moved there as the head of the department. Here again he started building a proper library and introduced many advanced subjects such as abstract algebra, etc. After reaching the age of 58, he retired in 1966 from the University of Mysore and settled down in Bangalore. Here he had the company of his earlier collaborators Prof. B. S. Madhava Rao and Prof. V. R. Tiruvenkatachar. For two years, he served as an emeritus professor at Madurai University, teaching advanced courses to the faculty there. The following is worth mentioning here. One of the faculty members told him that he was stuck on some problem connected with Krull algebra. KVI on the spot proved the necessary result, and this was submitted as a paper in a special volume of a journal in honour of Prof. Krull. Prof. Krull personally wrote a letter of appreciation.

During this time, Prof. KVI was also associated with the National Council for Educational Research and Training, and heading the geometry study group. He made geometry exciting for youngsters, by having plastic stick models of polyhedra made. These, and the wooden models of complex topological surfaces, made during his Central College days (1940–50's) by a master carpenter Giriappa, can be used for teaching at all levels. For example, counting can be taught to a child, trigonometry to middle and high school students, and graph colouring and group theory to Ph.D. level candidates. The topological surfaces reflect some of the deepest aspects of topology, including non-Euclidean relativistic geometries. This material has been donated to the International Institute of Information Technology, Bangalore, and is on display there.

When he was past 50, he noticed that mathematical books could be procured from the USSR through some specialized book stores specializing in USSR books. Since these could be obtained at rock bottom prices which he could afford, he started teaching himself enough Russian to understand mathematics in Russian.

At this time Prof. B. S. Madhava Rao was in his house in South Bangalore, and Prof. V. R. Tiruvenkatachar lived near KVI's house. KVI and Prof. V. R. Tiruvenkatachar used to go for walks in the mornings and evenings discussing mathematics as usual. They used to visit Prof. B. S. Madhava Rao around once every few months.

After his retirement, KVI decided to devote most of his energies on Ramanujan's life and works. This culminated in his beautiful book, *Development of Elliptic Functions According to Ramanujan*, which has been described in detail elsewhere in this volume.

In 1977, Bruce Berndt began to devote all of his research efforts toward proving the claims left behind by Ramanujan in notebooks that he compiled before heading to Cambridge. After the publication of Berndt's first volumes on Ramanujan, KVI was very happy, and told Berndt that this appeared to be the only example where all the work, including unpublished work, of a great mathematician is being critically examined.

Being a born-mathematician, KVI always breathed and lived mathematics. Whenever he sat down to dinner, he never failed to scribble mathematics on his dinner plate. His wife had to disturb him from this so that he could be served. This trait was with him throughout his life. More often than not, the scribbling included a triangle.

Around the 1960's, he was requested by the then pontiff of the Parkala Mutt (Hindu religious body) at Mysore, who was a very great scholar, to become the president of the Mutt for some administrative duties. Since the request came from a great seer, KVI accepted it, even though he had until then never taken part in any such organized cultural activity. Once he accepted this task, he devoted his energies to it and started learning certain portions of philosophy required by his duties. Due to his intellectual acumen, he succeeded eminently and served the Swamiji (seer) and the Mutt with utmost devotion.

The triangle remained his most endearing object spanning his entire life. He had a great liking to imbibe mathematics into any willing child or young person. The elliptic functions and lots of other mathematics were first taught to his grandson – of course it went well above the young child's (between 7–15 years of age) head. He also taught his grandson the "Morley-Trisector Theorem." His grandson was inspired enough and learnt enough about the construction to do it during one evening when he had gone for a walk, and showed it to him when he returned with great pride. KVI taught all this, plus trigonometry, Gauss polynomials, complex variables, especially using Hurwitz's double series approach, etc. Some of this material is with his grandson in India and the USA. He was especially happy when his great-granddaughter Anagha gave him a lecture on polyhedra!

Even though he walked the high roads of mathematics, he never in his whole life felt any 'pride'. He was always eager to teach any willing soul less endowed than himself in mathematics.

In closing, we include below views of Prof. KVI from some of his former students: Prof. Mahabala, a prominent Indian Computer Scientist; Prof. Prabhu, a control theorist and former Head of Electrical Engineering at IIT-Kanpur; and Dr. Ramamani, his Ph.D. student.

KV Fondly Remembered by Prof. H. N. Mahabala, IIITb

Prof. K. Venkatachala Iyengar was a young researcher's delight. He would often recall results of great mathematicians of the past. He did not like teaching routine math courses to undergraduate students, but was transferred to an engineering college due to departmental politics. His talent was wasted on disinterested engineering students. At any opportunity, he would talk about great mathematicians and their work. Students were perplexed. He would walk across to talk to graduate students where I was a student studying for B.Sc. (Hons.) in mathematics. We were fascinated by his obsession with mathematics. He would advise us to learn Russian, because Russian books on advanced mathematics were sold very cheap! We heard from him about Bourbaki. We were attracted to read Hardy's book on Ramanujan.

At that time (1950's) a scientific exposition was held, and who should organize the maths exhibits, but KV. There were wooden models of interesting solids, made by a carpenter under his close direction, which we volunteers had to explain. We found out to our great surprise that the circle was not the only curve with constant width. He got constructed a model where one could roll the curve to check that it had constant width. We had wire frames to demonstrate minimal surfaces using soap films. We had to prepare a soap solution using oxalic acid and some other ingredients. We enjoyed confounding the visitors with minimal surfaces, in particular using the cube frame. We found out how interesting mathematics can be. We wrote in big letters the digits of pi accurate to some 600 digits and fixed it along the wall for visitors to check that pi was a non-repeating irrational number. The lattice components were developed later. He would train us on how to explain mathematics and would frequently call us "Hopeless Fellows," all with a smirk on his face. He breathed mathematics and explained in an excited way about great results in mathematics. We set up a 5-inch telescope and showed the moon and Saturn to long queues of people. It is surprising that he was the only faculty member who helped us put up a good show. His complete disregard for every day etiquette and lack of interest in things other than mathematics made him a poor conversationalist though.

I am proud that IIITb celebrated KV's centenary, with many Ramanujan enthusiasts around the world participating.

Professor K. Venkatachaliengar by Prof. Prabhu

It was the academic year 1958–1959. I was a third year Electrical (Power) Engineering student in the Bachelor of Engineering course in the Government College of Engineering, Bangalore. It was a traditional course. It dealt mainly with engineering practice of those days. Much of it was descriptive. Engineering education was poised to undergo radical changes. It was already happening in the advanced countries. Our batch of students were lucky. The university had started taking a few tentative steps to usher in refreshing changes. We were offered a new curriculum which had incorporated some small, but significant, changes to the old curriculum. One aspect of it was to offer us two elective courses, one in electronics and radio engineering, and the other in mathematics. We were extraordinarily lucky in getting Professor Venkatachaliengar as our mathematics teacher. He taught us the theory of complex variables.

Mathematics was not given much importance earlier in engineering curricula. Now, of course, everybody knows the importance of a good mathematical foundation to build the superstructure of engineering knowledge and ability.

For Professor Venkatachaliengar, who was an outstanding mathematician, spending time teaching undergraduate engineering students, rather than guiding research work of gifted mathematics students, must have been a difficult experience. He was a mathematics teacher in a government run system, and had been posted to an engineering college. It must have been highly demoralizing and depressing for him. But his behavior did not indicate anything of this. He was one of the best teachers I have ever had. Always full of energy, he had a child-like ability to get thrilled with ideas and insights, and was ever ready to share his knowledge. It was just a blackboard and a piece of chalk; but what an experience they wrought through the hands of Professor Venkatachaliengar!

It was only later, much later, that I realized the value of what we had gained as students of Professor Venkatachaliengar. He made a big difference to us. He taught us deep things. He taught us that no matter what the circumstances in your life may be, as a teacher you should always be ever enthusiastic and dedicated towards your students.

A Brief Reminiscence by Dr. V. Ramamani

Scientist, Defence Research and Development Organization (Retd.)

My association with Prof. KV was a long one – for over 40 years. I was a postgraduate student of Prof. KV during 1961–63. Thereafter, I continued as a research scholar and did a Ph.D. under his able guidance. However, I continued to meet him. Whatever knowledge I gained in his association is a valuable one. Prof. KV was a versatile scholar both in pure and applied mathematics. But his love for S. Ramanujan was unique. He was devoting a lot his time in solving problems found in the note books of S. Ramanujan. His other favorite subjects were ancient Indian mathematics and astronomy. Prof. KV was a simple and a straightforward person and never hesitated to express his opinion. He expected hard work from his students and encouraged them whenever they did good work.

Prof. KV was a voracious reader and knowledgeable in varied fields. I had the opportunity and privilege to meet him on numerous occasions and during every visit I learned something from him. It was a treat to listen to him.

References

[1] B. S. Madhava Rao, V. R. Thiruvenkatachar, and K. Venkatachaliengar, Algebra related to elementary particles of spin $\frac{3}{2}$, *Proc. Roy. Soc. London, Ser. A.*, **187** (1946) 385–397.

K. Venkatachaliengar

Bruce C. Berndt

Department of Mathematics, University of Illinois, 1409 West Green Street, Urbana, IL 61801, USA
e-mail: berndt@illinois.edu

During the period from 1985 to 1994, the author received at least nine lengthy letters from Venkatachaliengar. The first, written on 15 July 1985, came in response to receiving a copy of the author's first book on Ramanujan's notebooks [4], which the author had earlier sent to Venkatachaliengar. In order to set the context of a long passage that we are going to quote from this letter, we need to recall a brief history of Ramanujan's *Quarterly Reports*.

On 26 February 1913, Sir Gilbert Walker, Head of the Meteorological Observatory in Madras, sent a letter to Francis Dewsbury, Registrar at the University of Madras, exhorting the University to provide Ramanujan with a scholarship. On 1 May 1913, Ramanujan was granted such a scholarship with the stipulation that he write *Quarterly Reports* on his research activity. Before departing for England in March of 1914, Ramanujan wrote three reports, dated 5 August 1913, 7 November 1913, and 9 March 1914. An account of these reports may be found in the author's book [4] or his papers [2,3]. In his first letter to the author, Venkatachaliengar provided an account of the deliberations that took place before the scholarship was approved. In the quote below, S. R. is an abbreviation for Srinivasa Ramanujan.

> The English professors in Madras headed by Littlehailes stoutly opposed the grant of a research scholarship to S. R. although the Board of Studies in Mathematics in Madras University consisting of a majority of Indian professors had recommend[ed] the award at the suggestion of the competent distinguished mathematician G. Walker. The Indian professors were apprehensive; the vice-chancellor was the chief justice of the Madras High Court who was appointed directly by the Secretary of State of India (India office, London). One of the members of the Syndicate of the Madras Univ. was Justice Sundaram Iyer of the Madras High Court who had been briefed by the Indian professors; at the meeting of the Syndicate all the English Profs. opposed the proposal with all their vehement speeches. After all of them had talked out their breath, Justice Sundaram Iyer stood up and read out the preamble to Madras University act passed by the British Parliament in London: The Madras University is founded to *promote learning* and *encourage research* .... The English Chief Justice was told that nothing had been

Research partially supported by grant H98230-07-1-0088 from the National Security Agency.

> done by the Univ. till that time to encourage research and he immediately stifled the opposition and made the recommendation to the Governor of Madras Lord Pentland, who was also a direct appointee of the British Govt. Even with this there was a hitch; and Sir Francis Spring wrote a letter to the private secretary of the Governor strongly recommending the case. After S. R. became famous, Prof. Littlehailes used to take S. R. in the sidecar of his motorbike. His Indian colleagues were having a hearty laugh. This was narrated to me by late Prof. S. R. Ranganathan, who had written a book on S. R.

Descriptions of events leading to Ramanujan's scholarship vary somewhat; see for example, [17] and [4, p. 295]. Venkatachaliengar's account highlights the objections of English mathematicians, in particular Littlehailes. In one of the two Presidential addresses [20] that Venkatachliengar gave before the Indian Mathematical Society, he pointed out that this was the first research scholarship awarded by the University some 50 years after the British Parliament had passed the Act. Sir Francis Spring was the Chairman of the Madras Port Trust Office where Ramanujan served as a clerk, and was one of Ramanujan's earliest and most devoted supporters. Ranganathan was Librarian at the University of Madras and wrote the first thorough biography of Ramanujan [17]. He is also famous in the field of library science and is considered to be one of the founders of modern library science. Littlehailes was Professor of Mathematics at Presidency College in Madras and later became Vice-Chancellor of the University of Madras; more information about him can be found in [6].

In his visits to India, the author can recall three long conversations with Venkatachaliengar. Although our conversations were chiefly about mathematics, in particular, about modular equations, he would often relate interesting facts about Indian mathematics and mathematicians, including Ramanujan, over a broad expanse of the twentieth century. It is unfortunate that portions of these conversations were not recorded for posterity.

An examination of the bibliography of Venkatachaliengar shows a broad stretch of interests. For roughly the first half of his career, Venkatachaliengar's interests were algebraic, covering a wide variety of topics. In the second half of his career, his interests took on a more analytic bent, and he became fascinated with the work of Ramanujan. Although he published few papers on Ramanujan's work, his passion for Ramanujan was evident in his two Presidential addresses to the Indian Mathematical Society [20], his highly original monograph on Ramanujan's discoveries in elliptic functions [21], and his unpublished, handwritten monograph on selected results from Ramanujan's work that he coauthored with his close friend V. R. Thiruvenkatachar [19]. We offer a few remarks about some of Venkatachaliengar's publications.

Venkatachaliengar's monograph [21] is thoroughly reviewed by S. Cooper elsewhere in this volume. A corrected, edited, and more widely distributed edition of this work is greatly desired.

A typed version of [19] has been made by M. D. Hirschhorn. This will be the starting point of an edited version that is being undertaken by the present author and several of his graduate students, and which will be published by the Ramanujan Mathematical Society.

Readers might be interested in the origin of the author's paper [7] with Venkatachaliengar. In early 1999, the author visited the University of Mysore, and one of his hosts, Professor S. Bhargava, accompanied him to the home of retired University of Mysore Professor T. S. Nanjundiah, who presented the author with a handwritten partial manuscript of a paper that he had once started to write with Venkatachaliengar. Although the Dedekind eta-function $\eta(\tau)$ was not mentioned in this partial manuscript, the intended authors had very cleverly derived the transformation formula for $\eta(\tau)$. Nanjundiah demurred about his own contributions to the paper, and so eventually with the gracious accedence of both Nanjundiah and Venkatachaliengar, the present author added a few minor contributions of his own and coauthored [7] with the latter. Readers might turn to the paper by the author, C. Gugg, S. Kongsiriwong, and J. Thiel [5] in this volume, where they will find Venkatachaliengar's ingenious idea generalized.

Venkatachaliengar supervised the doctoral dissertation of only one student, V. Ramamani [12]. In her beautiful thesis, she extended some of the ideas from Ramanujan's epic paper [14]. Two of the author's doctoral students, Heekyoung Hahn and Tim Huber, in turn used her work in their own research, [8,9] and [11], respectively.

Ramamani and Venkatachaliengar coauthored [13], which, for the present author, has been the latter's most influential paper, for he has lectured on its content in his classes and seminars. Recall first a famous theorem of Euler on partitions: The number of partitions of the positive integer n into odd parts is equal to the number of partitions of n into distinct parts. J. J. Sylvester [18] gave a beautiful extension of Euler's theorem in 1884, and Ramamani and Venkatachaliengar provided an elegant proof in [13], which one can find in the text by G. E. Andrews [1, pp. 24–25] and which motivated a further proof by M .D. Hirschhorn [10]. We state Sylvester's theorem.

Theorem 1. *Let $A_k(n)$ denote the number of partitions of the positive integer n into odd parts such that exactly k different parts occur. Let $B_k(n)$ denote the number of partitions $\lambda = (\lambda_1, \ldots, \lambda_r)$ of n, where $\lambda_i \geq \lambda_{i+1}$ and $1 \leq i \leq r-1$, such that the sequence $(\lambda_1, \ldots, \lambda_r)$ is composed of exactly k noncontiguous sequences of one or more consecutive integers. Then $A_k(n) = B_k(n)$, for all k and n.*

As intimated above, Venkatachaliengar served as President of the Indian Mathematical Society for two years. In his two addresses to the Society [20], he urged Indian mathematicians to study Ramanujan's (earlier) notebooks [15] and lost notebook [16]. In particular, he focused on particular examples from the lost notebook and his proofs of them. He also discussed the manuscripts of Ramanujan that were found in the library at Oxford University and published with Ramanujan's lost notebook [16].

References

[1] G. E. Andrews, The Theory of Partitions, Addison–Wesley, Reading, MA, 1976; reissued: Cambridge University Press, Cambridge (1998).

[2] B. C. Berndt, The quarterly reports of S. Ramanujan, *Amer. Math. Monthly*, **90** (1983) 505–516.

[3] B. C. Berndt, Ramanujan's quarterly reports, *Bull. London Math. Soc.*, **16** (1984) 449–489.

[4] B. C. Berndt, Ramanujan's Notebooks, Part I, Springer-Verlag, New York (1985).

[5] B. C. Berndt C. Gugg, S. Kongsiriwong and J. Thiel, A proof of the general theta transformation formula, this volume, pp. 53–62.

[6] B. C. Berndt and R. A. Rankin, Ramanujan: Letters and Commentary, American Mathematical Society, Providence, RI, 1995; London Mathematical Society, London (1995).

[7] B. C. Berndt and K. Venkatachaliengar, On the transformation formula for the Dedekind eta-function, in Symbolic Computation, Number Theory, Special Functions, Physics and Combinatorics, F. G. Garvan and M. E. H. Ismail, eds., Kluwer, Dordrecht (2001) pp. 73–77.

[8] H. Hahn, Convolution sums of some functions on divisors, *Rocky Mt. J. Math.*, **37** (2007) 315–344.

[9] H. Hahn, Eisenstein series associated with $\Gamma_0(2)$, *Ramanujan J.*, **15** (2008) 235–257.

[10] M. D. Hirschhorn, Sylvester's partition theorem and a related result, *Michigan Math. J.*, **21** (1975) 133–136.

[11] T. Huber, Zeros of Generalized Rogers–Ramanujan Series and Topics from Ramanujan's Theory of Elliptic Functions, Ph.D. thesis, University of Illinois at Urbana–Champaign, Urbana (2007).

[12] V. Ramamani, Some Identities Conjectured by Srinivasa Ramanujan Found in His Lithographed Notes Connected with Partition Theory and Elliptic Modular Functions – Their Proofs – Interconnection with Various Other Topics in the Theory of Numbers and Some Generalisations Thereon, Ph.D. thesis, University of Mysore, Mysore (1970).

[13] V. Ramamani and K. Venkatachaliengar, On a partition theorem of Sylvester, *Michigan Math. J.*, **19** (1972) 137–140.

[14] S. Ramanujan, On certain arithmetical functions, *Trans. Cambridge Philos. Soc.*, **22** (1916) 159–184.

[15] S. Ramanujan, Notebooks (2 volumes), Tata Institute of Fundamental Research, Bombay (1957).

[16] S. Ramanujan, The Lost Notebook and Other Unpublished Papers, Narosa, New Delhi (1988).

[17] S. R. Ranganathan, Ramanujan: The Man and the Mathematician, Asia Publishing House, Bombay (1967).

[18] J. J. Sylvester, A constructive theory of partitions arranged in three acts, an interact, and an exodion, *Amer. J. Math.*, **5** (1884) 251–330; **6** (1886) 334–336.

[19] V. R. Thiruvenkatachar and K. Venkatachaliengar, Ramanujan at Elementary Levels: Glimpses, Unpublished handwritten notes.

[20] K. Venkatachaliengar, Ramanujan Manuscripts, I, II, Presidential Addresses given in December 1979, 1980 to the Indian Mathematical Society, *Math. Student*, **52** (1984) 1–4; 215–233.

[21] K. Venkatachaliengar, Development of Elliptic Functions According to Ramanujan, Tech. Rep. 2, Madurai Kamaraj University, Madurai (1988).

A Review of Venkatachaliengar's Book on Elliptic Functions

Shaun Cooper

Institute of Information and Mathematical Sciences, Massey University-Albany,
Private Bag 102904, North Shore Mail Centre, Auckland, New Zealand
e-mail: s.cooper@massey.ac.nz

1. Introduction

One of the things Professor K. Venkatachaliengar is best remembered for is his book: "Development of Elliptic Functions according to Ramanujan", published by Madurai Kamaraj University in 1988[1]. In this review we shall survey several of the topics covered in the book and discuss some of the methods and ideas.

We will deliberately be brief and not go too heavily into the details. The exception is in Section 4, where a complete proof of Ramanujan's ${}_1\psi_1$ summation formula, due to Venkatachaliengar, is given.

Throughout this review, τ will always be a complex number with positive imaginary part, and $q = \exp(2\pi i \tau)$, so that $|q| < 1$. Occasionally, reference will be made to the Weierstrassian functions $\wp$, ζ, and σ, and the Weierstrassian invariants g_2 and g_3. Definitions and all of the main properties of the Weierstrassian functions can be found in Chapter 20 of the book by Whittaker and Watson [42]. For this review, it should be sufficient to refer to the equations (13)–(16), (49) and (50), below.

The page numbers that appear in the section headings refer to Venkatachaliengar's book [39].

2. A generalization of Ramanujan's identity [pp. 1–13]

The first part of Venkatachaliengar's book takes much of its inspiration from Ramanujan's paper "On Certain Arithmetical Functions" [32]. Two of the fundamental formulas in Ramanujan's paper are

[1]There is no publication date in Venkatachaliengar's book [39]. Dr. G. N. Srinivasa Prasanna has told me that according to Professor Soundararajan of Madurai University, the book went to press in Dec. 1987 and was released in Feb. 1988. This is consistent with information I have received from Professors R. A. Askey and B. C. Berndt. The article [38], published in vol. 52 of The Mathematics Student, reports that Venkatachaliengar's book was published in 1980. This is most likely an error. For, although this volume of The Mathematics Student has a publication year of 1984, MathSciNet indicates that it was not actually published until 1990. We conclude that Venkatachaliengar's book [39] must have been published in 1988.

$$\left(\frac{1}{4}\cot\frac{\theta}{2}+\sum_{j=1}^{\infty}\frac{q^j}{1-q^j}\sin j\theta\right)^2$$
$$=\left(\frac{1}{4}\cot\frac{\theta}{2}\right)^2+\sum_{j=1}^{\infty}\frac{q^j}{(1-q^j)^2}\cos j\theta+\frac{1}{2}\sum_{j=1}^{\infty}\frac{jq^j}{1-q^j}(1-\cos j\theta) \qquad (1)$$

and

$$\left(\frac{1}{8}\cot^2\frac{\theta}{2}+\frac{1}{12}+\sum_{j=1}^{\infty}\frac{jq^j}{1-q^j}(1-\cos j\theta)\right)^2$$
$$=\left(\frac{1}{8}\cot^2\frac{\theta}{2}+\frac{1}{12}\right)^2+\frac{1}{12}\sum_{j=1}^{\infty}\frac{j^3q^j}{1-q^j}(5+\cos j\theta). \qquad (2)$$

These identities are each valid in the horizontal strip $|\mathrm{Im}\,\theta| < 2\pi\ \mathrm{Im}\,\tau$, although all of the functions involved have analytic continuations to functions that are meromorphic on the complex plane.

The identities (1) and (2) can be expressed in terms of theta and elliptic functions, as follows. It is easy to check that the theta function

$$u(\theta,t)=-i\sum_{j=-\infty}^{\infty}q^{(2j+1)^2/8}e^{i(2j+1)\theta/2}, \quad \text{where} \quad q=e^{-t},$$

is a solution of the heat equation

$$2\frac{\partial u}{\partial t}=\frac{\partial^2 u}{\partial\theta^2}.$$

Therefore, $\log u$ satisfies the non-linear partial differential equation

$$\left(\frac{\partial\log u}{\partial\theta}\right)^2=2\frac{\partial\log u}{\partial t}-\frac{\partial^2\log u}{\partial\theta^2}. \qquad (3)$$

By Jacobi's triple product identity [5, p. 10] we have

$$u(\theta,t)=2q^{1/8}\sin\frac{\theta}{2}\prod_{j=1}^{\infty}(1-q^je^{i\theta})(1-q^je^{-i\theta})(1-q^j). \qquad (4)$$

In 1951, B. van der Pol [36] observed that if (4) is substituted into (3), the result is Ramanujan's identity (1). Moreover, if we apply the differential operator $\partial/\partial\theta$ to (3) and put $w=-\partial\log u/\partial\theta$, the result simplifies to Burgers' equation [14]

$$\frac{\partial w}{\partial t}+w\frac{\partial w}{\partial\theta}=\frac{1}{2}\frac{\partial^2 w}{\partial\theta^2}$$

which has applications in fluid mechanics.

Ramanujan's identity (2) may be shown to be equivalent to the identity

$$\wp^2(\theta) = \frac{1}{6}\wp''(\theta) + \frac{g_2}{12}$$

satisfied by the Weierstrass elliptic $\wp$ function, [42, p. 450].

Venkatachaliengar's book begins with a generalization of Ramanujan's identity (1). Let ρ_1 and ρ_2 be the functions defined by

$$\rho_1(x) = \frac{1}{2} + \sum{}' \frac{x^n}{1-q^n} \quad \text{and} \quad \rho_2(x) = -\frac{1}{12} + \sum{}' \frac{q^n x^n}{(1-q^n)^2}, \tag{5}$$

where the primes denote that the summations are over all non-zero integers n. These series converge in the annuli $|q| < |x| < 1$ and $|q| < |x| < |q|^{-1}$, respectively. Venkatachaliengar's identity is:

$$\rho_1(x)\rho_1(y) - \rho_1(xy)\{\rho_1(x) + \rho_1(y)\} = \rho_2(x) + \rho_2(y) + \rho_2(xy). \tag{6}$$

It may be proved by computing the coefficient of $x^m y^n$ on the left hand side and noting that it simplifies to zero unless $n = 0$, $m = 0$ or $m = n$. The non-zero terms give rise to the three functions that occur on the right hand side of the identity.

The analytic continuation for ρ_1 may be obtained as follows. We have

$$\begin{aligned}\rho_1(x) &= \frac{1}{2} + \sum_{j=1}^{\infty}\left(\frac{x^j}{1-q^j} + \frac{x^{-j}}{1-q^{-j}}\right)\\ &= \frac{1}{2} + \frac{x}{1-x} + \sum_{j=1}^{\infty}\left(\frac{q^j x^j}{1-q^j} - \frac{q^j x^{-j}}{1-q^j}\right).\end{aligned} \tag{7}$$

Expanding each term as a geometric series and interchanging the order of summation in the resulting double series gives

$$\begin{aligned}\rho_1(x) &= \frac{1+x}{2(1-x)} + \sum_{j=1}^{\infty}\sum_{k=1}^{\infty}(q^{jk}x^j - q^{jk}x^{-j})\\ &= \frac{1+x}{2(1-x)} + \sum_{k=1}^{\infty}\left(\frac{q^k x}{1-q^k x} - \frac{q^k x^{-1}}{1-q^k x^{-1}}\right).\end{aligned} \tag{8}$$

Thus $\rho_1(x)$ is analytic for $0 < |x| < \infty$ except for simple poles at $x = q^k$ for every integer k. In a similar way, we find that

$$\rho_2(x) = -\frac{1}{12} + \sum_{k=1}^{\infty}\left(\frac{kq^k x}{1-q^k x} + \frac{kq^k x^{-1}}{1-q^k x^{-1}}\right), \tag{9}$$

and therefore $\rho_2(x)$ is analytic for $0 < |x| < \infty$ except for simple poles at $x = q^k$ for every *non-zero* integer k.

From (8) and (9) we deduce the properties

$$\rho_1(x) = -\rho_1(x^{-1}) \quad \text{and} \quad \rho_2(x) = \rho_2(x^{-1}). \tag{10}$$

Combining (6) and (10) we obtain the symmetric form

$$\rho_1(x)\rho_1(y) + \rho_1(y)\rho_1(z) + \rho_1(z)\rho_1(x) = \rho_2(x) + \rho_2(y) + \rho_2(z), \tag{11}$$

which holds for all complex numbers x, y and z that satisfy $xyz = 1$.

Venkatachaliengar shows how (11) can be used to deduce two forms of the addition formula for the Weierstrass elliptic $\wp$ function [42, pp. 440–441]:

$$\wp(\alpha+\beta) = \frac{1}{4}\left(\frac{\wp'(\alpha) - \wp'(\beta)}{\wp(\alpha) - \wp(\beta)}\right)^2 - \wp(\alpha) - \wp(\beta) \tag{12}$$

and

$$\begin{vmatrix} 1 & 1 & 1 \\ \wp(\alpha) & \wp(\beta) & \wp(\gamma) \\ \wp'(\alpha) & \wp'(\beta) & \wp'(\gamma) \end{vmatrix} = 0, \quad \text{where } \alpha + \beta + \gamma = 0,$$

as well as the differential equation [42, p. 437]

$$(\wp'(\theta))^2 = 4\wp^3(\theta) - g_2\,\wp(\theta) - g_3.$$

Moreover, taking the limit as $y \to 1$ in (6) gives

$$-\rho_1^2(x) + x\rho_1'(x) = 2\rho_2(x) + \rho_2(1).$$

Putting $x = \exp(i\theta)$ in this and appealing to (7), we obtain Ramanujan's identity (1); that is, Venkatachaliengar's identities (6) and (11) are generalizations of Ramanujan's identity (1).

3. Ramanujan's Eisenstein series [pp. 15–18, 31–32]

Ramanujan's Eisenstein series P, Q and R are defined by

$$P = P(q) = 1 - 24\sum_{j=1}^{\infty} \frac{jq^j}{1-q^j}, \tag{13}$$

$$Q = Q(q) = 1 + 240\sum_{j=1}^{\infty} \frac{j^3q^j}{1-q^j}, \tag{14}$$

$$R = R(q) = 1 - 504\sum_{j=1}^{\infty} \frac{j^5q^j}{1-q^j}. \tag{15}$$

In terms of Weierstrassian parameters [42, pp. 437, 446], it turns out that

$$P(q) = \frac{12}{\pi}\eta_1, \quad Q(q) = 12g_2 \quad \text{and} \quad R(q) = 216g_3, \tag{16}$$

where the corresponding Weierstrass elliptic function has periods 2π and $2\pi\tau$.

The series P, Q and R were favorites of Ramanujan. They were studied extensively by him in the paper [32], in his second notebook [34] and in the lost notebook [35]. A survey of the Eisenstein series occurring in the lost notebook has been given by B. C. Berndt and A. J. Yee [8].

The Eisenstein series (13)–(15) arise as coefficients when the identities (1) and (2) are expanded in powers of θ. Ramanujan [32, p. 165] showed that by equating the coefficients of θ^2, θ^4 and θ^6 in the power series expansions of (1), and the coefficients of θ^4 in (2), the following differential equations may be deduced:

$$q\frac{dP}{dq} = \frac{P^2 - Q}{12}, \quad q\frac{dQ}{dq} = \frac{PQ - R}{3}, \quad \text{and} \quad q\frac{dR}{dq} = \frac{PR - Q^2}{2}. \tag{17}$$

These have come to be known as Ramanujan's differential equations, although they were known earlier to G.-H. Halphen in 1886 [23, p. 450]. Detailed proofs of Ramanujan's differential equations, following Ramanujan's method, are given at the beginning of Chapter 2 of Venkatachaliengar's book, and also in Chapter 4 of the introductory text by B. C. Berndt [5]. The reader is encouraged to work through the proof in Ramanujan's original paper [32], as well.

As noted by Ramanujan, the differential equations in (17) imply, by logarithmic differentiation, the result

$$Q^3 - R^2 = 1728q \prod_{j=1}^{\infty} (1 - q^j)^{24}.$$

Let U_n and V_n be defined by

$$U_n = \frac{1^{n+1} - 3^{n+1}q + 5^{n+1}q^3 - 7^{n+1}q^6 + \cdots}{1 - 3q + 5q^3 - 7q^6 + \cdots},$$

$$V_n = \frac{1^n - 5^n q - 7^n q^2 + 11^n q^5 + 13^n q^7 + \cdots}{1 - q - q^2 + q^5 + q^7 + \cdots},$$

or equivalently,

$$U_n = \frac{\sum j^{n+1} q^{j^2/8}}{\sum j q^{j^2/8}} \quad \text{and} \quad V_n = \frac{\sum (-1)^{(k-1)/6} k^n q^{k^2/24}}{\sum (-1)^{(k-1)/6} q^{k^2/24}}, \tag{18}$$

where the sums range over all integers j and k (positive as well as negative) that satisfy $j \equiv 1 \pmod 4$ and $k \equiv 1 \pmod 6$, respectively. Clearly, $U_0 = V_0 = 1$. In the lost notebook [35, p. 369], Ramanujan derived the recurrence relations

$$U_{n+2} = PU_n + 8q\frac{dU_n}{dq}, \tag{19}$$

$$V_{n+2} = PV_n + 24q\frac{dV_n}{dq}. \tag{20}$$

To prove these, recall that the denominators in (18) have factorizations given by [5, pp. 12, 14]

$$\sum_{j\equiv 1 \pmod 4} j q^{j^2/8} = q^{1/8}\prod_{\ell=1}^{\infty}(1-q^{\ell})^3 \tag{21}$$

and

$$\sum_{k\equiv 1 \pmod 6} (-1)^{(k-1)/6} q^{k^2/24} = q^{1/24}\prod_{\ell=1}^{\infty}(1-q^{\ell}). \tag{22}$$

Therefore,

$$\begin{aligned}
&8q\frac{d}{dq}\log U_n \\
&= 8q\frac{d}{dq}\log\left(\sum_{j\equiv 1 \pmod 4} j^{n+1}q^{j^2/8}\right) - 8q\frac{d}{dq}\log\left(q^{1/8}\prod_{\ell=1}^{\infty}(1-q^{\ell})^3\right) \\
&= \frac{\sum_{j\equiv 1 \pmod 4} j^{n+3}q^{j^2/8}}{\sum_{j\equiv 1 \pmod 4} j^{n+1}q^{j^2/8}} - \left(1-24\sum_{\ell=1}^{\infty}\frac{\ell q^{\ell}}{1-q^{\ell}}\right) \\
&= \frac{U_{n+2}}{U_n} - P.
\end{aligned}$$

The recurrence for U_n now follows, and the recurrence for V_n may be proved by the same method. Using the recurrence relations (19), (20) and Ramanujan's differential equations (17), we obtain

$$\begin{aligned}
&U_0 = 1, \quad U_2 = P, \quad U_4 = \frac{1}{3}(5P^2-2Q), \quad U_6 = \frac{1}{9}(35P^3-42PQ+16R), \\
&V_0 = 1, \quad V_2 = P, \quad V_4 = 3P^2-2Q, \quad V_6 = 15P^3-30PQ+16R.
\end{aligned}$$

By induction, it follows that for any positive integer n,

$$U_{2n} = \sum_{\substack{i,j,k\geq 0 \\ i+2j+3k=n}} \mu_{ijk}P^iQ^jR^k \quad \text{and} \quad V_{2n} = \sum_{\substack{i,j,k\geq 0 \\ i+2j+3k=n}} \nu_{ijk}P^iQ^jR^k, \tag{23}$$

where μ_{ijk} and ν_{ijk} are rational and integer constants, respectively.

In his presidential address delivered to the Indian Mathematical Society [38], Venkatachaliengar described the proof of (23) outlined by Ramanujan in the lost notebook [35, p. 369] as "disarmingly simple". The identities in (23) are discussed further in [2, pp. 355–364] and [39, pp. 31–32]. Analogues of (23) for which the products in (21) and (22) are replaced with $q^{r/24}\prod_{\ell=1}^{\infty}(1-q^{\ell})^r$ for $r \in \{2, 4, 6, 8, 10, 14, 26\}$ have been given in [16].

4. Ramanujan's ${}_1\psi_1$ summation formula [pp. 22–30]

One of Ramanujan's most important and well-known results is the ${}_1\psi_1$ summation formula. Venkatachaliengar gave two proofs, and we reproduce one of them here. This elegant proof makes use of symmetry and functional equations. First, recall Jacobi's triple product identity, which in one form is [5, p. 10]

$$\prod_{n=1}^{\infty}(1+q^{2n-1}z)(1+q^{2n-1}z^{-1})(1-q^{2n}) = \sum_{n=-\infty}^{\infty} q^{n^2}z^n. \tag{24}$$

It may be regarded as the Laurent series expansion of the product

$$\prod_{n=1}^{\infty}(1+q^{2n-1}z)(1+q^{2n-1}z^{-1}) \tag{25}$$

in the annulus $0 < |z| < \infty$. Ramanujan's ${}_1\psi_1$ summation formula involves a generalization of this product to include a denominator which consists of two infinite products. Specifically, let

$$\phi(z,\alpha,\beta) = \prod_{n=1}^{\infty}\frac{(1+q^{2n-1}z)(1+q^{2n-1}z^{-1})}{(1+\alpha q^{2n-1}z)(1+\beta q^{2n-1}z^{-1})} \tag{26}$$

and suppose in addition that

$$|\beta q| < |z| < |\alpha q|^{-1}. \tag{27}$$

The poles of ϕ are given by

$$z \in \{-\beta q, -\beta q^3, -\beta q^5, \ldots\} \cup \{-(\alpha q)^{-1}, -(\alpha q^3)^{-1}, -(\alpha q^5)^{-1}, \ldots\},$$

and the purpose of the condition (27) is to ensure that ϕ is analytic in an annulus that separates the two families of poles. By Laurent's theorem, ϕ has an expansion of the form

$$\phi(z,\alpha,\beta) = \sum_{n=-\infty}^{\infty} c_n(\alpha,\beta)z^n \tag{28}$$

in the annulus (27). It is easy to check that

$$(1+\alpha qz)\phi(z,\alpha,\beta) = (\beta+qz)\phi(q^2z,\alpha,\beta). \tag{29}$$

If we put

$$\Psi(z) = (1+\alpha qz)\phi(z,\alpha,\beta),$$

we see that $\Psi(z)$ has a Laurent expansion valid in the larger annulus

$$|\beta q| < |z| < |\alpha q^3|^{-1}, \tag{30}$$

because the pole of $\phi(z,\alpha,\beta)$ at $z = -(\alpha q)^{-1}$ has been eliminated. Now

$$\Psi(z) = (1+\alpha qz)\sum_{n=-\infty}^{\infty} c_n(\alpha,\beta)z^n = \sum_{n=-\infty}^{\infty}(c_n(\alpha,\beta)+\alpha q c_{n-1}(\alpha,\beta))z^n$$

and by (29),

$$\Psi(z) = (\beta + qz) \sum_{n=-\infty}^{\infty} c_n(\alpha, \beta) q^{2n} z^n$$

$$= \sum_{n=-\infty}^{\infty} (\beta c_n(\alpha, \beta) q^{2n} + c_{n-1}(\alpha, \beta) q^{2n-1}) z^n.$$

Both of these expansions for Ψ are valid in the larger annulus given by (30). Comparing coefficients of z^n gives

$$c_n(\alpha, \beta) + \alpha q c_{n-1}(\alpha, \beta) = \beta c_n(\alpha, \beta) q^{2n} + c_{n-1}(\alpha, \beta) q^{2n-1},$$

hence, we obtain the recurrence relation

$$c_n(\alpha, \beta) = \frac{q^{2n-2} - \alpha}{1 - q^{2n}\beta} q c_{n-1}(\alpha, \beta).$$

Iterating, gives

$$c_n(\alpha, \beta) = \frac{(1-\alpha)(q^2-\alpha)\cdots(q^{2n-2}-\alpha)}{(1-\beta q^2)(1-\beta q^4)\cdots(1-\beta q^{2n})} q^n c_0(\alpha, \beta), \tag{31}$$

for any positive integer n. From the definition (26), it is clear that

$$\phi(z, \alpha, \beta) = \phi(z^{-1}, \beta, \alpha)$$

and comparing coefficients of z^{-n} on each side gives

$$c_{-n}(\alpha, \beta) = c_n(\beta, \alpha). \tag{32}$$

Combining (31) and (32) it follows that for any positive integer n,

$$\begin{aligned} c_{-n}(\alpha, \beta) &= c_n(\beta, \alpha) \qquad (33) \\ &= \frac{(1-\beta)(q^2-\beta)\cdots(q^{2n-2}-\beta)}{(1-\alpha q^2)(1-\alpha q^4)\cdots(1-\alpha q^{2n})} q^n c_0(\beta, \alpha) \\ &= \frac{(1-\beta)(q^2-\beta)\cdots(q^{2n-2}-\beta)}{(1-\alpha q^2)(1-\alpha q^4)\cdots(1-\alpha q^{2n})} q^n c_0(\alpha, \beta), \end{aligned}$$

where the last step follows by taking $n = 0$ in (32). It remains to evaluate $c_0(\alpha, \beta)$. Again, it is easy to check from the definition (26) that

$$\phi(z, \alpha q^2, \beta) = (1 + \alpha q z)\phi(z, \alpha, \beta),$$

and both sides represent analytic functions of z for $|\beta q| < |z| < |\alpha q^3|^{-1}$. Equating the constant coefficients and using (33) we obtain

$$c_0(\alpha q^2, \beta) = c_0(\alpha, \beta) + \alpha q c_{-1}(\alpha, \beta) = \frac{1 - \alpha\beta q^2}{1 - \alpha q^2} c_0(\alpha, \beta).$$

Iterating gives

$$c_0(\alpha,\beta)=c_0(\alpha q^{2n},\beta)\prod_{j=1}^{n}\frac{1-\alpha q^{2j}}{1-\alpha\beta q^{2j}},$$

and taking the limit as $n\to\infty$ gives

$$c_0(\alpha,\beta)=c_0(0,\beta)\prod_{j=1}^{\infty}\frac{1-\alpha q^{2j}}{1-\alpha\beta q^{2j}}. \tag{34}$$

By (32) and (34), we deduce further that

$$c_0(0,\beta)=c_0(\beta,0)=c_0(0,0)\prod_{j=1}^{\infty}(1-\beta q^{2j}), \tag{35}$$

and substituting (35) into (34) gives

$$c_0(\alpha,\beta)=c_0(0,0)\prod_{j=1}^{\infty}\frac{(1-\alpha q^{2j})(1-\beta q^{2j})}{1-\alpha\beta q^{2j}}. \tag{36}$$

Now take $\alpha=\beta=1$ and observe that $\phi(z,1,1)=1$ identically. Hence, $c_0(1,1)=1$. Using this in (36) gives

$$1=c_0(0,0)\prod_{j=1}^{\infty}(1-q^{2j})$$

and so

$$c_0(0,0)=\prod_{j=1}^{\infty}\frac{1}{(1-q^{2j})},$$

therefore (36) becomes

$$c_0(\alpha,\beta)=\prod_{j=1}^{\infty}\frac{(1-\alpha q^{2j})(1-\beta q^{2j})}{(1-\alpha\beta q^{2j})(1-q^{2j})}. \tag{37}$$

If we divide both sides of (28) by $c_0(\alpha,\beta)$ and use (31), (33) and (37), the final result is Ramanujan's ${}_1\psi_1$ summation formula:

$$\prod_{n=1}^{\infty}\frac{(1+q^{2n-1}z)(1+q^{2n-1}z^{-1})(1-\alpha\beta q^{2j})(1-q^{2j})}{(1+\alpha q^{2n-1}z)(1+\beta q^{2n-1}z^{-1})(1-\alpha q^{2j})(1-\beta q^{2j})}$$

$$=1+\sum_{n=1}^{\infty}\frac{(1-\alpha)(q^2-\alpha)\cdots(q^{2n-2}-\alpha)}{(1-\beta q^2)(1-\beta q^4)\cdots(1-\beta q^{2n})}q^nz^n$$

$$+\sum_{n=1}^{\infty}\frac{(1-\beta)(q^2-\beta)\cdots(q^{2n-2}-\beta)}{(1-\alpha q^2)(1-\alpha q^4)\cdots(1-\alpha q^{2n})}q^nz^{-n}. \tag{38}$$

By the ratio test, the series involving positive powers of z in (38) converges for $|\alpha q z| < 1$, while the series involving negative powers of z converges for $|\beta q/z| < 1$. The two series will converge simultaneously and represent an analytic function of z when both conditions are satisfied, and this is precisely the condition given by (27).

It should be noted that the Jacobi triple product identity (24) is the limiting case $\alpha, \beta \to 0$ of (38). In this case, the annulus of convergence becomes $0 < |z| < \infty$.

For some historical background and references to other proofs of Ramanujan's ${}_1\psi_1$ summation formula (38), see [2, pp. 54–56] and [28].

5. The Jordan–Kronecker function [pp. 37–43]

The Jordan-Kronecker function is defined by

$$F(x, y) = \prod_{j=1}^{\infty} \frac{(1-q^{j-1}xy)(1-q^j x^{-1}y^{-1})(1-q^j)^2}{(1-q^{j-1}x)(1-q^j x^{-1})(1-q^{j-1}y)(1-q^j y^{-1})}. \tag{39}$$

It has two important properties. The first is its expansion as a series:

$$F(x, y) = \sum_{j=-\infty}^{\infty} \frac{x^j}{1-yq^j}, \quad \text{provided } |q| < |x| < 1. \tag{40}$$

This is an immediate consequence of Ramanujan's ${}_1\psi_1$ summation formula: set $(z, \alpha, \beta) = (-xy/q, 1/y, y)$ in (38), divide both sides by $1-y$, then replace q^2 with q.

The infinite product (39) is symmetric in x and y, but the series in (40) is not. Symmetry in the series can be restored by a calculation that is similar to (7). That is, consider the terms for which j is positive, negative or zero, and expand into a double series, to get

$$F(x, y) = \frac{1-xy}{(1-x)(1-y)} + \sum_{j=1}^{\infty}\sum_{k=1}^{\infty} q^{jk}(x^j y^k - x^{-j}y^{-k}). \tag{41}$$

This converges for $|q| < |x|, |y| < |q|^{-1}$, apart from simple poles at $x = 1$, $y = 1$. The series (41) can be manipulated to yield a series that converges for all x and y, apart from at the poles, by separating the terms in (41) into the three cases $j = k$, $j > k$ and $j < k$, and summing the geometric series. The final result is

$$\begin{aligned} F(x, y) = {} & \frac{1-xy}{(1-x)(1-y)} \\ & + \sum_{j=1}^{\infty} q^{j^2} x^j y^j \left(1 + \frac{q^j x}{1-q^j x} + \frac{q^j y}{1-q^j y}\right) \\ & - \sum_{j=1}^{\infty} q^{j^2} x^{-j} y^{-j} \left(1 + \frac{q^j x^{-1}}{1-q^j x^{-1}} + \frac{q^j y^{-1}}{1-q^j y^{-1}}\right). \end{aligned} \tag{42}$$

The formulas (39) and (41) were given by Kronecker[2] in lectures in July 1876, and then in a paper published in 1881 [30]. The formulas (39), (40) and (42) appear in Jordan's Cours d'Analyse [29, p. 507–511]. For these reasons, Venkatachaliengar refers to the function $F(x, y)$ as the Jordan-Kronecker function.

Kroencker's paper has a reference to the 1850 paper of Jacobi "Sur la rotation d'un corps", [25]. If eq. (3) in [25, p. 297] is multiplied by $i = \sqrt{-1}$ and the result subtracted from eq. (4), ibid., the result eventually simplifies to the identity (40). Thus, although (40) is implicit in Jacobi's work, it would be a bit of a stretch to attribute the identity to him. However, Jacobi's work is noteworthy because the genesis of the identity may be seen there, and because the identity arises in a physical application.

Kronecker's analysis has been examined and simplified by Weil [41, pp. 70–71]. Both Kronecker and Weil expand the product in (39) as a Laurent series and use Cauchy's theorem to compute the coefficients.

The second important property is the result which Venkatachaliengar calls the fundamental multiplicative identity:

$$F(x,t)F(y,t) - t\frac{\partial}{\partial t}F(xy,t) + F(xy,t)\left(\rho_1(x) + \rho_1(y)\right), \tag{43}$$

where ρ_1 is defined in (5). This can be proved by writing

$$F(x,t)F(y,t) = \sum_{n=-\infty}^{\infty}\sum_{m=-\infty}^{\infty}\frac{x^m y^n}{(1-tq^m)(1-tq^n)},$$

breaking the sum into the cases $m = n$ and $m \neq n$, and using partial fractions in the latter case. The details have been reproduced in [18] and [21].

Two special cases of (43) deserve special mention. First, if we set $t = e^{\upsilon}$ in (43), expand both sides in powers of υ, then equate coefficients of υ^0, the result is Venkatachaliengar's identity (11). See [18, p. 75] for the details. Thus, (43) may be regarded as a generalization of Ramanujan's identity (1) that involves two extra parameters.

The other noteworthy special case is to take the limit as $y \to 1/x$ in (43). The result simplifies to

$$F(x,t)F(x^{-1},t) = t\frac{d}{dt}\rho_1(t) - x\frac{d}{dx}\rho_1(x). \tag{44}$$

This can be shown to be equivalent to the Weierstrassian identity [42, p. 451]

$$\frac{\sigma(\alpha+\beta)\sigma(\alpha-\beta)}{\sigma^2(\alpha)\sigma^2(\beta)} = \wp(\beta) - \wp(\alpha).$$

[2]Kronecker gives the identity in the form

$$\frac{\sum(-1)^{(\mu-1)/2}\mu q^{\mu^2/4}\sum(-1)^{(\nu-1)/2}q^{\nu^2/4}(x^\nu y^\nu - x^{-\nu}y^{-\nu})}{\sum(-q)^{m^2}x^{2m}\sum(-q)^{n^2}y^{2n}} = \sum\sum q^{\mu\nu/2}(x^\mu y^\nu - x^{-\mu}y^{-\nu})$$

where the sums are over positive odd integers μ and ν, and all integers m and n. Kronecker's form can be manipulated into the form we have given by converting each of the four sums in the quotient into infinite products via Jacobi's triple product identity and making a change of variable.

The identity (43) should be regarded as classical. It has been rediscovered and reproved many times. For, dividing by $F(xy,t)$ gives

$$\frac{F(x,t)F(y,t)}{F(xy,t)} = t\frac{\partial}{\partial t}\log F(xy,t) + \rho_1(x) + \rho_1(y)$$
$$= \rho_1(x) + \rho_1(y) + \rho_1(t) - \rho_1(xyt),$$

and this can be shown to be equivalent to the Weierstrassian identity

$$\frac{\sigma(\alpha+\beta)\sigma(\alpha-\beta)\sigma(2\gamma)}{\sigma(\alpha+\gamma)\sigma(\alpha-\gamma)\sigma(\beta+\gamma)\sigma(\beta-\gamma)} = \zeta(\alpha+\gamma)-\zeta(\alpha-\gamma)+\zeta(\beta-\gamma)-\zeta(\beta+\gamma).$$

This was given by G.-H. Halphen [23, p. 187]. Further historical background for the identity (43) has been given in [2, p. 60].

6. Inversion for elliptic functions, and theories for alternative bases [pp. 52–56, 89–95]

At the end of Chapter 3, a novel approach to the connection between theta functions and hypergeometric functions is proposed. For $|q| < 1$, define $x = x(q)$ by

$$x = \left(\sum_{n=-\infty}^{\infty} q^{\left(n+\frac{1}{2}\right)^2} \Big/ \sum_{n=-\infty}^{\infty} q^{n^2} \right)^4. \tag{45}$$

It can be shown that as q increases from 0 to 1, x also increases from 0 to 1. The basic problem is to find a formula for the inverse function $q = q(x)$. The result is

$$q = \exp\left(-\pi \frac{{}_2F_1\left(\frac{1}{2},\frac{1}{2};1;1-x\right)}{{}_2F_1\left(\frac{1}{2},\frac{1}{2};1;x\right)} \right), \tag{46}$$

where

$${}_2F_1(a,b;c;x) = 1 + \sum_{n=1}^{\infty} \frac{(a)_n(b)_n}{(c)_n n!}x^n$$

is the hypergeometric function, and

$$(a)_n = \frac{\Gamma(a+n)}{\Gamma(a)} = a(a+1)\cdots(a+n-1)$$

is the shifted factorial. These results were known to Jacobi in 1828, see, e.g., [24].

Venkatachaliengar's approach is to disregard (45) and begin instead by defining z and x implicitly by

$$Q(q) = z^4(1-x+x^2) \quad \text{and} \quad R(q) = z^6(1+x)(1-2x)(1-x/2). \tag{47}$$

The goal is to solve the equations in (47) for q to obtain (46). The main step in the argument is to use Ramanujan's differential equations (17) to show that z satisfies the following hypergeometric differential equation with respect to x:

$$\frac{d}{dx}\left(x(1-x)\frac{dz}{dx}\right) = \frac{z}{4}.$$

Venkatachaliengar's proof has been explained in more detail by S. Bhargava [9, pp. 315–318].

In Chapter 5, [pp. 89–95] Venkatachaliengar proposes to use the same approach to prove the analogous results for (what are now known as) Ramanujan's theories of elliptic functions to alternative bases. These are commonly referred to by their signatures, which are 3, 4, or 6. Jacobi's classical theory involving (45) and (46) is said to belong to the theory of signature 2. In the signature 3 theory Venkatachaliengar begins by defining z_3 and x_3 implicitly by

$$Q = z_3^4(1 + 8x_3), \qquad R = z_3^6(1 - 20x_3 - 8x_3^2). \tag{48}$$

There are similar definitions in each of the theories for signatures 4 and 6.

Let us discuss some of the merits and objections to Venkatachaliengar's approach just described. The utility of Venkatachaliengar's method has been demonstrated by H. H. Chan and Y. L. Ong [17] in establishing a septic theory. The most obvious objection is that the formulas (47) and (48) are unmotivated and have to be known in advance. Also, z and x are multivalued and a selection has to be made. Moreover, this is almost certainly not the method Ramanujan would have used. Ramanujan in fact outlined his approach to this inversion problem for elliptic functions in the first few entries of Chapter 17 in his second notebook [34]. Full details of Ramanujan's method have been worked out by B. C. Berndt [3, pp. 87–102]. Ramanujan's method is original, and Berndt makes the comment [3, p. 99]: "Proofs in the latter half of the second notebook are very rare indeed". In contrast, Ramanujan left no clues as to how he discovered the results for signatures 3, 4 and 6. The reader is referred to [4, Chapter 33], [6,7,10–13,15], and [20] for more information about Ramanujan's theories of elliptic functions to alternative bases.

The reviewer [20] has proposed the following alternative starting point, which leads to a unified treatment of the classical and alternative theories. Let $r = 1, 2, 3$ or 4. Let $z_r = z_r(q)$ be defined by

$$z_r = \begin{cases} (Q(q))^{1/4} & \text{if } r = 1, \\ \left(\frac{rP(q^r) - P(q)}{r-1}\right)^{1/2} & \text{if } r = 2, 3 \text{ or } 4. \end{cases}$$

Let $x_r = x_r(q)$ be defined to be the solution of the initial value problem

$$q\frac{dx_r}{dq} = z_r^2 x_r(1 - x_r), \qquad x_r(e^{-2\pi/\sqrt{r}}) = \frac{1}{2}.$$

It can be shown that as q increases from 0 to 1, x_r increases from 0 to 1. Therefore, the inverse function $q = q_r(x_r)$ exists. It is given explicitly by the formula

$$q = \exp\left(-\frac{2\pi}{\sqrt{r}} \frac{{}_2F_1(c_r, 1-c_r; 1; 1-x_r)}{{}_2F_1(c_r, 1-c_r; 1; x_r)}\right),$$

where $c_r = 1/6, 1/4, 1/3$ or $1/2$, according to whether $r = 1, 2, 3$ or 4, respectively. The quantity $1/c_r$ has come to be known as the signature. In the theory of modular forms, the parameter r is called the level. Thus, levels $r = 1, 2, 3$ and 4 correspond to the signatures 6, 4, 3 and 2, respectively. The formulas (45) and (46) are the results for level 4 (signature 2).

7. Halphen's differential equations [pp. 57–76]

Let

$$\Omega(\theta) = i\frac{d}{d\theta}\rho_1(e^{i\theta}).$$

From (7), we may obtain the explicit formula

$$\Omega(\theta) = \frac{1}{4}\csc^2\frac{\theta}{2} - 2\sum_{j=1}^{\infty}\frac{jq^j}{1-q^j}\cos j\theta. \tag{49}$$

It can be shown, e.g., [18, p. 69], that

$$\Omega(\theta) = \wp(\theta) + \frac{P}{12}, \tag{50}$$

where $\wp(z)$ is the Weierstrass elliptic function with periods 2π and $2\pi\tau$ and P is Ramanujan's Eisenstein series defined by (13).

Venkatachaliengar defines three functions u_1, u_2 and u_3 by

$$\begin{aligned} u_1 &= 4\Omega(\pi), \\ u_2 &= 4\Omega(\pi\tau), \\ u_3 &= 4\Omega(\pi + \pi\tau). \end{aligned}$$

Using (49), and setting $h = e^{i\pi\tau}$ so that $h^2 = q$, we find that

$$u_1 = 1 - 8\sum_{n=1}^{\infty}\frac{(-1)^n nh^{2n}}{1-h^{2n}}, \tag{51}$$

$$u_2 = -8\sum_{n=1}^{\infty}\frac{nh^n}{1-h^{2n}}, \tag{52}$$

$$u_3 = -8\sum_{n=1}^{\infty}\frac{(-1)^n nh^n}{1-h^{2n}}. \tag{53}$$

G.-H. Halphen [23, p. 330] proved that the functions given by (51)–(53) satisfy the system of differential equations

$$h\frac{d}{dh}(u_1+u_2)=u_1u_2, \tag{54}$$

$$h\frac{d}{dh}(u_2+u_3)=u_2u_3, \tag{55}$$

$$h\frac{d}{dh}(u_3+u_1)=u_3u_1. \tag{56}$$

Halphen noted that the system (54)–(56) is a special case of the differential equations

$$C(dA+dB)=B(dA+dC)=A(dB+dC),$$

posed by G. Darboux [22, p. 149] in 1878. More recently, the system (54)–(56) and other related systems of nonlinear differential equations have been studied by W. Zudilin [43].

By ingenious use of (11), Venkatachaliengar gave a simple proof that the functions defined by (51)–(53) are solutions of the system (54)–(56).

By series manipulations using (13) and (51)–(53), we find that

$$u_1+u_2+u_3=P.$$

Applying the differential operator $h\frac{d}{dh}=2q\frac{d}{dq}$, and using (54)–(56) on the left and (17) on the right, we find that

$$u_1^2+u_2^2+u_3^2-u_1u_2-u_2u_3-u_3u_1=Q.$$

Differentiating again using (54)–(56) and (17), and simplifying, it can be shown that

$$\frac{1}{2}(2u_1-u_2-u_3)(2u_2-u_3-u_1)(2u_3-u_1-u_2)=R.$$

The differences u_j-u_k have simple representations as infinite products. For example, subtracting (55) from (56) gives

$$h\frac{d}{dh}(u_1-u_2)=(u_1-u_2)u_3.$$

Divide by $(u_1-u_2)h$ and integrate with respect to h, to get

$$\begin{aligned}\log(u_1-u_2)&=-8\sum_{n=1}^{\infty}\int\frac{(-1)^n nh^{n-1}}{1-h^{2n}}\,dh\\&=4\sum_{n=1}^{\infty}(-1)^n\log\left(\frac{1-h^n}{1+h^n}\right)+c.\end{aligned}$$

Letting $h=0$ we find that the constant of integration is given by $c=0$. It follows that

$$u_1-u_2=\prod_{n=1}^{\infty}\frac{(1-h^{2n})^4(1+h^{2n-1})^4}{(1-h^{2n-1})^4(1+h^{2n})^4}.$$

A formula, for the number of representations of an integer as a sum of four squares as a divisor sum, can be deduced from this result. Infinite products for the differences $u_1 - u_3$ and $u_3 - u_2$ can be obtained by the same method.

Lastly, we note that by expanding each of (51)–(53) as double series and interchanging the order of summation, if follows that $u_1 > 0$, $u_2 < 0$ and $u_3 > 0$ for $0 < h < 1$.

8. Jacobian elliptic functions [pp. 97–114]

In Chapter 6 of [39], Venkatachaliengar uses the Jordan-Kronecker function (39) and the fundamental multiplicative identity (43) to give a new and efficient development of the main properties of Jacobian elliptic functions. An outline of this theory goes as follows. Let

$$f_1(\theta) = -iF(e^{i\pi}, e^{i\theta}), \tag{57}$$

$$f_2(\theta) = -ie^{i\theta/2}F(e^{i\pi\tau}, e^{i\theta}), \tag{58}$$

$$f_3(\theta) = -ie^{i\theta/2}F(e^{i\pi+i\pi\tau}, e^{i\theta}). \tag{59}$$

The factors $-i$ and $-ie^{i\theta/2}$ are included so that f_1, f_2 and f_3 will be real valued when θ is real. Infinite product expansions follow directly from the definition (39). The results are

$$f_1(\theta) = \frac{1}{2}\cot\frac{\theta}{2}\prod_{n=1}^{\infty}\frac{(1+2q^n\cos\theta+q^{2n})(1-q^n)^2}{(1-2q^n\cos\theta+q^{2n})(1+q^n)^2}, \tag{60}$$

$$f_2(\theta) = \frac{1}{2}\csc\frac{\theta}{2}\prod_{n=1}^{\infty}\frac{(1-2q^{n-\frac{1}{2}}\cos\theta+q^{2n-1})(1-q^n)^2}{(1-2q^n\cos\theta+q^{2n})(1-q^{n-\frac{1}{2}})^2}, \tag{61}$$

$$f_3(\theta) = \frac{1}{2}\csc\frac{\theta}{2}\prod_{n=1}^{\infty}\frac{(1+2q^{n-\frac{1}{2}}\cos\theta+q^{2n-1})(1-q^n)^2}{(1-2q^n\cos\theta+q^{2n})(1+q^{n-\frac{1}{2}})^2}. \tag{62}$$

The Fourier expansions follow directly from (40) or (41), and using the symmetry property $F(x, y) = F(y, x)$. After simplification, we find that

$$f_1(\theta) = \frac{1}{2}\cot\frac{\theta}{2} - 2\sum_{n=1}^{\infty}\frac{q^n}{1+q^n}\sin n\theta, \tag{63}$$

$$f_2(\theta) = \frac{1}{2}\csc\frac{\theta}{2} + 2\sum_{n=1}^{\infty}\frac{q^{n-\frac{1}{2}}}{1-q^{n-\frac{1}{2}}}\sin\left(n-\frac{1}{2}\right)\theta, \tag{64}$$

$$f_3(\theta) = \frac{1}{2}\csc\frac{\theta}{2} - 2\sum_{n=1}^{\infty}\frac{q^{n-\frac{1}{2}}}{1+q^{n-\frac{1}{2}}}\sin\left(n-\frac{1}{2}\right)\theta. \tag{65}$$

The Fourier expansions for the squares of each function can be obtained from (44). For, taking $t = e^{i\theta}$ and successively setting $x = e^{i\pi}$, $e^{i\pi\tau}$ and $e^{i\pi+i\pi\tau}$ in (44) gives, respectively,

$$f_1^2(\theta) = \left[i \frac{d}{d\alpha} \rho_1(e^{i\alpha}) \right]_{\pi}^{\theta}, \tag{66}$$

$$f_2^2(\theta) = \left[i \frac{d}{d\alpha} \rho_1(e^{i\alpha}) \right]_{\pi\tau}^{\theta}, \tag{67}$$

$$f_3^2(\theta) = \left[i \frac{d}{d\alpha} \rho_1(e^{i\alpha}) \right]_{\pi+\pi\tau}^{\theta}, \tag{68}$$

and the Fourier expansions now follow by applying (49). For future reference, we note that (66)–(68) imply

$$\left[i \frac{d}{dz} \rho_1(e^{iz}) \right]_{\alpha}^{\beta} = [f_1^2(z)]_{\alpha}^{\beta} = [f_2^2(z)]_{\alpha}^{\beta} = [f_3^2(z)]_{\alpha}^{\beta}. \tag{69}$$

Comparison of the products (60)–(62) and series (63)–(65) with those in [42, pp. 508, 511–512], reveals that the functions f_1, f_2 and f_3 are, up to rescaling, the Jacobian elliptic functions cs, ns and ds, respectively.

From the infinite products, it is easy to show that

$$f_1(\theta + 2\pi m + 2\pi\tau n) = (-1)^n f_1(\theta), \tag{70}$$

$$f_2(\theta + 2\pi m + 2\pi\tau n) = (-1)^m f_2(\theta), \tag{71}$$

$$f_3(\theta + 2\pi m + 2\pi\tau n) = (-1)^{m+n} f_3(\theta). \tag{72}$$

Here m and n are integers. Thus f_1 is doubly periodic with periods 2π and $4\pi\tau$, f_2 is doubly periodic with periods 4π and $2\pi\tau$, while f_3 is doubly periodic with periods 4π and $2\pi + 2\pi\tau$.

The zeros and poles may be determined from the infinite products (60)–(62). The zeros of f_1, f_2 and f_3 are at $\theta = (2m+1)\pi + 2n\pi\tau$, $\theta = 2m\pi + (2n+1)\pi\tau$ and $\theta = (2m+1)\pi + (2n+1)\pi\tau$, respectively, where m and n are any integers. The poles of f_1, f_2 and f_3 all occur when $\theta = 2m\pi + 2n\pi\tau$.

The fundamental multiplicative identity (43) may be used to compute the derivatives of each of f_1, f_2 and f_3, as well as their addition formulas. We begin with the derivatives. In (43), let $t = e^{i\theta}$ to get

$$F(a, e^{i\theta})F(b, e^{i\theta}) = \frac{1}{i} \frac{\partial}{\partial\theta} F(ab, e^{i\theta}) + F(ab, e^{i\theta})(\rho_1(a) + \rho_1(b)). \tag{73}$$

Now let $a = e^{i\pi}$ and $b = e^{i\pi\tau}$. Noting that $\rho_1(e^{i\pi}) = 0$ and $\rho_1(e^{i\pi\tau}) = \frac{1}{2}$, the result (73) eventually simplifies to

$$f_3'(\theta) = -f_1(\theta) f_2(\theta). \tag{74}$$

Similarly, letting $a = e^{i\pi\tau}$, $b = e^{i\pi + i\pi\tau}$ and $a = e^{i\pi+i\pi\tau}$, $b = e^{i\pi}$ in (73) leads, respectively, to

$$f_1'(\theta) = -f_2(\theta)f_3(\theta), \tag{75}$$

$$f_2'(\theta) = -f_3(\theta)f_1(\theta). \tag{76}$$

In order to deduce the addition formulas, write the fundamental multiplicative identity (43) in the form

$$F(e^{i\alpha}, e^{i\theta})F(e^{i\beta}, e^{i\theta}) = \frac{1}{i}F(e^{i(\alpha+\beta)}, e^{i\theta}) + F(e^{i(\alpha+\beta)}, e^{i\theta})(\rho_1(e^{i\alpha}) + \rho_1(e^{i\beta})).$$

Apply $\partial/\partial\alpha - \partial/\partial\beta$ to both sides. The result is

$$\begin{aligned}&\frac{\partial}{\partial\alpha}F(e^{i\alpha}, e^{i\theta})F(e^{i\beta}, e^{i\theta}) - F(e^{i\alpha}, e^{i\theta})\frac{\partial}{\partial\beta}F(e^{i\beta}, e^{i\theta})\\ &\quad = F(e^{i(\alpha+\beta)}, e^{i\theta})\left(\frac{d}{d\alpha}\rho_1(e^{i\alpha}) - \frac{d}{d\beta}\rho_1(e^{i\beta})\right).\end{aligned}$$

Rearranging this and using (69) gives

$$F(e^{i(\alpha+\beta)}, e^{i\theta}) = \frac{i\left[\frac{\partial}{\partial\alpha}F(e^{i\alpha}, e^{i\theta})F(e^{i\beta}, e^{i\theta}) - F(e^{i\alpha}, e^{i\theta})\frac{\partial}{\partial\beta}F(e^{i\beta}, e^{i\theta})\right]}{f_j^2(\alpha) - f_j^2(\beta)}, \tag{77}$$

which holds for $j = 1, 2$ or 3. Let $\theta = \pi$, take $j = 1$ and apply (75) to get the addition formula for f_1:

$$f_1(\alpha+\beta) = \frac{f_1(\alpha)f_2(\beta)f_3(\beta) - f_1(\beta)f_2(\alpha)f_3(\alpha)}{f_1^2(\beta) - f_1^2(\alpha)}.$$

Similarly, letting $\theta = \pi\tau$ and $\theta = \pi + \pi\tau$ in (77) leads to

$$f_2(\alpha+\beta) = \frac{f_2(\alpha)f_3(\beta)f_1(\beta) - f_2(\beta)f_3(\alpha)f_1(\alpha)}{f_2^2(\beta) - f_2^2(\alpha)},$$

$$f_3(\alpha+\beta) = \frac{f_3(\alpha)f_1(\beta)f_2(\beta) - f_3(\beta)f_1(\alpha)f_2(\alpha)}{f_3^2(\beta) - f_3^2(\alpha)}.$$

In summary, this efficient approach to the Jacobian elliptic functions, due to Venkatachaliengar, is quite different from any other theory that has been presented before.

9. Conclusions

Many of Ramanujan's results relate to, or involve, elliptic functions. The usual way of studying these is to use the theory of complex variables. Yet, it is generally agreed that Ramanujan did not use the theory of complex variables. An interesting discussion on how Ramanujan may have studied elliptic functions has been given by B. C. Berndt [3, p. 2]. The fact will always remain, that because Ramanujan's notebooks generally do not contain proofs, we will never know for sure what his methods were.

The book by A. Weil [41] shows how the theory of elliptic functions may be developed, without the theory of complex variables, by a process called Eisenstein summation. Weil's book is based on the pioneering works of Eisenstein and Kronecker. Venkatachaliengar's book shows that it is also possible to use q-series to develop the theory of elliptic functions in an efficient, logical and systematic way. Thus, there are now two comprehensive answers, elucidated by Weil [41] and Venkatachaliengar [39], to the question of how the theory of elliptic functions can be developed by series manipulations and without using the theory of complex variables as the major tool.

Venkatachaliengar's methods frequently make maximal use of symmetry. The proof of Ramanujan's ${}_1\psi_1$ summation formula (38) is one example of this. Moreover, the statement of the ${}_1\psi_1$ identity is the same as the symmetric form given by Ramanujan in his second notebook. Most proofs of this identity prove the non-symmetric version in terms of the ${}_1\psi_1$ basic hypergeometric series, from which the symmetric form can be deduced by a change of variable. The quintuple product identity in [39, Appendix B] (not discussed in this review, but see [19]) is similarly presented in its most symmetric form.

Many proofs of Ramanujan's ${}_1\psi_1$ summation formula rely on functional equations like (29). Since $\phi(z, \alpha, \beta)$ and $\phi(q^2z, \alpha, \beta)$ converge in different annuli, the usual approach is to make an additional assumption on the parameters to ensure that the two annuli of convergence overlap; the extra assumption is then removed at the end of the proof by analytic continuation. In Venkatachlaliengar's proof, the introduction of the function Ψ eliminates the need for any analytic continuation argument. The series manipulations in the proof hold for all values of the parameters for which the identity (38) is valid.

Unfortunately, the book was not carefully edited. There are many typographical errors. Sometimes topics are repeated unnecessarily. Some of the proofs are incomplete, but most can be (and have been) fixed.

Due to space limitations, several other topics in Venkatachaliengar's book have not been discussed in this review. These include the modular transformation $\tau \to -1/\tau$, Landen's transformation $\tau \to 2\tau$, Picard's theorem, modular equations, the quintuple product identity, and the addition theorem for elliptic integrals.

In summary, a high degree of originality is present throughout the book. The successive generalizations of Ramanujan's identity (1) given by (11) and then by (43) are wonderful achievements. The use of the Jordan-Kronecker function (39) along with the fundamental multiplicative identity (43) to study the Jacobian elliptic functions is also quite different from any other method that has been used to study these functions.

We end this review by mentioning three other published works on elliptic functions by Venkatachaliengar: the research article [37], the presidential address [38], and the survey [40].

Acknowledgements

The reviewer thanks Drs. B. C. Berndt, M. Schlosser, W. L. Sweatman and P. C. Toh for helpful comments and suggestions. The reviewer gratefully acknowledges the help of Dr. G. N. Srinivasa Prasanna in confirming the year of publication of [39].

References

[1] R. P. Agarwal, Resonance of Ramanujan's mathematics, Vol. III, New Age International Publishers Limited, New Delhi (1999).

[2] G. E. Andrews and B. C. Berndt, Ramanujan's Lost Notebook, Part II, Springer, New York (2009).

[3] B. C. Berndt, Ramanujan's Notebooks, Part III, Springer-Verlag, New York (1991).

[4] B. C. Berndt, Ramanujan's Notebooks, Part V, Springer-Verlag, New York (1998).

[5] B. C. Berndt, Number theory in the spirit of Ramanujan, American Mathematical Society, Providence, RI (2006).

[6] B. C. Berndt, S. Bhargava and F. G. Garvan, Ramanujan's theories of elliptic functions to alternative bases, *Trans. Amer. Math. Soc.*, **347** (1995) 4163–4244.

[7] B. C. Berndt, H. H. Chan and W.-C. Liaw, On Ramanujan's quartic theory of elliptic functions, *J. Number Theory*, **88** (2001) 129–156.

[8] B. C. Berndt and A. J. Yee, Ramanujan's contributions to Eisenstein series, especially in his lost notebook. In: Number theoretic methods (Iizuka, 2001), 31–53, *Dev. Math.*, **8** Kluwer, Dordrecht (2002).

[9] S. Bhargava, A look at Ramanujan's work in elliptic function theory and further developments, *Proc. Nat. Acad. Sci. India, Sect. A*, **68** (1998) 299–347. Reprinted in [1, pp. 168–216].

[10] J. M. Borwein and P. B. Borwein, A remarkable cubic mean iteration. In: Computational methods and function theory (Valparaíso, 1989), 27–31, Lecture Notes in Math., 1435, Springer, Berlin (1990).

[11] J. M. Borwein and P. B. Borwein, A cubic counterpart of Jacobi's identity and the AGM, *Trans. Amer. Math. Soc.*, **323** (1991) 691–701.

[12] J. M. Borwein, P. B. Borwein and F. G. Garvan, Hypergeometric analogues of the arithmetic-geometric mean iteration, *Constr. Approx.*, **9** (1993) 509–523.

[13] J. M. Borwein, P. B. Borwein and F. G. Garvan, Some cubic modular identities of Ramanujan, *Trans. Amer. Math. Soc.*, **343** (1994) 35–47.

[14] J. M. Burgers, Application of a model system to illustrate some points of the statistical theory of free turbulence, *Nederl. Akad. Wetensch., Proc.*, **43** (1940) 2–12.

[15] H. H. Chan, On Ramanujan's cubic transformation for ${}_2F_1\left(\frac{1}{3}, \frac{2}{3}; 1; z\right)$, *Math. Proc. Cambridge Phil. Soc.*, **124** (1998) 193–204.

[16] H. H. Chan, S. Cooper and P. C. Toh, Ramanujan's Eisenstein series and powers of Dedekind's eta-function, *J. London Math. Soc. (2)*, **75** (2007) 225–242.

[17] H. H. Chan and Y. L. Ong, On Eisenstein series and $\sum_{m,n=-\infty}^{\infty} q^{m^2+mn+2n^2}$, *Proc. Amer. Math. Soc.*, **127** (1999) 1735–1744.

[18] S. Cooper, The Development of Elliptic Functions According to Ramanujan and Venkatachaliengar, *Research Letters in the Information and Mathematical Sciences*, **1** (2000) 65–78. http://iims.massey.ac.nz/research/letters/volume1number1/index.html

[19] S. Cooper, The quintuple product identity, *Int. J. Number Theory*, **2** (2006) 115–161.

[20] S. Cooper, Inversion formulas for elliptic functions, *Proc. London Math. Soc.*, **99** (2009) 461–483.

[21] S. Cooper and H. Y. Lam, Sums of two, four, six and eight squares and triangular numbers: an elementary approach, *Indian J. Math.*, **44** (2002) 21–40.

[22] G. Darboux, Mémoire sur la théorie des coordonnées curvilignes, et des systèmes orthogonaux, *Annales scientifiques de l'École Normale Supérieure, Sér. 2*, **7** (1878) 101–150.
[23] G.-H. Halphen, Traité des fonctions elliptiques et de leurs applications, Part 1, Gauthier-Villars, Paris (1886).
[24] C. G. J. Jacobi, Note sur les fonctions elliptiques, *Journal für die reine und angewandte Mathematik*, **3** (1828) 192–195. Reprinted in [26, pp. 251–254].
[25] C. G. J. Jacobi, Sur la rotation d'un corps. Extrait d'une lettre adressée à l'académie des sciences de Paris, *Journal für die reine und angewandte Mathematik*, **39** (1850) 293–350. Reprinted in [27, pp. 289–352].
[26] C. G. J. Jacobi, Gesammelte Werke, vol. 1, Chelsea, New York (1969).
[27] C. G. J. Jacobi, Gesammelte Werke, vol. 2, Chelsea, New York (1969).
[28] W. P. Johnson, How Cauchy missed Ramanujan's ${}_1\psi_1$ summation, *Amer. Math. Monthly*, **111** (2004) 791–800.
[29] C. Jordan, Cours d'Analyse, vol. 2, 3rd edition, Gauthier-Villars, Paris, 1913. Reprinted by Jacques Gabay, Sceaux (1991).
[30] L. Kronecker, Zur Theorie der elliptischen Functionen, Monatsberichte der Königlich Preussischen Akademie der Wissenschaften zu Berlin vom Jahre 1881, 1165–1172. Reprinted in [31, 311–318].
[31] L. Kronecker, Leopold Kronecker's Werke, vol. 4, Teubner, Leipzig, 1929. Reprinted by Chelsea, New York (1968).
[32] S. Ramanujan, On certain arithmetical functions, *Trans. Cambridge Philos. Soc.*, **22** (1916) 159–184. Reprinted in [33, pp. 136–162].
[33] S. Ramanujan, Collected Papers, Cambridge, University Press, 1927. Reprinted by AMS Chelsea, Providence, RI (2000).
[34] S. Ramanujan, Notebooks (2 volumes), Tata Institute of Fundamental Research, Bombay (1957).
[35] S. Ramanujan, The lost notebook and other unpublished papers, Narosa, New Delhi (1988).
[36] B. van der Pol, On a non-linear partial differential equation satisfied by the logarithm of the Jacobian theta-functions, with arithmetical applications. I, II, *Nederl. Akad. Wetensch. Proc. Ser. A*, **54** (Same as: Indagationes Math., **13** (1951)) 261–271, 272–284.
[37] K. Venkatachaliengar, Elliptic modular functions and Picard's theorem, Hyperbolic complex analysis (Proc. All India Sem., Ramanujan Inst., Univ. Madras, Madras, 1977) pp. 21–41, Publ. Ramanujan Inst., 4, Univ. Madras, Madras (1979).
[38] K. Venkatachaliengar, Ramanujan manuscripts, II, *Math. Student*, **52** (1984) 215–233.
[39] K. Venkatachaliengar, Development of elliptic functions according to Ramanujan, Madurai Kamaraj University, Madurai (1988).
[40] K. Venkatachaliengar, Ramanujan's contributions to the theory of elliptic functions. In: Wazir Hasan Abdi (ed.), Toils and triumphs of Srinivasa Ramanujan the man and the mathematician, pp. 182–206. National Publishing House, Chaura Rasta, Jaipur (1992).
[41] A. Weil, Elliptic functions according to Eisenstein and Kronecker, Springer-Verlag, Berlin-New York (1976).
[42] E. T. Whittaker and G. N. Watson, A Course of Modern Analysis, Cambridge University Press, 4th edition (1927).
[43] W. Zudilin, The hypergeometric equation and Ramanujan functions, *Ramanujan J.*, **7** (2003) 435–447.

List of Publications of K. Venkatachaliengar

[1] *Weierstrass' non-differentiable function*, J. Indian Math. Soc. **19** (1931), 54–58.

[2] *On the validity of the Raman-Banerjee analysis of the pianoforte hammer problem*, Proc. Indian Acad. **1** (1934), 60–66.

[3] *On series whose terms as well as the sumfunction are continuous in an inteval, and which converges non-uniformly in every sub-interval*, J. Indian Math. Soc. **20** (1934), 173–175.

[4] *Generalisation of Jacobi's θ-function formulae*, J. Indian Math. Soc. **1** (1934), 23–28.

[5] *On the reducibility of the general elliptic integral into logarithms*, J. Indian Math. Soc. **1** (1934), 59–65.

[6] *A note on Julia's Lemma*, Math. Student **2** (1934), 149–150.

[7] *The method of finding the class-number and the structure of the class group of any algebraic field*, Proc. Indian Acad. Sci., Sect. A, **1** (1935), 446–450.

[8] *A simple proof of the Jordan-curve theorem*, Quart. J. Math. (Oxford) **8** (1937), 241–244.

[9] *On a problem of the tetrahedron*, Proc. Indian Acad. Sci., Sect. A **7** (1938), 257–260.

[10] (with K. S. K. Iyengar) *On a problem relating to a tetrahedron*, Proc. Indian Acad. Sci. A **7** (1938), 305–311.

[11] (with B. S. Madhava Rao) *On an inequality concerning lattice sums*, Proc. Indian Acad. Sci. A **8** (1938), 524–528.

[12] *A note on Poncelet's problem*, J. Mysore Univ., Sect. B **3** (1942), 131–134.

[13] *A deepening of the binomial inequality*, J. Mysore Univ., Sect. B **3** (1942), 135–138.

[14] *Pairs of symmetric and skew matrices in an arbitrary field. I*, Proc. Indian Acad. Sci., Sect. A. **22** (1945), 243–264.

[15] (with B. S. Madhava Rao and V. R. Thiruvenkatachar) *Algebra related to elementary particles of spin $\frac{3}{2}$*, Proc. Roy. Soc. London, Ser. A. **187** (1946), 385–397.

[16] (with C. N. Srinivasiengar, H. Gupta, S. C. Chakrabarty, and V. R. Tiruvenkatachar) *On Kemmer's identity in combinatory functions*, Math. Student **15** (1947), 93–100.

[17] (with K. N. Srinivasa Rao) *The irreducible representations of the Lie-algebra of the orthogonal group with spin* 3/2, Proc. Nat. Inst. Sci. India **20** (1954), 509–523.

[18] (with S. K. Lakshmana Rao) *On Turán's inequality for ultraspherical polynomials*, Proc. Amer. Math. Soc. **8** (1957), 1075–1087.

[19] *An elementary proof of Morley's theorem*, Amer. Math. Monthly **65** (1958), 612–613.

[20] *Elementary proofs of the infinite product for Sin Z and allied formulae*, Amer. Math. Monthly **69** (1962), 541–545.

[21] *The congruent reduction of a pencil of two skew symmetric matrices over a commutative field of characteristic 2, and the criteria for congruence of two such given pencils*, Math. Student **33** (1965), 33–35.

[22] (with T. A. Soundararajan) *A covering theorem for skew fields*, Nederl. Akad. Wetensch. Proc. Ser. A **72** = Indag. Math. **31** (1969), 441–442.

[23] (with T. A. Soundararajan) *Topologies for Galois groups of extension fields*, J. Reine Angew. Math. **242** (1970), 1–7.

[24] (with T. A. Soundararajan) *On the Krull Galois theory for non-algebraic extension fields*, Bull. Austral. Math. Soc. **4** (1971), 367–387.

[25] (with V. Ramamani) *On a partition theorem of Sylvester*, Michigan Math. J. **19** (1972), 137–140.

[26] *Elliptic modular functions and Picard's theorem*, in Hyperbolic Complex Analysis (Proc. All India Sem., Ramanujan Inst., Univ. Madras, Madras, 1977), pp. 21–41; Publ. Ramanujan Inst., 4, Univ. Madras, Madras, 1979.

[27] *Ramanujan manuscripts. II*, Math. Student **52** (1984), 215–233.

[28] *Development of Elliptic Functions According to Ramanujan*, Technical Report, No. 2, Department of Mathematics, Madurai Kamaraj University, Madurai, 1988.

[29] *Ramanujan's contributions to the theory of elliptic functions*, in *Toils and Triumphs of Srinivasa Ramanujan the Man and the Mathematician*, Wazir Hasan Abdi, ed., National Publishing House Chaura Rasta, Jaipur, 1992, pp. 182–206.

[30] (with T. Soundararajan) *A variety of applications of a theorem of B. H. Neumann on groups*, Acta Math. Vietnam **19** (1994), 55–60.

[31] (with B. C. Berndt) *On the transformation formula for the Dedekind eta-function*, in *Symbolic Computation, Number Theory, Special Functions, Physics and Combinatorics*, F. G. Garvan and M. E. H. Ismail, eds., Kluwer, Dordrecht, 2001, pp. 73–77.

[32] *Some features of Brahmagupta's work in geometry*, bibliographic data unknown.

[33] *The development of mathematics in ancient India: role of Brahmagupta*, bibliographic data unknown.

Contents

Ramanujan Rediscovered No. 14, 2010, pp. 1–8.

On a Generalized Basic Series and New Combinatorial Interpretations of Göllnitz–Gordon Identities

S. Anand and A. K. Agarwal*

Centre for Advanced Study in Mathematics, Panjab University, Chandigarh 160 014, India
e-mail: shivani.bedi.maths@gmail.com; aka@pu.ac.in

Abstract. We interpret a generalized basic series as a generating function for certain restricted Bender–Knuth matrices. This in combination with recent work of Agarwal and Rana leads to an infinite class of 5-way combinatorial identities. In some cases we get even 6-way combinatorial identities. We discuss three such particular cases and obtain five new combinatorial interpretations for each of the two Göllnitz–Gordon partition identities and an identity due to Göllnitz.

Keywords. Basic series, Göllnitz–Gordon identities, n-color partitions, lattice paths, Frobenius partitions, Bender–Knuth matrices.

2000 Subject Classification: Primary: 05A15, 05A17, 05A19; Secondary: 11P81

1. Introduction

The following two analytical identities from Slater's Compendium [13, (36), p. 155; (34), p. 155]

$$\prod_{n=1}^{\infty}\frac{(1-q^{2n})}{(1+q^{2n-1})}\sum_{n=0}^{\infty}\frac{(-q;q^2)_n q^{n^2}}{(q^2;q^2)_n}=\prod_{n=1}^{\infty}(1-q^{8n-3})(1-q^{8n-5})(1-q^{8n}), \quad (1.1)$$

and

$$\prod_{n=1}^{\infty}\frac{(1-q^{2n})}{(1+q^{2n-1})}\sum_{n=0}^{\infty}\frac{(-q;q^2)_n q^{n(n+2)}}{(q^2;q^2)_n}=\prod_{n=1}^{\infty}(1-q^{8n-1})(1-q^{8n-7})(1-q^{8n}), \quad (1.2)$$

were interpreted combinatorially by Göllnitz [10] and Gordon [12] independently as follows:

Theorem 1.1. *The number of partitions of n into parts differing by atleast 2 among which no two consecutive even numbers appear is equal to the number of partitions of n into parts congruent to* 1, 4 *or* 7 (mod 8),

---

*Supported by CSIR Research Grant No. 25(0158)/07/EMR-II.

and

Theorem 1.2. *The number of partitions of n into parts differing by atleast 2 among which no two consecutive even numbers appear and with each part being at least equal to 3 is equal to the number of partitions of n into parts which are congruent to 3, 4 or 5 (mod 8).*

Another analytical identity [7, Corollary 2.7, p. 21, with q replaced by q^2 and a by $-q$]

$$\sum_{n=0}^{\infty} \frac{(-q;q^2)_n q^{n(n+1)}}{(q^2;q^2)_n} = (-q^2;q^2)_\infty(-q^3;q^4)_\infty, \tag{1.3}$$

was interpreted combinatorially by Göllnitz [11] in the form

Theorem 1.3. *The number of partitions of n into parts differing by atleast 2 among which no two consecutive odd numbers appear and with each part being atleast equal to 2 is equal to the number of partitions of n into parts which are congruent to 2, 3 or 7 (mod 8).*

In an attempt to unify the partition functions with certain difference conditions appearing in Theorems 1.1–1.3, Agarwal [1] defined a more generalized partition function $A_k(\nu)$ as the number of partitions of ν in which each part $\geq k$, minimal difference ≥ 2 between the parts, consecutive odd integers are not allowed if k is even and consecutive even integers are not allowed if k is odd. Obviously, $A_1(\nu)$ and $A_3(\nu)$ are Göllnitz–Gordon partition functions of Theorems 1.1 and 1.2, respectively and $A_2(\nu)$ is the Göllnitz function appearing in Theorem 1.3. Agarwal [1] also proved the following:

$$\sum_{\nu=0}^{\infty} A_k(\nu)q^\nu = \sum_{n=0}^{\infty} \frac{(-q;q^2)_n q^{n(n+k-1)}}{(q^2;q^2)_n}, \quad |q|<1. \tag{1.4}$$

Recently, Agarwal and Rana [5] interpreted the right-hand side of (1.4) combinatorially, in two different ways (i) as generating function for certain n-colour partition function and (ii) as generating function for a weighted lattice paths function. This resulted in an infinite class of 3-way combinatorial identities. Later in [6] they extended their result to a 4-way infinite class of combinatorial identities by using Frobenius partitions. In this paper, we further extend the result of [6] to a 5-way infinite class of combinatorial identities using Bender–Knuth matrices. In some particular cases we get even 6-way combinatorial identities. Our each 6-way identity yields 15 combinatorial identities in the usual sense. We shall discuss three such cases and obtain five new combinatorial versions of each of Theorems 1.1–1.3. Before we state our main result we recall the following definitions which will be used in the sequel.

Definition 1 (Agarwal and Andrews [3]). *An n-colour partition of a positive integer ν is a partition in which a part of size n, $n \geq 0$ can come in n different colours denoted by subscripts: $n_1, n_2, \ldots, n_n$ and the parts satisfy the order $1_1 < 2_1 < 2_2 < 3_1 < 3_2 < 3_3 < \cdots$.*

Thus, for example, n-colour partitions of 3 are: $3_1, 3_2, 3_3, 2_1+1_1, 2_2+1_1, 1_1+1_1+1_1$.

Definition 2 (Agarwal and Andrews [3]). *The weighted difference of two parts* $m_i, n_j, m \geq n$ *is defined by* $m - n - i - j$ *and is denoted by* $((m_i - n_j))$.

Definition 3. *We reproduce the following definition of lattice paths from* [4]*:*

All paths will be of finite length lying in the first quadrant. They will begin on the y-axis and terminate on the x-axis. Only three moves are allowed at each step:

northeast: *from* (i, j) *to* $(i + 1, j + 1)$,

southeast: *from* (i, j) *to* $(i + 1, j - 1)$, *only allowed if* $j > 0$,

horizontal: *from* $(i, 0)$ *to* $(i + 1, 0)$, *only allowed along x-axis.*

The following terminology will be used in describing the lattice paths:

Peak. Either a vertex on the y-axis which is followed by a southeast step or a vertex preceded by a northeast step and followed by a southeast step.

Valley. A vertex preceded by a southeast step and followed by a northeast step. Note that a southeast step followed by a horizontal step followed by a northeast step does not constitute a valley.

Plain. A section of the path consisting of only horizontal steps which starts either on the y-axis or at a vertex preceded by a southeast step and ends at a vertex followed by a northeast step.

The **Height** of a vertex is its y-coordinate. The **Weight** of a vertex is its x-coordinate. The **Weight** of a path is the sum of the weights of its peaks.

Definition 4 ([8]). *A two-rowed array of non-negative integers*

$$\begin{pmatrix} a_1 & a_2 & \cdots & a_r \\ b_1 & b_2 & \cdots & b_r \end{pmatrix}, \qquad a_1 > a_2 > \cdots > a_r \geq 0,$$

$b_1 > b_2 > \cdots > b_r \geq 0$ *is known as a Frobenius partition of n if*

$$n = r + \sum_{i=1}^{r} a_i + \sum_{i=1}^{r} b_i.$$

For example, if $n = 28 = 4 + (6+5+2+0) + (5+3+2+1)$, *then the corresponding Frobenius notation is* $\left(\begin{smallmatrix} 6 & 5 & 2 & 0 \\ 5 & 3 & 2 & 1 \end{smallmatrix}\right)$.

Definition 5. *A plane partition* π *of a positive integer* ν *is an array of non-negative integers*

$$\begin{array}{|llll} \hline n_{1,1} & n_{1,2} & n_{1,3} & \cdots \\ n_{2,1} & n_{2,2} & n_{2,3} & \cdots \\ \vdots & \vdots & \vdots & \end{array}$$

for which $\sum_{i,j} n_{i,j} = \nu$ *and rows and columns are arranged in non-increasing order. The non-zero entries* $n_{i,j}$ *are called the parts of* π.

E. A. Bender and D. E. Knuth [9] proved the following:

Theorem (Bender and Knuth). *There is a one-to-one correspondence between plane partitions of* ν*, on the one hand, and infinite matrices* $a_{i,j}(i, j \geq 1)$ *of non-negative integer entries which satisfy*

$$\sum_{r\geq 1} r \left\{ \sum_{i+j=r+1} a_{i,j} \right\} = \nu,$$

on the other.

Note. For the definition and other details of the one-to-one correspondence of this theorem which is denoted by ϕ the reader is referred to [9].

Corresponding to every non-negative integer ν we shall call the matrices of the above theorem BK_ν-matrices (BK for Bender and Knuth). These are infinite matrices but will be represented in the sequel by the largest possible square matrices whose last row (column) is non-zero. Thus, for example, we will represent six BK_3 matrices by

$$(3), \begin{pmatrix} 1 & 0 \\ 1 & 0 \end{pmatrix}, \begin{pmatrix} 1 & 1 \\ 0 & 0 \end{pmatrix}, \begin{pmatrix} 0 & 0 \\ 0 & 1 \end{pmatrix}, \begin{pmatrix} 0 & 0 & 1 \\ 0 & 0 & 0 \\ 0 & 0 & 0 \end{pmatrix}, \begin{pmatrix} 0 & 0 & 0 \\ 0 & 0 & 0 \\ 1 & 0 & 0 \end{pmatrix}.$$

We give here four more definitions:

Definition 6. *We define a matrix* $E_{i,j}$ *as an infinite matrix whose* $(i, j)^{th}$ *entry is* 1 *and the other entries are all zeros. We call* $E_{i,j}$ *distinct units of a* BK_ν *matrix.*

Definition 7. *In the set of all units the order is defined as follows:*

If $r + s < p + q$ *then* $E_{r,s} < E_{p,q}$ *and if* $r + s = p + q$*, then* $E_{r,s} < E_{p,q}$ *where* $r < p$*. Thus, the units satisfy the order:*

$$E_{1,1} < E_{1,2} < E_{2,1} < E_{1,3} < E_{2,2} < E_{3,1} < E_{1,4} < E_{2,3} < E_{3,2} < \cdots$$

Definition 8. *The order difference of two units* $E_{p,q}$*,* $E_{r,s}(p + q \geq r + s)$ *is defined by* $q - s - 2r$ *and is denoted by* $\{\{E_{p,q} - E_{r,s}\}\}$.

Agarwal and Rana [5,6] proved the following result:

Theorem 1.4 (Agarwal and Rana). *For a positive integer* k*, let* $B_k(\nu)$ *denote the number of n-colour partitions of* ν *such that*

(1.4 a) *the parts are greater than or equal to* k*,*

(1.4 b) *the parts used are of the type* $(2l - 1)_1$ *or* $(2l)_2$ *if* k *is odd, and* $(2l - 1)_2$ *or* $(2l)_1$ *if* k *is even, and*

(1.4 c) *the weighted difference between any two parts is non-negative and even.*

Let $C_k(\nu)$ *denote the number of lattice paths of weight* ν *which start at* $(0, 0)$ *such that*

(1.4 d) *there is no valley above height* 0*,*

(1.4 e) *there is a plain of length* $(k-1)+2s$, $s \geq 0$, *in the beginning of the path, other plains, if any, are of even lengths, and*

(1.4 f) *the height of each peak of odd (resp. even) weight is* 1 *(resp.* 2*) if* k *is odd and* 2 *(resp.* 1*) if* k *is even.*

Let $D_k(\nu)$ *denote the number of Frobenius partitions* $\binom{a_1\ a_2\ \cdots\ a_r}{b_1\ b_2\ \cdots\ b_r}$ *of* ν *such that*

(1.4 g) $a_i = \begin{cases} b_i & \text{if } a_i \text{ and } b_i \text{ are of the same parity and } k \text{ odd} \\ b_i+2 & \text{if } a_i \text{ and } b_i \text{ are of the same parity and } k \text{ even} \\ b_i+1 & \text{if } a_i \text{ and } b_i \text{ are of opposite parity} \end{cases}$

(1.4 h) $b_i > a_{i+1}$, $1 \leq i \leq r-1$, *and*

(1.4 i) $a_i \geq k - b_i - 1$.

Then

$$A_k(\nu) = B_k(\nu) = C_k(\nu) = D_k(\nu), \quad \textit{for all } \nu. \tag{1.5}$$

In this paper we shall prove the following result:

Theorem 1.5. *For* $k, \nu \geq 1$, *let* $E_k(\nu)$ *denote the number of* BK_ν*-matrices* Δ *such that*

(1.5 a) *if* k *is odd (resp. even), then even (resp. odd) columns are zero,*

(1.5 b) *all rows after the second row are zero,*

(1.5 c) *the order difference of any two units of* Δ *is non-negative and even,*

(1.5 d) *for odd* $k > 1$, *the first* $\frac{(k-1)}{2}$ *odd columns are zero, and*

(1.5 e) *for even* $k > 2$, *the first* $\frac{(k-2)}{2}$ *even columns are zero.*

Then, $B_k(\nu) = E_k(\nu)$, for all k and ν.

Note. In view of (1.5 c) the entries in Δ cannot exceed 1.

Remark. Theorem 1.5 extends the identity (1.5) to a five-way identity

$$A_k(\nu) = B_k(\nu) = C_k(\nu) = D_k(\nu) = E_k(\nu). \tag{1.6}$$

Using Agarwal's bijection [2] between BK_ν-matrices and n-colour partitions of ν, we shall prove Theorem 1.5 in our next section and discuss three particular cases in our last section which provide us five new combinatorial versions of each of the Theorems 1.1–1.3.

Here we recall Agarwal's bijection from [2] for clarity.

$$\text{Let} \quad \Delta = a_{1,1}E_{1,1} + a_{1,2}E_{1,2} + \cdots + a_{2,1}E_{2,1} + a_{2,2}E_{2,2} + \cdots \tag{1.7}$$

be a BK_ν-matrix, where a_{ij} are non-negative integers which denote the multiplicities of $E_{i,j}$.

We map each unit $E_{p,q}$ of Δ to a single part m_i of an n-colour partition of ν. The mapping denoted by f is defined as

$$f : E_{p,q} \to (p+q-1)_p, \tag{1.8}$$

and the inverse mapping f^{-1} is easily seen to be

$$f^{-1} : m_i \to E_{i,m-i+1}. \tag{1.9}$$

For $\nu = 3$, this bijection is illustrated in the following table:

K_3-matrices Δ	$f(\Delta)$
$(3) = 3E_{1,1}$	$3 1_1 = 1_1 + 1_1 + 1_1$
$\begin{pmatrix} 0 & 1 \\ 0 & 1 \end{pmatrix} = E_{1,1} + E_{1,2}$	$1_1 + 2_1$
$\begin{pmatrix} 1 & 1 \\ 0 & 0 \end{pmatrix} = E_{1,1} + E_{2,1}$	$1_1 + 2_2$
$\begin{pmatrix} 0 & 0 & 0 \\ 0 & 0 & 0 \\ 0 & 0 & 1 \end{pmatrix} = E_{1,3}$	3_1
$\begin{pmatrix} 0 & 0 \\ 0 & 1 \end{pmatrix} = E_{2,2}$	3_2
$\begin{pmatrix} 0 & 0 & 1 \\ 0 & 0 & 0 \\ 0 & 0 & 0 \end{pmatrix} = E_{3,1}$	3_3

2. Proof of theorem 1.5

We shall prove that if Δ is a matrix enumerated by $E_k(\nu)$ then the n-colour partition $f(\Delta)$ is enumerated by $B_k(\nu)$, and conversely, if π is an n-colour partition enumerated by $B_k(\nu)$, then the BK_ν matrix $f^{-1}(\pi)$ is enumerated by $E_k(\nu)$.

Let the matrix Δ enumerated by $E_k(\nu)$ have the representation (1.7). Clearly, in view of the note given after Theorem 1.5, each $a_{i,j} = 1$ or 0. Let $E_{p,q}$, $E_{r,s}$ $(p + q \geq r + s)$ be two units of Δ which correspond to two n-colour parts m_i, n_j of $f(\Delta)$. Then $m_i = (p + q - 1)_p$ and $n_j = (r + s - 1)_r$ by (1.8). Since $p + q \geq r + s$, therefore $m \geq n$ and

$$((m_i - n_j)) = (p + q - 1) - p - (r + s - 1) - r = q - s - 2r = \{\{E_{p,q} - E_{r,s}\}\}$$

which is non-negative and even.

This shows that (1.5 c) implies (1.4 c).

Next, we see that if k is odd, then by (1.5 a), even columns in Δ are zero which means that in $E_{p,q}$, q is odd. Further, since $p \leq 2$ by (1.5 b), we conclude that

$$f(E_{p,q}) = \begin{cases} q_1 & \text{if } p = 1 \\ (q+1)_2 & \text{if } p = 2 \end{cases}$$

This shows that in $f(\triangle)$, the parts are of the form $(2l-1)_1$ or $(2l)_2$. Similarly, we can show that if k is even, then in $f(\triangle)$, the parts are of the form $(2l-1)_2$ or $(2l)_1$. Thus (1.5 a) and (1.5 b) imply (1.4 b). Finally, when k is odd say, $(2l-1)$, the first $l-1$ odd columns, that is, 1st, 3rd, ..., $(2l-3)$th are zero by (1.5 d) and since $E_{1,2l-3} = (2l-3)_1$ and $E_{2,2l-3} = (2l-2)_2$, we see that in $f(\triangle)$ the parts are $\geq k$. Thus, (1.5 d) implies (1.4 a) when k is odd. Similarly, we can show that (1.5 e) implies (1.4. a) when k is even. Thus $f(\triangle)$ is enumerated by $B_k(\nu)$.

To see the reverse implication, let π be an n-colour partition of ν enumerated by $B_k(\nu)$. We shall prove that the BK_ν-matrix $f^{-1}(\pi)$ is enumerated by $E_k(\nu)$. Let $m_i, n_j, m \geq n$ be two parts of π such that $f^{-1}(m_i) = E_{p,q}$ and $f^{-1}(n_j) = E_{r,s}$. Then $E_{p,q} = E_{i,m-i+1}$ and $E_{r,s} = E_{j,n-j+1}$ by (1.9). Since $(m \geq n)$, we have $p+q = m+1 \geq n+1 = r+s$, and

$$\begin{aligned}\{\{E_{p,q} - E_{r,s}\}\} &= \{\{E_{i,m-i+1} - E_{j,n-j+1}\}\}\\ &= (m-i+1) - (n-j+1) - 2j\\ &= m-n-i-j\\ &= ((m_i - n_j)), \quad \text{which is even and non-negative.}\end{aligned}$$

Thus (1.4 c) implies (1.5 c).

To prove (1.5 a), (1.5 b), (1.5 d), and (1.5 e), we first consider the case when k is odd.

Since $f^{-1}((2l-1)_1) = E_{1,2l-1}$ and $f^{-1}((2l)_2) = E_{2,2l-1}$, we see that in $f^{-1}(\pi)$ even columns are zero and all rows after the second row are zero. This proves (1.5 a) and (1.5 b). Furthermore, by (1.4 a) we see that in $f^{-1}((2l-1)_1) = E_{1,2l-1}$, $(2l-1) \geq k$ and in $f^{-1}((2l)_2) = E_{2,2l-1}$, $(21) \geq k$, that is (1.5 d) and (1.5 e) are satisfied. Similarly, we can prove the case when k is even. This completes the proof of Theorem 1.5.

3. Particular cases

When $k = 1$, Theorem 1.5 in view of (1.1) yields the following 6-way identity

$$A_1(\nu) = B_1(\nu) = C_1(\nu) = D_1(\nu) = E_1(\nu) = F_1(\nu), \quad \text{for all } \nu \tag{3.1}$$

where $F_1(\nu)$ is the number of partitions of ν into parts $\equiv 1, 4$ or $7 \pmod 8$. Identity (3.1) induces 15 identities in the usual sense of which the identity $A_1(\nu) = F_1(\nu)$ is Göllnitz–Gordon first identity, that is, Theorem 1.1, 9 are contained in [5,6] and the remaining five involving $E_1(\nu)$ are new. Similarly, the particular cases $k = 3$ and $k = 2$ of (1.6) yield 5 new combinatorial versions of Theorems 1.3 and 1.2 respectively.

References

[1] A. K. Agarwal, On a generalized partition theorem, *J. Indian Math. Soc.*, **50** (1986) 185–190.

[2] A. K. Agarwal, n-color analogues of Gaussian polynomials, *ARS Combinatoria*, **61** (2001) 97–117.

[3] A. K. Agarwal and G. E. Andrews, Rogers-Ramanujan identities for partitions with "N copies of N", *J. Combin. Theory Ser. A*, **45(1)** (1987) 40–49.

[4] A. K. Agarwal and D. M. Bressoud, Lattice paths and multiple basic hypergeometric series, *Pacific J. Math.*, **136(2)** (1989) 209–228.

[5] A. K. Agarwal and M. Rana, New combinatorial versions of Göllnitz–Gordon identities, *Utilitas Mathematica*, **79** (2009) 145–155.

[6] M. Rana and A. K. Agarwal, On an extension of a combinatorial identity, *Proc. Indian Acad. Sci. (Math. Sci.)*, **119(1)** (2009) 1–7.

[7] G. E. Andrews, The Theory of Partitions, Encyclopedia of Mathematics and its Applications, Addison-Wesley (1976).

[8] G. E. Andrews, Generalized Frobenius Partitions, *Mem. Amer. Math. Soc.*, **49** (1984) 301, IV+ 44 pp.

[9] E. A. Bender and D. E. Knuth, Enumeration of Plane Partitions, *J. Combin. Theory (A)*, **13** (1972) 40–54.

[10] H. Göllnitz, Einfache Partitionen (unpublished), Diplomarbeit, W. S. (1960) (Göttingen) 65 pp.

[11] H. Göllnitz, Partitionen mit Differenzenbedingungen, *J. Reine Angew. Math.*, **225** (1967) 154–190.

[12] B. Gordon, Some continued fractions of the Rogers-Ramanujan type, *Duke J. Math.*, **32** (1965) 741–748.

[13] L. J. Slater, Further identities of the Rogers-Ramanujan type, *Proc. London Math. Soc.*, **2(54)** (1952) 147–167.

Ramanujan Rediscovered No. 14, 2010, pp. 9–30.

Series for $1/\pi$ Arising from Certain Representations for Eisenstein Series in Ramanujan's Second Notebook

Nayandeep Deka Baruah and Narayan Nayak

Department of Mathematical Sciences, Tezpur University, Sonitpur 784 028, India
e-mail: nayan@tezu.ernet.in; narayan.nayak05@gmail.com

Abstract. Recently, N. D. Baruah and B. C. Berndt found many old and new Ramanujan-type series for $1/\pi$ by using certain representations for Eisenstein series recorded mostly by Ramanujan. In this paper, we prove several known series for $1/\pi$ by using different kinds of representations for Eisenstein series recorded by Ramanujan in his second notebook. In certain cases, we have found proofs of some series for $1/\pi$ which are simpler/shorter than those given by previous authors.

Keywords. Eisenstein series, modular equations, theta functions, hypergeometric series.

2000 Subject Classification: 33E05, 33C05, 11F11, 11R29

Dedicated to the memory of Professor K. Venkatachaliengar

1. Introduction

Define Ramanujan's Eisenstein series

$$P(q) := 1 - 24\sum_{k=1}^{\infty} \frac{kq^k}{1-q^k}, \qquad |q| < 1. \tag{1.1}$$

In Section 10 of his paper [33], [34, pp. 33–34], Ramanujan recorded without proofs twelve representations for

$$f_n(q) := nP(q^{2n}) - P(q^2), \tag{1.2}$$

corresponding to twelve values of n, namely, $n = 2, 3, 4, 5, 7, 11, 15, 17, 19, 23, 31$, and 35. He also recorded the representations for $n = 2$ and 4 in Chapter 17 and for the remaining ten values and for $n = 9$ and $n = 25$ in Chapter 21 of his second notebook [35]. In Sections 13 and 14 of the same paper [33], [34, pp. 36–38], Ramanujan recorded 17 hypergeometric-like series representations for $1/\pi$. Proofs of the first three series representations were briefly sketched by him [34, p. 36]. These three series belong to the classical theory of elliptic functions, while the later fourteen series depend on Ramanujan's theories of elliptic functions to alternative bases. It was not until 1987 that proofs of all 17 formulas were found by J. M. and P. B. Borwein [15]. They also discovered many further series for $1/\pi$ [16], [17], [18],

[19], [20], where D. H. Bailey also coauthored [18]. At about the same time, D. V. and G. V. Chudnovsky [26] also independently proved several of Ramanujan's series representations for $1/\pi$ and established new ones as well. Further particular series representations for $1/\pi$ as well as some general formulas have subsequently been derived by B. C. Berndt and H. H. Chan [12], Berndt, Chan, and W.-C. Liaw [13], H. H. Chan, S. H. Chan, and Z.-G. Liu [21], H. H. Chan and Liaw [22], H. H. Chan and K. P. Loo [24], H. H. Chan, Liaw, and V. Tan [23], H. H. Chan and H. Verrill [25], and S. Cooper [27]. J. Guillera [28]–[32] discovered "experimentally" some beautiful series for $1/\pi$ as well as for $1/\pi^2$ and proved some of his series by using the WZ-method. Further work has been accomplished by W. Zudilin [38]–[42]. More recently, N. D. Baruah and Berndt [3], returned to Ramanujan's ideas expressed in Section 13 of his fundamental paper [33], [34, p. 36]. By employing the aforementioned representations for $f_n(q)$ with $q = e^{-\pi/\sqrt{n}}$, a transformation formula for $P(q^2)$ and certain hypergeometric identities, they established 13 of Ramanujan's 17 identities from [33] as well as derived many new series representations for $1/\pi$. In particular, they relied on Ramanujan's initial ideas more so than previous authors. In [2], they used analogous formulas in the alternative theories to derive several series for $1/\pi$ in cubic and quartic theories. Furthermore, Baruah and Berndt [4] and Baruah and N. Nayak [5], [6] found many new hypergeometric-like series for $1/\pi^2$. We refer to [7] for a recent survey on Ramanujan-type series for $1/\pi$.

In Chapter 21 of his second notebook, Ramanujan [35, Vol. II] also recorded without proofs three representations for

$$h_n(q) := nP(-q^n) - P(-q), \tag{1.3}$$

corresponding to $n = 3$, 5, and 7. In the same chapter, he also recorded the representations for $h_9(-q)$ and $h_{25}(-q)$. Proofs of all these results can be found in Berndt's book [9, pp. 473–480]. Trivially, we have $P(-q) = -P(q) + 6P(q^2) - 4P(q^4)$. But there is no trivial way to transform the representations for $f_n(q)$ into $h_n(q)$. One might speculate why Ramanujan was interested in such representations for $h_n(\pm q)$. One reason could be, as in the cases of $f_n(q)$, Ramanujan might have tried to find series for $1/\pi$ arising from these representations as well. The purpose of this paper is to prove several known series for $1/\pi$ by employing the above mentioned representations for $h_n(\pm q)$ along with a transformation formula for $P(-q)$ and certain hypergeometric series identities. In certain cases, we have found proofs of some series for $1/\pi$ which are simpler/shorter than those in [3]. In the process, we have also rediscovered an explicit well-known value of Eisenstein series (see (4.10)) and certain quotients of theta functions (see (7.12) and (8.10)) recorded by Ramanujan in his first notebook [35]. Further remarks on these can be found in appropriate places of Sections 4–8.

2. Definitions, preliminary results, and notation

We use the standard shifted or rising factorial notation

$$(a)_0 := 1, \quad (a)_k := a(a+1)(a+2)\dots(a+k-1), \quad k \geq 1.$$

The hypergeometric functions ${}_pF_{p-1}$, $p \geq 1$, are defined by

$$ {}_pF_{p-1}(a_1, \ldots, a_p; b_1, \ldots, b_{p-1}; x) := \sum_{k=o}^{\infty} \frac{(a_1)_k \cdots (a_p)_k}{(b_1)_k \cdots (b_{p-1})_k} \frac{x^n}{n!}, \qquad |x| < 1. $$

If

$$ q = \exp\left(-\pi \frac{{}_2F_1\left(\frac{1}{2}, \frac{1}{2}; 1; 1-x\right)}{{}_2F_1\left(\frac{1}{2}, \frac{1}{2}; 1; x\right)} \right), \tag{2.1} $$

then one of the fundamental results in the theory of elliptic functions [9, p. 101, Entry 6] is given by

$$ z(q) := \phi^2(q) = {}_2F_1\left(\frac{1}{2}, \frac{1}{2}; 1; x\right), \tag{2.2} $$

where Ramanujan's theta function $\phi(q)$ is defined by

$$ \phi(q) = \sum_{k=-\infty}^{\infty} q^{k^2}, \qquad |q| < 1. \tag{2.3} $$

It is also well-known, for example, from Entries 10(i) and 11(iii) of [9, pp. 122–123], that

$$ x(q) = 16q \frac{\psi^4(q^2)}{\phi^4(q)}, \tag{2.4} $$

where $\phi(q)$ is as defined in (2.3) and Ramanujan's special theta function $\psi(q)$ is defined by

$$ \psi(q) = \sum_{k=0}^{\infty} q^{k(k+1)/2}, \qquad |q| < 1. \tag{2.5} $$

In the sequel, we often write x and z for $x(q)$ and $z(q)$, respectively.
We also need Ramanujan's theta function

$$ f(-q) := \sum_{k=-\infty}^{\infty} (-1)^k q^{k(3k-1)/2} = \prod_{j=1}^{\infty} (1 - q^j), \tag{2.6} $$

where the later equality in (2.6) is Euler's pentagonal number theorem. Note that, if $q = \exp(2\pi i \tau)$, with $\Im \tau > 0$, then $f(-q) = q^{-1/24}\eta(\tau)$, where $\eta(\tau)$ is the classical Dedekind eta-function.

In his second notebook, Ramanujan [35], [9, p. 124, Entry 12(i)] recorded the identity

$$ f(q) = \sqrt{z} \left\{ \frac{x(1-x)}{16q} \right\}^{1/24}. \tag{2.7} $$

If we replace q by q^n in the above above, then the same representation also holds with x replaced by $\beta = x(q^n)$ and z replaced by $z(q^n) = {}_2F_1(\frac{1}{2}, \frac{1}{2}; 1; \beta)$. We will use such representations in Sections 7 and 8.

We end this introduction by defining a modular equation of degree n. Let n denote a fixed natural number, and assume that

$$n\frac{{}_2F_1\left(\frac{1}{2}, \frac{1}{2}; 1; 1-\alpha\right)}{{}_2F_1\left(\frac{1}{2}, \frac{1}{2}; 1; \alpha\right)} = \frac{{}_2F_1\left(\frac{1}{2}, \frac{1}{2}; 1; 1-\beta\right)}{{}_2F_1\left(\frac{1}{2}, \frac{1}{2}; 1; \beta\right)}. \tag{2.8}$$

Then a modular equation of degree n is a relation between α and β induced by (2.8). The corresponding multiplier $m(\alpha, \beta)$ is defined by

$$m(\alpha, \beta) = \frac{{}_2F_1\left(\frac{1}{2}, \frac{1}{2}; 1; \alpha\right)}{{}_2F_1\left(\frac{1}{2}, \frac{1}{2}; 1; \beta\right)}. \tag{2.9}$$

3. Main ideas

In brief, we combine two different relations between $P(-q)$ and $P(-q^n)$, for certain positive integers n, along with a Clausen's formula, to obtain the series representations for $1/\pi$. The development of the method for deriving series for $1/\pi$ described in this section is analogous to that in [3].

First we recall the following Clausen's formulas from Theorems 5.7(i)–(vi) and Formula (5.5.9) in [15, pp. 180–181]. Let

$$A_k := \frac{\left(\frac{1}{2}\right)_k^3}{k!^3}, \quad B_k := \frac{\left(\frac{1}{4}\right)_k\left(\frac{1}{2}\right)_k\left(\frac{3}{4}\right)_k}{k!^3}, \quad C_k := \frac{\left(\frac{1}{6}\right)_k\left(\frac{1}{2}\right)_k\left(\frac{5}{6}\right)_k}{k!^3}, \tag{3.1}$$

and

$$X := 4x(1-x), \quad Y := \frac{4x}{(1-x)^2}, \quad U := \frac{x^2}{4(1-x)}, \quad V := \frac{4\sqrt{x}\,(1-x)}{(1+x)^2},$$

$$W := \frac{2\sqrt{X}}{1-X}, \quad L := \frac{27X^2}{(4-X)^3}, \quad \text{and} \quad M := \frac{27X}{(1-4X)^3}.$$

Then

$$z^2 = {}_3F_2\left(\frac{1}{2}, \frac{1}{2}, \frac{1}{2}; 1, 1; X\right) = \sum_{k=0}^{\infty} A_k X^k, \quad 0 \le x \le \frac{1}{2}, \tag{3.2}$$

$$= \frac{1}{1-x}\,{}_3F_2\left(\frac{1}{2}, \frac{1}{2}, \frac{1}{2}; 1, 1; -Y\right) = \frac{1}{1-x}\sum_{k=0}^{\infty}(-1)^k A_k Y^k,$$

$$0 \le x \le 3-2\sqrt{2}, \tag{3.3}$$

$$= \frac{1}{\sqrt{1-x}}\, {}_3F_2\left(\frac{1}{2}, \frac{1}{2}, \frac{1}{2}; 1, 1; -U\right) = \frac{1}{\sqrt{1-x}} \sum_{k=0}^{\infty} (-1)^k A_k U^k,$$

$$0 \le x \le 2\sqrt{2}-2, \tag{3.4}$$

$$= \frac{1}{1+x}\, {}_3F_2\left(\frac{1}{4}, \frac{1}{2}, \frac{3}{4}; 1, 1; V^2\right) = \frac{1}{1+x} \sum_{k=0}^{\infty} B_k V^{2k}, \quad 0 \le x \le 3-2\sqrt{2}, \tag{3.5}$$

$$= \frac{1}{1-2x}\, {}_3F_2\left(\frac{1}{4}, \frac{1}{2}, \frac{3}{4}; 1, 1; -W^2\right) = \frac{1}{1-2x} \sum_{k=0}^{\infty} (-1)^k B_k W^{2k},$$

$$0 \le x \le \frac{1 - 2^{1/4}\sqrt{2-\sqrt{2}}}{2}, \tag{3.6}$$

$$= \frac{2}{\sqrt{4-X}}\, {}_3F_2\left(\frac{1}{6}, \frac{1}{2}, \frac{5}{6}; 1, 1; L\right) = \frac{2}{\sqrt{4-X}} \sum_{k=0}^{\infty} C_k L^k, \quad 0 \le x \le \frac{1}{2}, \tag{3.7}$$

$$= \frac{1}{\sqrt{1-4X}}\, {}_3F_2\left(\frac{1}{6}, \frac{1}{2}, \frac{5}{6}; 1, 1; -M\right) = \frac{1}{\sqrt{1-4X}} \sum_{k=0}^{\infty} (-1)^k C_k M^k,$$

$$0 \le x \le \frac{1}{2}. \tag{3.8}$$

Next, we find a representation for $P(-q)$ in terms of x and z.
Let

$$y = \pi \frac{{}_2F_1\left(\frac{1}{2}, \frac{1}{2}; 1; 1-x\right)}{{}_2F_1\left(\frac{1}{2}, \frac{1}{2}; 1; x\right)},$$

so that, by (2.1), $q = e^{-y}$. Thus, from (1.1), we obtain

$$\begin{aligned} P(-q) &= 1 - 24 \sum_{k=1}^{\infty} \frac{(-1)^k k q^k}{1-(-1)^k q^k} \\ &= 1 - 24 \sum_{k=1}^{\infty} \frac{k}{(-1)^{-k} e^{ky} - 1} \\ &= 1 - 24 \frac{d}{dy} \sum_{k=1}^{\infty} \operatorname{Log}(1-(-1)^k e^{-ky}) \\ &= 1 - 24 \frac{d}{dy} \operatorname{Log} \prod_{k=1}^{\infty} (1-(-1)^k e^{-ky}). \end{aligned} \tag{3.9}$$

Using (2.6) in (3.9), we find that

$$P(-q) = 1 - 24\frac{d}{dy}\operatorname{Log} f(e^{-y})$$

$$= -8\frac{d}{dy}\operatorname{Log}\{e^{-y/8} f^3(e^{-y})\}. \tag{3.10}$$

Transcribing the expression inside the curly brackets in (3.10) with the aid of (2.7), we arrive at

$$P(-q) = -8\frac{d}{dy}\operatorname{Log}\{z^{3/2}2^{-1/2}(x(1-x))^{1/8}\}$$

$$= -\frac{d}{dy}\operatorname{Log}\{x(1-x)z^{12}\}. \tag{3.11}$$

Now, recall from Entry 9(i) of [9, p. 120] that

$$\frac{dy}{dx} = -\frac{1}{x(1-x)z^2}. \tag{3.12}$$

From (3.11) and (3.12), we deduce that

$$P(-q) = x(1-x)z^2\frac{d}{dx}\operatorname{Log}\{x(1-x)z^{12}\}$$

$$= (1-2x)z^2 + 12x(1-x)z\frac{dz}{dx}, \tag{3.13}$$

which is the desired representation for $P(-q)$ in terms of x and z.

Now, differentiating (3.2) with respect to x, we find that

$$2z\frac{dz}{dx} = \sum_{k=0}^{\infty} A_k k X^{k-1} \cdot 4(1-2x). \tag{3.14}$$

Employing (3.2) and (3.14) in (3.13), we deduce that

$$P(-q) = (1-2x)\sum_{k=0}^{\infty}(6k+1)A_k X^k. \tag{3.15}$$

Now set

$$x_n := x(e^{-\pi\sqrt{n}}) \quad \text{and} \quad z_n := z(e^{-\pi\sqrt{n}}). \tag{3.16}$$

The numbers x_n are the so-called *singular moduli*. Ramanujan recorded the values of many singular moduli in his notebooks [35]. We refer to [14] and [10, Chapter 34] for the proofs and commentaries on Ramanujan's singular moduli. It can be shown that [1, pp. 378] x_n and z_n satisfy the transformation formulas

$$1 - x_n = x_{1/n} \quad \text{and} \quad z_{1/n} = \sqrt{n}z_n. \tag{3.17}$$

Setting $q = e^{-\pi\sqrt{n}}$ in (3.15), we deduce that

$$P(-e^{-\pi\sqrt{n}}) = (1-2x_n)\sum_{k=0}^{\infty}(6k+1)A_k X_n^k, \tag{3.18}$$

where $X_n = 4x_n(1-x_n)$.

Similarly, differentiating each of (3.3)–(3.8) with respect to x, and proceeding as above, we can find that

$$P(-e^{-\pi\sqrt{n}}) = \frac{1}{1-x_n}\sum_{k=0}^{\infty}\{6(1+x_n)k+1+4x_n\}(-1)^k A_k Y_n^k, \tag{3.19}$$

$$= \frac{1}{\sqrt{1-x_n}}\sum_{k=0}^{\infty}\{6(2-x_n)k+1+x_n\}(-1)^k A_k U_n^k, \tag{3.20}$$

$$= \frac{1}{4(1+x_n)^2}\sum_{k=0}^{\infty}\{24(x_n^2-6x_n+1)k-6X_n$$

$$+4(1+x_n)\sqrt{1-X_n}\}B_k V_n^{2k}, \tag{3.21}$$

$$= \frac{1}{1-X_n}\sum_{k=0}^{\infty}\{6(1+X_n)k+1+2X_n\}(-1)^k B_k W_n^{2k}, \tag{3.22}$$

$$= \frac{4\sqrt{1-X_n}}{(4-X_n)\sqrt{4-X_n}}\sum_{k=0}^{\infty}\{3(8+X_n)k+2+X_n\}C_k L_n^k, \tag{3.23}$$

$$= \frac{(1+8X_n)\sqrt{1-X_n}}{(1-4X_n)\sqrt{1-4X_n}}\sum_{k=0}^{\infty}(6k+1)(-1)^k C_k M_n^k, \tag{3.24}$$

where

$$Y_n := \frac{4x_n}{(1-x_n)^2}, \quad U_n := \frac{x_n^2}{4(1-x_n)}, \quad V_n := \frac{4\sqrt{x_n}(1-x_n)}{(1+x_n)^2}, \tag{3.25}$$

$$W_n := \frac{2\sqrt{X_n}}{1-X_n}, \quad L_n := \frac{27X_n^2}{(4-X_n)^3}, \quad \text{and} \quad M_n := \frac{27X_n}{(1-4X_n)^3}. \tag{3.26}$$

Next, we determine a transformation formula for $P(-q)$. If $\alpha, \beta > 1$ with $\alpha\beta = \pi^2$, then Ramanujan's function $f(q)$ defined in (2.6) satisfies the transformation formula [9, p. 43, Entry 27 (iv)]

$$e^{-\alpha/24}\alpha^{1/4}f(e^{-\alpha}) = e^{-\beta/24}\beta^{1/4}f(e^{-\beta}). \tag{3.27}$$

Taking logarithms of both sides of (3.27), we find that

$$-\frac{\alpha}{24}+\frac{\log\alpha}{4}+\sum_{k=0}^{\infty}\log(1+(-1)^k e^{-(k+1)\alpha})$$
$$=-\frac{\beta}{24}+\frac{\log\beta}{4}+\sum_{k=0}^{\infty}\log(1+(-1)^k e^{-(k+1)\beta}). \tag{3.28}$$

Differentiating both sides of (3.28) with respect to α, we obtain

$$-\frac{1}{24}+\frac{1}{4\alpha}+\sum_{k=1}^{\infty}\frac{k(-e^{-\alpha})^k}{1-(-e^{-\alpha})^k}=\frac{\beta}{24\alpha}-\frac{1}{4\alpha}-\sum_{k=1}^{\infty}\frac{(k\beta/\alpha)(-e^{-\beta})^k}{1-(-e^{-\beta})^k}. \tag{3.29}$$

Multiplying both sides of (3.29) by 24α, rearranging, and then employing the definition of $P(q)$ from (1.1), we deduce that

$$12-\alpha P(-e^{-\alpha})=\beta P(-e^{-\beta}). \tag{3.30}$$

Setting $\alpha=\pi/\sqrt{n}$, so that $\beta=\pi\sqrt{n}$, in (3.30), we arrive at

$$nP(-e^{-\pi\sqrt{n}})+P(-e^{-\pi/\sqrt{n}})=\frac{12\sqrt{n}}{\pi}, \tag{3.31}$$

which is the desired transformation formula for $P(-q)$.

Note that, if we set $n=1$ in (3.31), then we arrive at the evaluation

$$P(-e^{-\pi})=\frac{6}{\pi}. \tag{3.32}$$

By combining (3.31) and the representations for $h_n(q):=nP(-q^n)-P(-q)$ with $q=e^{-\pi/\sqrt{n}}$, for each $n=3,\ 5,\ 7,\ 9$, and 25 along with the identities (3.2)–(3.8), we can derive certain known series representations for $1/\pi$. We present these in the remaining sections. In the cases for $n=5,\ 9$, and 25, one can also get a few more series for $1/\pi$, that are not so elegant. So we do not record them.

4. Series corresponding to x_3

Theorem 4.1. *If A_k, B_k, and C_k, $k\geq 0$, are defined by* (3.1), *then*

$$\frac{4}{\pi}=\sum_{k=0}^{\infty}(6k+1)A_k\frac{1}{4^k}, \tag{4.1}$$

$$\frac{1}{\pi}=\sum_{k=0}^{\infty}(-1)^k\{(15\sqrt{3}-24)k+6\sqrt{3}-10\}A_k 2^k(\sqrt{3}-1)^{6k}, \tag{4.2}$$

$$\frac{4\sqrt{2}}{\pi}=\sum_{k=0}^{\infty}(-1)^k\{(30-6\sqrt{3})k+7-3\sqrt{3}\}A_k\frac{(2-\sqrt{3})^{3k}}{2^{4k}}, \tag{4.3}$$

$$\frac{8\sqrt{2}}{\pi} = \sum_{k=0}^{\infty}\{(85\sqrt{3}-135)k+8\sqrt{3}-12\}B_k\left(\frac{(8\sqrt{2})}{51\sqrt{3}-75}\right)^{2k+1}, \tag{4.4}$$

$$\frac{5\sqrt{5}}{2\pi\sqrt{3}} = \sum_{k=0}^{\infty}(11k+1)C_k\left(\frac{4}{125}\right)^{k}. \tag{4.5}$$

The identities (4.1) and (4.5) are due to Ramanujan [33, Eqs. (28), (33), resp.], [34, pp. 36–37], [35, pp. 355, 378], [36, p. 370]. The remaining identities are due to Baruah and Berndt [3]. To prove Theorem 4.1, Baruah and Berndt [3] calculated and used the identity $f_3(e^{-\pi/\sqrt{3}}) = 3P(e^{-2\pi\sqrt{3}}) - P(e^{-2\pi/\sqrt{3}}) = \frac{3\sqrt{3}}{2}z_3^2$. In the following proof, we calculated and used (4.8), which makes our calculations somewhat shorter.

Proof. First of all, we derive an expression for

$$h_3(e^{-\pi/\sqrt{3}}) = 3P(-e^{-\pi\sqrt{3}}) - P(-e^{-\pi/\sqrt{3}}).$$

Recall from [9, p. 473, Entry 6(i)], that

$$1+12\sum_{k=1}^{\infty}\frac{k(-q)^k}{1-(-q)^k} - 36\sum_{k=1}^{\infty}\frac{k(-q)^{3k}}{1-(-q)^{3k}}$$
$$= \phi^2(q)\phi^2(q^3)((x(q)x(q^3))^{1/4} - \{(1-x(q))(1-x(q^3))\}^{1/4})^2. \tag{4.6}$$

With the aid of (1.1) and (2.2) we rewrite (4.6) in the form

$$h_3(q) = 3P(-q^3) - P(-q)$$
$$= 2z(q)z(q^3)((x(q)x(q^3))^{1/4} - \{(1-x(q))(1-x(q^3))\}^{1/4})^2. \tag{4.7}$$

Now we set $q = e^{-\pi/\sqrt{3}}$, so that, by (3.17), $x(q) = x(e^{-\pi/\sqrt{3}}) = 1 - x(e^{-\pi\sqrt{3}}) = 1 - x_3$, $x(q^3) = x_3$ and $z(q) = z(e^{-\pi/\sqrt{3}}) = \sqrt{3}z(e^{-\pi\sqrt{3}}) = \sqrt{3}z_3$. We therefore deduce from (4.7) that

$$h_3(e^{-\pi/\sqrt{3}}) = 3P(-e^{-\pi\sqrt{3}}) - P(-e^{-\pi/\sqrt{3}}) = 0. \tag{4.8}$$

Next, setting $n = 3$ in (3.31), we obtain

$$3P(-e^{-\pi\sqrt{3}}) + P(-e^{-\pi/\sqrt{3}}) = \frac{12\sqrt{3}}{\pi}. \tag{4.9}$$

From (4.8) and (4.9), we readily deduce

$$P(-e^{-\pi\sqrt{3}}) = \frac{2\sqrt{3}}{\pi}, \tag{4.10}$$

which is a well-known result (See, for example, Berndt's [8, Proposition 2.13]). Chan, Liaw and Tan [23] also established (4.10) and wrongly noted this value to be new.

Now, setting $n = 3$ in (3.18), we find that

$$P(-e^{-\pi\sqrt{3}}) = (1 - 2x_3)\sum_{k=0}^{\infty}(6k + 1)A_k X_3^k. \tag{4.11}$$

From (4.10) and (4.11), we obtain

$$\frac{2\sqrt{3}}{\pi} = (1 - 2x_3)\sum_{k=0}^{\infty}(6k + 1)A_k X_3^k. \tag{4.12}$$

But, from [10, p. 290, Theorem 9.9], we note that

$$x_3 = \frac{2 - \sqrt{3}}{4}, \tag{4.13}$$

so that

$$X_3 = 4x_3(1 - x_3) = \frac{1}{4}. \tag{4.14}$$

Employing (4.13) and (4.14) in (4.12), we complete the proof of (4.1).

Proof of (4.2). Setting $n = 3$ in (3.19), we find that

$$P(-e^{-\pi\sqrt{3}}) = \frac{1}{1 - x_3}\sum_{k=0}^{\infty}\{6(1 + x_3)k + 1 + 4x_3\}(-1)^k A_k Y_3^k, \tag{4.15}$$

where, by (3.25),

$$Y_3 = \frac{4x_3}{(1 - x_3)^2}. \tag{4.16}$$

Using (4.13) in (4.16) and (4.15), we obtain

$$P(-e^{-\pi\sqrt{3}}) = \sum_{k=0}^{\infty}\{(90 - 48\sqrt{3})k + 36 - 20\sqrt{3}\}(-1)^k A_k (2(\sqrt{3} - 1)^6)^k. \tag{4.17}$$

From (4.17) and (4.10), we easily arrive at (4.2).

Proof of (4.3). Setting $n = 3$ in (3.20), we find that

$$P(-e^{-\pi\sqrt{3}}) = \frac{1}{\sqrt{1 - x_3}}\sum_{k=0}^{\infty}\{6(2 - x_3)k + 1 + x_3\}(-1)^k A_k U_3^k, \tag{4.18}$$

where, by (3.25),

$$U_3 := \frac{x_3^2}{4(1 - x_3)}. \tag{4.19}$$

Using (4.13) in (4.19) and (4.18), we arrive at

$$P(-e^{-\pi\sqrt{3}}) = \sum_{k=0}^{\infty}\left\{\frac{3}{2}(\sqrt{6} - \sqrt{2})(6 + \sqrt{3})k + \frac{1}{4}(\sqrt{6} - \sqrt{2})(6 - \sqrt{3})\right\}$$
$$\times (-1)^k A_k \frac{(2 - \sqrt{3})^{3k}}{2^{4k}}. \tag{4.20}$$

We readily deduce (4.3) from (4.20) and (4.10).

Proof of (4.4). Setting $n = 3$ in (3.21), we obtain

$$P(-e^{-\pi\sqrt{3}}) = \frac{1}{4(1+x_3)^2}\sum_{k=0}^{\infty}\{24(x_3^2 - 6x_3 + 1)k - 6X_3 + 4(1+x_3)\sqrt{1-X_3}\}B_k V_3^{2k}, \tag{4.21}$$

where, again by (3.25),

$$V_3 := \frac{4\sqrt{x_3}(1-x_3)}{(1+x_3)^2}. \tag{4.22}$$

Using (4.13) in (4.22) and (4.21), we obtain

$$P(-e^{-\pi\sqrt{3}}) = \frac{1}{121}\sum_{k=0}^{\infty}\{(320\sqrt{3}-170)k + 36\sqrt{3}-4\}B_k\left(\frac{8\sqrt{2}}{51\sqrt{3}-75}\right)^{2k}, \tag{4.23}$$

From (4.23) and (4.10), we derive (4.4).

Proof of (4.5). Setting $n = 3$ in (3.23), we find that

$$P(-e^{-\pi\sqrt{3}}) = \frac{4\sqrt{1-X_3}}{(4-X_3)(\sqrt{4-X_3})}\sum_{k=0}^{\infty}\{3(8+X_3)k + 2 + X_3\}C_k L_3^k, \tag{4.24}$$

where, by (3.26),

$$L_3 = \frac{27X_3^2}{(4-X_3)^3}. \tag{4.25}$$

Employing (4.14) in (4.25) and (4.24), we obtain

$$P(-e^{-\pi\sqrt{3}}) = \frac{12}{5\sqrt{5}}\sum_{k=0}^{\infty}(11k+1)C_k\left(\frac{4}{125}\right)^k. \tag{4.26}$$

From (4.26) and (4.10), we deduce (4.4) to complete the proof of the theorem.

5. Series corresponding to x_7

Theorem 5.1. *If* A_k, B_k, *and* C_k, $k \geq 0$, *are defined by* (3.1), *then*

$$\frac{16}{\pi} = \sum_{k=0}^{\infty}(42k+5)A_k\frac{1}{2^{6k}}, \tag{5.1}$$

$$\frac{1}{\pi} = \sum_{k=0}^{\infty}(-1)^k\{(255\sqrt{7}-672)k + 112\sqrt{7} - 296\}A_k(32-12\sqrt{7})^{3k}, \tag{5.2}$$

$$\frac{8\sqrt{2}}{\pi} = \sum_{k=0}^{\infty}(-1)^k\{(102\sqrt{7}-210)k + 35\sqrt{7} - 89\}A_k\left(\frac{8-3\sqrt{7}}{4}\right)^{3k}, \tag{5.3}$$

$$\frac{29241}{\pi}=\sum_{k=0}^{\infty}\{(76160-455\sqrt{7})k+784\sqrt{7}+6728\}B_k\left(\frac{8\sqrt{2}(325+119\sqrt{7})}{29241}\right)^{2k}, \tag{5.4}$$

$$\frac{9\sqrt{7}}{\pi}=\sum_{k=0}^{\infty}(-1)^k(65k+8)B_k\left(\frac{16}{63}\right)^{2k}, \tag{5.5}$$

$$\frac{85\sqrt{85}}{18\pi\sqrt{3}}=\sum_{k=0}^{\infty}(133k+8)C_k\left(\frac{4}{85}\right)^{3k+1}, \tag{5.6}$$

$$\frac{5\sqrt{15}}{\pi}=\sum_{k=0}^{\infty}(-1)^k(63k+8)C_k\left(\frac{4}{5}\right)^{3k}. \tag{5.7}$$

The identities (5.1) and (5.6) are due to Ramanujan [33, Eqs. (29), (34), resp.], [34, pp. 36–37], [35, pp. 355, 378], [36, p. 370] and (5.5) is due to Berndt, Chan, and Liaw [13]. The other four identities are due to Baruah and Berndt [3]. The level of difficulty in the following proof is the same as in [3].

Proof. At first, we derive an expression for $h_7(e^{-\pi/\sqrt{3}}) = 7P(-e^{-\pi\sqrt{3}}) - P(-e^{-\pi/\sqrt{3}})$. Recall from [9, p. 473, Entry 6(iii)], that

$$1+4\sum_{k=1}^{\infty}\frac{k(-q)^k}{1-(-q)^k}-28\sum_{k=1}^{\infty}\frac{k(-q)^{7k}}{1-(-q)^{7k}}$$
$$=\phi^2(q)\phi^2(q^7)(\{x(q)x(q^7)\}^{1/4}+\{(1-x(q))(1-x(q^7))\}^{1/4})^2. \tag{5.8}$$

Using (1.1) and (2.2) in (5.8), we find that

$$h_7(q)=7P(-q^7)-P(-q)$$
$$=6z(q)z(q^7)((x(q)x(q^7))^{1/4}+\{(1-x(q))(1-x(q^7))\}^{1/4})^2. \tag{5.9}$$

Now we set $q=e^{-\pi/\sqrt{7}}$, so that, by (3.17), $x(q)=x(e^{-\pi/\sqrt{7}})=1-x(e^{-\pi\sqrt{7}})=1-x_7$, $x(q^7)=x_7$, $z(q)=z(e^{-\pi/\sqrt{7}})=\sqrt{7}z(e^{-\pi\sqrt{7}})=\sqrt{7}z_7$, and $z(q^7)=z(e^{-\pi\sqrt{7}})=z_7$. Thus, we deduce from (5.9) that

$$h_7(e^{-\pi/\sqrt{7}})=7P(-e^{-\pi\sqrt{7}})-P(-e^{-\pi/\sqrt{7}})=24\sqrt{7}\sqrt{x_7(1-x_7)}z_7^2. \tag{5.10}$$

But, from [10, p. 290, Theorem 9.9], we have

$$x_7=\frac{8-3\sqrt{7}}{16}, \tag{5.11}$$

and consequently,

$$X_7=4x_7(1-x_7)=\frac{1}{64}. \tag{5.12}$$

Using (5.11) in (5.10), we find that

$$7P(-e^{-\pi\sqrt{7}}) - P(-e^{-\pi/\sqrt{7}}) = \frac{3}{2}\sqrt{7}z_7^2. \tag{5.13}$$

Again, setting $n = 7$, in (3.31), we obtain

$$7P(-e^{-\pi\sqrt{7}}) + P(-e^{-\pi/\sqrt{7}}) = \frac{12\sqrt{7}}{\pi}. \tag{5.14}$$

From (5.13) and (5.14), we arrive at

$$P(-e^{-\pi\sqrt{7}}) = \frac{12\sqrt{7}}{14\pi} + \frac{3\sqrt{7}}{28}z_7^2. \tag{5.15}$$

Next, setting $n = 7$ in (3.18), we find that

$$P(-e^{-\pi\sqrt{7}}) = (1 - 2x_7)\sum_{k=0}^{\infty}(6k + 1)A_k X_7^k. \tag{5.16}$$

Using (5.11) and (5.12) in (5.16), we obtain

$$P(-e^{-\pi\sqrt{7}}) = \frac{3\sqrt{7}}{8}\sum_{k=0}^{\infty}(6k + 1)A_k \frac{1}{2^{6k}}. \tag{5.17}$$

From (5.15), (5.17), and (3.2), we easily deduce (5.1).

Proof of (5.2). Setting $n = 7$ in (3.19), we find that

$$P(-e^{-\pi\sqrt{7}}) = \frac{1}{1 - x_7}\sum_{k=0}^{\infty}\{6(1 + x_7)k + 1 + 4x_7\}(-1)^k A_k Y_7^k. \tag{5.18}$$

Using (5.11) in (5.18), we obtain

$$P(-e^{-\pi\sqrt{7}}) = \sum_{k=0}^{\infty}\{6(255 - 96\sqrt{7})k + 4(159 - 60\sqrt{7})\}(-1)^k A_k Y_7^k, \tag{5.19}$$

where $Y_7 = (32 - 12\sqrt{7})^3$.

Again, using (3.3) and (5.11) in (5.15), we obtain

$$P(-e^{-\pi\sqrt{7}}) = \frac{12\sqrt{7}}{14\pi} + \frac{12\sqrt{7}}{7}\sum_{k=0}^{\infty}(-1)^k A_k Y_7^k. \tag{5.20}$$

From (5.19) and (5.20), we arrive at (5.2).

Proof of (5.3). Setting $n = 7$ in (3.20), we find that

$$P(-e^{-\pi\sqrt{7}}) = \frac{1}{\sqrt{1 - x_7}}\sum_{k=0}^{\infty}\{6(2 - x_7)k + 1 + x_7\}(-1)^k A_k U_7^k. \tag{5.21}$$

Using (5.11) in (5.21), we obtain

$$P(-e^{-\pi\sqrt{7}}) = \frac{3}{4\sqrt{2}}\sum_{k=0}^{\infty}\{6(17-5\sqrt{7})k+31-11\sqrt{7}\}(-1)^k A_k U_7^k, \tag{5.22}$$

where $U_7 = \left(\frac{8-3\sqrt{7}}{4}\right)^3$.

But, employing (3.4) and (5.11) in (5.15), we also obtain

$$P(-e^{-\pi\sqrt{7}}) = \frac{12\sqrt{7}}{14\pi} + \frac{3\sqrt{7}}{7\sqrt{2}}(3-\sqrt{7})\sum_{k=0}^{\infty}(-1)^k A_k U_7^k. \tag{5.23}$$

From (5.22) and (5.23), we easily deduce (5.3).

Proof of (5.4). Setting $n = 7$ in (3.21), we find that

$$P(-e^{-\pi\sqrt{7}}) = \frac{1}{4(1+x_7)^2}\sum_{k=0}^{\infty}\{24(x_7^2-6x_7+1)k-6X_7 + 4(1+x_7)\sqrt{1-X_7}\}B_k V_7^{2k}. \tag{5.24}$$

Using (5.11) in (5.24), we obtain

$$P(-e^{-\pi\sqrt{7}}) = \frac{1}{29241}\sum_{k=0}^{\infty}\{6(10880\sqrt{7}-455)k+6756+8112\sqrt{7}\}B_k V_7^{2k}, \tag{5.25}$$

where $V_7 = \frac{8\sqrt{2}(325+119\sqrt{7})}{29241}$.

Again, using (3.5) and (5.11) in (5.15), we obtain

$$P(-e^{-\pi\sqrt{7}}) = \frac{12\sqrt{7}}{14\pi} + \frac{4\sqrt{7}}{399}(8+\sqrt{7})\sum_{k=0}^{\infty}(-1)^k B_k V_7^{2k}. \tag{5.26}$$

From (5.25) and (5.26), we arrive at (5.4).

Proof of (5.5). Setting $n = 7$, in (3.22), we find that

$$P(-e^{-\pi\sqrt{7}}) = \frac{1}{1-X_7}\sum_{k=0}^{\infty}\{6(1+X_7)k+1+2X_7\}(-1)^k B_k W_7^{2k}. \tag{5.27}$$

Using (5.11) in (5.27), we obtain

$$P(-e^{-\pi\sqrt{7}}) = \frac{1}{21}\sum_{k=0}^{\infty}\{130k+22\}(-1)^k B_k W_7^{2k}, \tag{5.28}$$

where $W_7 = \frac{16}{63}$.

But, with the aid of (3.6) and (5.11) we can rewrite (5.15) in the form

$$P(-e^{-\pi\sqrt{7}}) = \frac{12\sqrt{7}}{14\pi} + \frac{2}{7}\sum_{k=0}^{\infty}(-1)^k B_k V_7^{2k}. \tag{5.29}$$

From (5.28) and (5.29), we arrive at (5.5).

Proof of (5.6). Setting $n = 7$, in (3.23), we have

$$P(-e^{-\pi\sqrt{7}}) = \frac{4\sqrt{1-X_7}}{(4-X_7)(\sqrt{4-X_7})}\sum_{k=0}^{\infty}\{3(8+X_7)k+2+X_7\}C_k L_7^k. \tag{5.30}$$

Using (5.11) in (5.30), we obtain

$$P(-e^{-\pi\sqrt{7}}) = \frac{12\sqrt{7}}{85\sqrt{85}\sqrt{3}}\sum_{k=0}^{\infty}(513k+43)C_k L_7^k, \tag{5.31}$$

where $L_7 = \left(\frac{4}{85}\right)^3$.

On the other hand, using (3.7) and (5.11) in (5.15), we obtain

$$P(-e^{-\pi\sqrt{7}}) = \frac{12\sqrt{7}}{14\pi} + \frac{24\sqrt{7}}{14\sqrt{255}}\sum_{k=0}^{\infty}(-1)^k B_k L_7^{2k}. \tag{5.32}$$

From (5.31) and (5.32), we arrive at (5.6).

Proof of (5.7). Setting $n = 7$, in (3.24), we find that

$$P(-e^{-\pi\sqrt{7}}) = \frac{(1+8X_7)\sqrt{1-X_7}}{(1-4X_7)(\sqrt{1-4X_7})}\sum_{k=0}^{\infty}(6k+1)(-1)^k C_k M_7^k. \tag{5.33}$$

Using (5.11) in (5.33), we obtain

$$P(-e^{-\pi\sqrt{7}}) = \frac{9\sqrt{7}}{5\sqrt{15}}\sum_{k=0}^{\infty}(6k+1)(-1)^k C_k M_7^k, \tag{5.34}$$

where $M_7 = \left(\frac{4}{5}\right)^3$.

Using (3.8) and (5.11) in (5.15), we obtain

$$P(-e^{-\pi\sqrt{7}}) = \frac{12\sqrt{7}}{14\pi} + \frac{6\sqrt{7}}{14\sqrt{15}}\sum_{k=0}^{\infty}(-1)^k C_k M_7^{2k}. \tag{5.35}$$

From (5.34) and (5.35), we arrive at (5.7).

6. Series corresponding to x_5

Theorem 6.1. *If A_k, and B_k, $k \geq 0$, are defined by* (3.1), *then*

$$\frac{2}{\pi} = \sum_{k=0}^{\infty} \{4\sqrt{5}k + \sqrt{5} - 1\} A_k (\sqrt{5} - 2)^{2k+1/2}, \tag{6.1}$$

$$\frac{8}{\pi} = \sum_{k=0}^{\infty} (-1)^k [2\{(15 + 5\sqrt{5})\sqrt{\sqrt{5}+1} - 7\sqrt{10} - 5\sqrt{2}\}k + (9 + 3\sqrt{5})\sqrt{\sqrt{5}+1}$$
$$- 7\sqrt{2} - 5\sqrt{10}] A_k \left(\frac{\sqrt{5}-1}{4}\right)^{3k} \left(\frac{\sqrt{5}+1}{2} - \sqrt{\frac{\sqrt{5}+1}{2}}\right)^{6k}, \tag{6.2}$$

$$\frac{8}{\pi} = \sum_{k=0}^{\infty} (-1)^k (20k + 3) B_k \frac{1}{4^k}. \tag{6.3}$$

The last identity was recorded by Ramanujan [33, Eq. (35)], [34, p. 38]. The remaining two identities are due to Baruah and Berndt [3]. Because the proofs are similar to those in previous sections, we do not give them. The proofs require the identity

$$h_5(e^{-\pi/\sqrt{5}}) = 5P(-e^{-\pi\sqrt{5}}) - P(-e^{-\pi/\sqrt{5}}) = 2\sqrt{10}(\sqrt{5}-2)(\sqrt{5}-1)^{1/2} z_5^2, \tag{6.4}$$

which can be derived, with $q = e^{-\pi/\sqrt{5}}$, from Ramanujan's representation [9, p. 473, Entry 6(ii)]

$$1 + 6\sum_{k=1}^{\infty} \frac{k(-q)^k}{1-(-q)^k} - 30\sum_{k=1}^{\infty} \frac{k(-q)^{5k}}{1-(-q)^{5k}}$$
$$= \phi^2(q)\phi^2(q^5)(\sqrt{x(q)x(q^5)} + \sqrt{(1-x(q))(1-x(q^5))})$$
$$\times \left(\frac{1}{2}\{1 + \sqrt{x(q)x(q^5)} + \sqrt{(1-x(q))(1-x(q^5))}\}\right), \tag{6.5}$$

and using the value of Ramanujan's singular modulus x_5 from [10, p. 290]. The level of difficulty in proving Theorem 6.1 is the same as in [3].

7. Series corresponding to x_9

Theorem 7.1. *If A_k, and B_k, $k \geq 0$, are defined by* (3.1), *then*

$$\frac{12^{1/4}}{\pi} = \sum_{k=0}^{\infty} \{(24\sqrt{3} - 36)k + 9\sqrt{3} - 15\} A_k (2 - \sqrt{3})^{4k}, \tag{7.1}$$

$$\frac{4}{\pi}=\sum_{k=0}^{\infty}[6\{9\sqrt{2}+7\sqrt{6}-3^{1/4}(5\sqrt{3}+11)\}k+15\sqrt{6}+21\sqrt{2}$$

$$-3^{1/4}(13\sqrt{3}+27)]\times(-1)^k A_k\left(\frac{(\sqrt{3}+1)^{1/3}(\sqrt{2}-3^{1/4})}{2^{2/3}}\right)^{6k}, \quad (7.2)$$

$$\frac{16}{\pi\sqrt{3}}=\sum_{k=0}^{\infty}(-1)^k(28k+3)B_k\frac{1}{48^k}. \quad (7.3)$$

The identity (7.3) was recorded by Ramanujan [33, Eq. (36)], [34, p. 38]. The first identity is due to the Borweins [15], [16] and the second is due to Baruah and Berndt [3]. One of the main differences of our following proof of Theorem 7.1 with that of Baruah and Berndt [3] (though details are not given there) is the fact that we make use of the explicit value of $\phi(e^{-\pi})/\phi(e^{-3\pi})$ recorded by Ramanujan in his first notebook [35].

Proof. We find an expression for $h_9(e^{-\pi/3})=9P(-e^{-3\pi})-P(-e^{-\pi/3})$.

At first we recall the following identity from [9, p. 475, Entry 7(i)]. If $f(-q)$ is defined by (2.6), then

$$1+3\sum_{k=1}^{\infty}\frac{kq^k}{1-q^k}-27\sum_{k=1}^{\infty}\frac{kq^{9k}}{1-q^{9k}}$$

$$=\frac{f^6(-q^3)}{f^2(-q)f^2(-q^9)}\{f^6(-q)+9qf^3(-q)f^3(-q^9)+27q^2f^6(-q^9)\}^{1/3}. \quad (7.4)$$

Multiplying both sides of (7.4) by 8, replacing q by $-q$, and then using the definition of $P(q)$ from (1.1), we find that

$$9P(-q^9)-P(-q)=\frac{8f^6(q^3)}{f^2(q)f^2(q^9)}\{f^6(q)-9qf^3(q)f^3(q^9)+27q^2f^6(q^9)\}^{1/3}. \quad (7.5)$$

Transcribing the right side of (7.5) with the aid of (2.7), we obtain

$$9P(-q^9)-P(-q)$$

$$=\frac{2^{8/3}z^3(q^3)\{x(q^3)(1-x(q^3))\}^{1/4}}{z(q)z(q^9)\{x(q)(1-x(q))x(q^9)(1-x(q^9))\}^{1/12}}\{z^3(q)\{x(q)(1-x(q))\}^{1/4}$$

$$-9z^{3/2}(q)z^{3/2}(q^9)\{x(q)(1-x(q))x(q^9)(1-x(q^9))\}^{1/8}$$

$$+27z^3(q^9)\{x(q^9)(1-x(q^9))\}^{1/4}\}^{1/3}. \quad (7.6)$$

Now we set $q=e^{-\pi/3}$ in (7.6), so that, by (3.17), $x(q)=x(e^{-\pi/3})=1-x(e^{-3\pi})=1-x_9$, $x(q^3)=x(e^{-\pi})=x_1=1/2$, $x(q^9)=x_9$, $z(q)=z(e^{-\pi/3})=3z(e^{-3\pi})=3z_9$,

$z(q^3) = z(e^{-\pi}) = z_1$, and $z(q^9) = z(e^{-3\pi}) = z_9$. Accordingly, we deduce that

$$9P(-e^{-3\pi}) - P(-e^{-\pi/3}) = \frac{4z_1^3(2-\sqrt{3})^{1/3}}{2^{1/6}z_9\{x_9(1-x_9)\}^{1/12}}$$

$$= \frac{4(2-\sqrt{3})^{1/3}}{2^{1/6}\{x_9(1-x_9)\}^{1/12}}\left(\frac{z_1}{z_9}\right)^3 z_9^2. \tag{7.7}$$

But, from Theorem 9.9 of [10, pp. 290–291], we have

$$x_9 = \frac{1}{2}\left(\frac{\sqrt{3}-1}{\sqrt{2}}\right)^4(\sqrt{4+2\sqrt{3}}-\sqrt{3+2\sqrt{3}})^2,$$

so that

$$x_9(1-x_9) = \frac{1}{4}\left(\frac{\sqrt{3}-1}{\sqrt{2}}\right)^8 = \frac{1}{4}(2-\sqrt{3})^4. \tag{7.8}$$

Employing (7.8) in (7.7), we find that

$$9P(-e^{-3\pi}) - P(-e^{-\pi/3}) = 4\left(\frac{z_1}{z_9}\right)^3 z_9^2. \tag{7.9}$$

On p. 284 of his first notebook, Ramanujan [35] recorded an explicit value for $\phi(e^{-\pi})/\phi(e^{-3\pi})$, i.e., by (2.2) and (3.16), for $(z_1/z_9)^{1/2}$. Berndt and Chan [11] gave three proofs of Ramanujan's evaluation, which were also reproduced in Chapter 35 of Berndt's book [10]. J. Yi [37] also established this value among many others. Here we find an alternative proof of the explicit value for the same. To do so, we first recast Entry 5(v) of [9, p. 230] by replacing q by q^3 to obtain

$$\frac{z(q^3)}{z(q^9)} = \left\{1+4\left(\frac{x^3(q^9)(1-x(q^9))^3}{x(q^3)(1-x(q^3))}\right)^{1/8}\right\}^{1/2}. \tag{7.10}$$

Setting $q = e^{-\pi/3}$ in (7.10) and noting that $x(q^3) = x(e^{-\pi}) = x_1 = 1/2, x(q^9) = x_9$, $z(q^3) = z(e^{-\pi}) = z_1$, and $z(q^9) = z(e^{-3\pi}) = z_9$, we deduce that

$$\frac{z_1}{z_9} = \{1+4\cdot 2^{1/4}(x_9(1-x_9))^{3/8}\}^{1/2}. \tag{7.11}$$

With the aid of (7.8), we deduce from (7.11) that

$$\frac{z_1}{z_9} = \{1+2^{3/2}(2-\sqrt{3})^{3/2}\}^{1/2} = \{1+(\sqrt{3}-1)^3\}^{1/2} = 3^{1/4}\cdot\frac{3-\sqrt{3}}{\sqrt{2}}. \tag{7.12}$$

Now, employing (7.12) in (7.9), we arrive at

$$9P(-e^{-3\pi}) - P(-e^{-\pi/3}) = 6\sqrt{2}(9-5\sqrt{3})\cdot 3^{3/4}z_9^2. \tag{7.13}$$

The remainder of the proof is similar to those in Sections 4 and 5. □

8. Series corresponding to x_{25}

Theorem 8.1. *If A_k, and B_k, $k \geq 0$, are defined by* (3.1), *then*

$$\frac{5^{1/4}}{\pi} = \sum_{k=0}^{\infty} \{(540\sqrt{5} - 1200)k + 235\sqrt{5} - 525\} A_k (\sqrt{5} - 2)^{8k}, \tag{8.1}$$

$$\frac{4}{5\pi} = \sum_{k=0}^{\infty} (-1)^k (644k + 41) B_k \frac{1}{(72\sqrt{5})^{2k+1}}. \tag{8.2}$$

The first series is due to Borwein and Borwein [16]. The second was recorded by Ramanujan [33, Eq. (38)], [34, p. 38]. As in Section 7, in the following proof of Theorem 8.1, we also reprove and use the explicit value of $\phi^2(e^{-\pi})/\phi^2(e^{-5\pi})$ recorded by Ramanujan in his first notebook [35].

Proof. The crucial part of the proof is finding an expression for $h_{25}(e^{-\pi/5}) = 25P(-e^{-5\pi}) - P(-e^{-\pi/5})$. To this end, we recall from [9, p. 475, Entry 7(iii)] that

$$1 + \sum_{k=1}^{\infty} \frac{kq^k}{1 - q^k} - 25 \sum_{k=1}^{\infty} \frac{kq^{25k}}{1 - q^{25k}}$$

$$= \frac{f^5(-q^5)}{f(-q)f(-q^{25})} \{f^2(-q) + 2qf(-q)f(-q^{25}) + 5q^2 f^2(-q^{25})\}^{1/2}. \tag{8.3}$$

Multiplying both sides of (8.3) by 24, replacing q by $-q$, and then using the definition of $P(q)$ from (1.1), we find that

$$25P(-q^{25}) - P(-q) = \frac{24 f^5(q^5)}{f(q)f(q^{25})} \{f^2(q) - 2qf(q)f(q^{25}) + 5q^2 f^2(q^{25})\}^{1/2}. \tag{8.4}$$

Transcribing the right side of (8.4) with the aid of (2.7), and then following a similar method as in the derivation of (7.7) from (7.5), but this time with $q = e^{-\pi/5}$, we arrive at

$$25P(-e^{-5\pi}) - P(-e^{-\pi/5}) = \frac{24 \cdot 5^{1/4}(\sqrt{5} - 1)^{1/2}}{2^{7/12}\sqrt{5}\{x_{25}(1 - x_{25})\}^{1/24}} \left(\frac{z_1}{z_{25}}\right)^{5/2} z_{25}^2. \tag{8.5}$$

But, from Theorem 9.9 of [10, p. 291], we note that

$$x_{25} = \frac{1}{2}(161 - 72\sqrt{5}) \left(\sqrt{\frac{5 + \sqrt{5}}{4}} - \sqrt{\frac{1 + \sqrt{5}}{4}}\right)^8,$$

so that

$$x_{25}(1 - x_{25}) = \frac{1}{4}(161 - 72\sqrt{5})^2 = \frac{1}{4}(\sqrt{5} - 2)^8 = \frac{1}{4}\left(\frac{\sqrt{5} - 1}{2}\right)^{24}. \tag{8.6}$$

Employing (8.6) in (8.5), we find that

$$25P(-e^{-5\pi}) - P(-e^{-\pi/5}) = \frac{24\sqrt{2}\cdot 5^{1/4}}{\sqrt{5}(\sqrt{5}-1)^{1/2}}\left(\frac{z_1}{z_{25}}\right)^{5/2} z_{25}^2. \tag{8.7}$$

Next, we need to know the explicit value of the expression z_1/z_{25}. But, on p. 285 of his first notebook, Ramanujan [35] recorded, without proof, an explicit value for z_1/z_{25}, equivalently, by (2.2) and (3.16), for $\phi^2(e^{-\pi})/\phi^2(e^{-5\pi})$. Berndt and Chan [11] found a proof of this value by applying Ramanujan's modular equations and an explicit value of his class invariant. The same proof can be found in [10, pp. 328–329]. A different proof was found by Yi [37]. Now we present a method of explicit evaluation of z_1/z_{25}, which is slightly different from that in [11] and [10]. To this end, we recall from Entry 13(iv) of [9, p. 281], with q replaced by q^5, that

$$\frac{z(q^5)}{z(q^{25})} = 1 + 2^{4/3}\left(\frac{x^5(q^{25})(1-x(q^{25}))^5}{x(q^5)(1-x(q^5))}\right)^{1/24}. \tag{8.8}$$

Setting $q = e^{-\pi/5}$ in (8.8) and noting that $x(q^5) = x(e^{-\pi}) = x_1 = 1/2, x(q^{25}) = x_{25}$, $z(q^{25}) = z(e^{-\pi}) = z_1$, and $z(q^{25}) = z(e^{-5\pi}) = z_{25}$, we obtain

$$\frac{z_1}{z_{25}} = 1 + 2\cdot 2^{5/12}(x_{25}(1-x_{25}))^{5/24}, \tag{8.9}$$

which, by (8.6) reduces to

$$\frac{z_1}{z_{25}} = 5(\sqrt{5}-2). \tag{8.10}$$

Using (8.10) in (8.7), we find that

$$25P(-e^{-5\pi}) - P(-e^{-\pi/5}) = 300\cdot 5^{1/4}(13\sqrt{5}-29)z_{25}^2. \tag{8.11}$$

Because the remainder of the proof is similar to those in the Sections 4 and 5, we do not record it.

Remark 8.2. Proceeding as in [4], but working with $h_n(q)$ instead of $f_n(q)$, we can also derive all the series for $1/\pi^2$ in [4] that correspond to singular moduli x_3, x_5, x_7, x_9, and x_{25}.

References

[1] G. E. Andrews and B. C. Berndt, Ramanujan's Lost Notebook, Part II, Springer, New York (2009).
[2] N. D. Baruah and B. C. Berndt, Ramanujan's series for $1/\pi$ arising from his cubic and quartic theories of elliptic functions, *J. Math. Anal. Appl.*, **341** (2008) 357–371.
[3] N. D. Baruah and B. C. Berndt, Eisenstein series and Ramanujan-type series for $1/\pi$, *Ramanujan J.*, to appear.
[4] N. D. Baruah and B. C. Berndt, Ramanujan's Eisenstein series and new hypergeometric-like series for $1/\pi^2$, *J. Approx. Theory*, **160** (2009) 135–153.

[5] N. D. Baruah and N. Nayak, New hypergeometric-like series for $1/\pi^2$ arising from Ramanujan's theory of elliptic functions to alternative base 3, *Trans. Amer. Math. Soc.*, **363** (2011) 887–900.

[6] N. D. Baruah and N. Nayak, Some new series for $1/\pi^2$ arising from Ramanujan's class invariant λ_n, Submitted for publication.

[7] N. D. Baruah, B. C. Berndt and H. H. Chan, Ramanujan's series for $1/\pi$: A survey, *Amer. Math. Monthly*, **116** (2009) 567–587.

[8] B. C. Berndt, Modular transformations and generalizations of several formulae of Ramanujan, *Rocky Mountain J. Math.*, **7** (1977) 147–189.

[9] B. C. Berndt, Ramanujan's Notebooks, Part III, Springer-Verlag, New York (1991).

[10] B. C. Berndt, Ramanujan's Notebooks, Part V, Springer-Verlag, New York (1998).

[11] B. C. Berndt and H. H. Chan, *Ramanujan's explicit values for the classical theta-function, Mathematika*, **42** (1995) 278–294.

[12] B. C. Berndt and H. H. Chan, Eisenstein series and approximations to π, *Illinois J. Math.*, **45** (2001) 75–90.

[13] B. C. Berndt, H. H. Chan and W.-C. Liaw, On Ramanujan's quartic theory of elliptic functions, *J. Number Theory*, **88** (2001) 129–156.

[14] B. C. Berndt, H. H. Chan and L.-C. Zhang, Ramanujan's singular moduli, *Ramanujan J.*, **1** (1997) 53–74.

[15] J. M. Borwein and P. B. Borwein, *Pi and the AGM; A Study in Analytic Number Theory and Computational Complexity*, Wiley, New York (1987).

[16] J. M. Borwein and P. B. Borwein, Ramanujan's rational and algebraic series for $1/\pi$, *J. Indian Math. Soc.*, **51** (1987) 147–160.

[17] J. M. Borwein and P. B. Borwein, More Ramanujan-type series for $1/\pi$, in Ramanujan Revisited, (G. E. Andrews, R. A. Askey, B. C. Berndt, K. G. Ramanathan and R. A. Rankin, eds.), Academic Press, Boston (1988) 359–374.

[18] J. M. Borwein, P. B. Borwein and D. H. Bailey, Ramanujan, modular equations, and approximations to pi or how to compute one billion digits of pi, *Amer. Math. Monthly*, **96** (1989) 201–219.

[19] J. M. Borwein and P. B. Borwein, Some observations on computer aided analysis, *Notices Amer. Math. Soc.*, **39** (1992) 825–829.

[20] J. M. Borwein and P. B. Borwein, Class number three Ramanujan type series for $1/\pi$, *J. Comput. Appl. Math.*, **46** (1993) 281–290.

[21] H. H. Chan, S. H. Chan and Z.-G. Liu, Domb's numbers and Ramanujan-Sato type series for $1/\pi$, *Adv. in Math.*, **186** (2004) 396–410.

[22] H. H. Chan and W.-C. Liaw, Cubic modular equations and new Ramanujan-type series for $1/\pi$, *Pacific J. Math.*, **343** (2000) 219–238.

[23] H. H. Chan, W.-C. Liaw and V. Tan, Ramanujan's class invariant λ_n and a new class of series for $1/\pi$, *London Math. Soc.*, (2) **64** (2001) 93–106.

[24] H. H. Chan and K. P. Loo, Ramanujan's cubic continued fraction revisited, *Acta Arith.*, **126** (2007) 305–313.

[25] H. H. Chan and H. Verrill, The Apéry numbers, the Almkvist-Zudilin numbers and new series for $1/\pi$, *Math. Res. Lett.*, **16** (2009) 405–420.

[26] D. V. Chudnovsky and G. V. Chudnovsky, Approximation and complex multiplication according to Ramanujan, in Ramanujan Revisited, (G. E. Andrews, R. A. Askey, B. C. Berndt, K. G. Ramanathan and R. A. Rankin, eds.), Academic Press, Boston (1988) 375–472.

[27] S. Cooper, Series and iterations for $1/\pi$, *Acta Arith.*, **141** (2010) 33–58.

[28] J. Guillera, Some binomial series obtained by the WZ-method, *Adv. Appl. Math.*, **29** (2002) 599–603.

[29] J. Guillera, About a new kind of Ramanujan-type series, *Experiment. Math.*, **12** (2003) 507–510.

[30] J. Guillera, Generators of some Ramanujan formulas, *Ramanujan J.*, **11** (2006) 41–48.

[31] J. Guillera, A new method to obtain series for $1/\pi$ and $1/\pi^2$, *Experiment. Math.*, **15** (2006) 83–89.

[32] J. Guillera, Hypergeometric identities for 10 extended Ramanujan-type series, *Ramanujan J.*, **15** (2008) 219–234.

[33] S. Ramanujan, Modular equations and approximations to π, *Quart. J. Math. (Oxford)*, **45** (1914) 350–372.

[34] S. Ramanujan, Collected Papers, Cambridge Univ. Press, Cambridge (1927); reprinted by Chelsea, New York (1962); reprinted by Amer. Math. Soc., Providence, RI (2000).

[35] S. Ramanujan, Notebooks (2 volumes), Tata Institute of Fundamental Research, Bombay (1957).

[36] S. Ramanujan, The Lost Notebook and Other Unpublished Papers, Narosa, New Delhi (1988).

[37] J. Yi, Theta function identities and the explicit formulas for theta-function and their applications, *J. Math. Anal. Appl.*, **292** (2004) 381–400.

[38] W. Zudilin, Ramanujan-type formulae and irrationality measures of certain multiples of π, *Mat. Sb.*, **196** (2005) 51–66.

[39] W. Zudilin, Quadratic transformations and Guillera's formulae for $1/\pi^2$, *Mat. Zametki.*, **81** (2007) 335–340 (Russian); *Math. Notes.*, **81** (2007) 297–301.

[40] W. Zudilin, More Ramanujan-type formulae for $1/\pi^2$, *Russian Math. Surveys*, **62** (2007) 634–636.

[41] W. Zudilin, Ramanujan-type formulae for $1/\pi$: A second wind, in Modular Forms and String Duality, (N. Yui, H. Verril and C. F. Doran, eds.), Fields Institute Communications, vol. 54, American mathematical Society and the Fields Institute for Research in Mathematical Sciences, Providence, RI (2008) pp. 179–188.

[42] W. Zudilin, Ramanujan-type supercongruences, *J. Number Theory*, **129** (2009) 1848–1857.

Ramanujan Rediscovered No. 14, 2010, pp. 31–51.

What is a q-Series?

Bruce C. Berndt

Department of Mathematics, University of Illinois at Urbana-Champaign,
1409 West Green Street, Urbana, IL 61801, USA
e-mail: berndt@illinois.edu

Abstract. Historically, research in q-series has not always been appreciated, and those who accomplished it even less acknowledged. "The problem is very much that of unscrambling an egg; we have to reverse the substitution $q = 1$, and this involves us, especially with series, in inserting, variously, various unexpected q^N ...." "...though everything he wrote was marked by a certain distinction, nothing else of first-rate importance was discovered. ... Yet of what the world counts success he achieved practically nothing." Despite the unkind remarks about two of the founders of the theory of q-series, upon which we shall elaborate later, the beautiful and useful theorems in q-series are now cherished, and many of their progenitors greatly admired. It is our fondest wish that the introduction to q-series that follows will inspire formerly uninitiated readers to willingly contract the "q-disease," in the words of Richard Askey, who has "suffered" from this disease for several decades.

Keywords. q-series, basic hypergeometric series, theta functions, partitions.

In Memory of K. Venkatachaliengar

1. Introduction

What is a q-series? The simplest and most manifestly useless definition would be a series with q's in the summands. Having begun this essay with meaningless drivel, we next admit that there does not exist a "good" definition of a q-series. We might define a q-series to be one with summands containing expressions of the type

$$(a)_n := (a; q)_n := (1-a)(1-aq)\cdots(1-aq^{n-1}), \quad n \geq 0, \tag{1.1}$$

where we interpret $(a; q)_0 = 1$. If the *base* q is understood, we often use the notation at the far left-hand side of (1.1), but in this paper, since we are going to use this notation for *rising factorials* or *shifted factorials*, to avoid confusion, we always write $(a; q)_n$. This definition of a q-series is also not completely satisfactory, because often in the theory of q-series, we let parameters in the summands tend to 0 or to ∞, and consequently it may happen that no factors of the type $(a; q)_n$ remain in the summands. In some cases, what may remain is a theta function. Following the lead of S. Ramanujan, we define a general theta function $f(a, b)$ by

$$f(a,b) := \sum_{n=-\infty}^{\infty} a^{n(n+1)/2} b^{n(n-1)/2}, \quad |ab| < 1. \tag{1.2}$$

In the theory of q-series, theta functions also frequently arise in identities satisfied by series with products (1.1) in their summands. Thus, for these reasons, theta functions are an integral part of the theory and are also considered to be q-series. As we shall see in the sequel, infinite q-products

$$(a;q)_\infty := \lim_{n\to\infty}(a;q)_n, \qquad |q|<1,$$

arise both in product representations of theta functions and, more generally, in identities for q-series.

In the theory of q-series, there is an important class of series that are defined for $|z|<1$ by

$$ {}_r\phi_s\left(\begin{matrix} a_1, a_2, \ldots, a_r \\ b_1, b_2, \ldots, b_s \end{matrix}; q, z\right)$$

$$:= \sum_{n=0}^{\infty} \frac{(a_1;q)_n(a_2;q)_n\cdots(a_r;q)_n}{(b_1;q)_n(b_2;q)_n\cdots(b_s;q)_n(q;q)_n}\left((-1)^n q^{n(n-1)/2}\right)^{1+s-r} z^n, \qquad (1.3)$$

which are called *basic hypergeometric series*. Observe that the quotient of successive terms in (1.3) is a rational function of q^n, and conversely every such series with this property is of the form (1.3). The parameters $a_1, a_2, \ldots, a_r, b_1, b_2, \ldots, b_s$ are arbitrary complex numbers, except for the restrictions $b_j \neq q^{-n}$, $1 \leq j \leq s$, $n \geq 0$. If the number of parameters is "small," then we write them on one line, instead of two; e.g., we usually write ${}_2\phi_1(a, b; c; q, t)$.

In Section 2, we give a brief history of q-series indicating a few of the prominent theorems in the subject and describing some of the mathematicians who played leading roles in its development. In the following Section 3, we indicate some of the primary tools that are used to prove theorems about q-series. We concentrate on the use of combinatorial methods, because of exciting recent activity and because they often give new, fascinating insights into the combinatorial nature of q-series identities. Theorems about basic hypergeometric series are often *q-analogues* of results about ordinary or generalized hypergeometric series, and so in Section 4, we describe this symbiosis and offer a few examples in illustration. In case readers think that theta functions might arise only trivially or coincidentally in the theory of q-series, in Section 5, we provide a few examples where theta functions arise in some of Ramanujan's beautiful theorems, in particular, on mock theta functions. Lastly, by the time readers reach the end of Section 5, if we have been successful in stimulating them to action, they will want to dig more deeply into the subject, and so we indicate in Section 6 some of the primary sources from which one can study and learn more about the properties and applications of q-series.

2. Important theorems and figures in the history of q-series

One could claim that the theory of q-series began with certain famous theorems of L. Euler [36, Chapter 16] and C. F. Gauss [40]. We begin non-chronologically with Gauss, who is generally considered to be the founder of the theory of theta functions.

Recall first that the general theta function $f(a, b)$ is defined by (1.2), and recall second that the famous Jacobi triple product identity [3, p. 21, Theorem 2.8], [21, p. 35, Entry 19] is given, for $|ab| < 1$, by

$$f(a, b) = (-a; ab)_\infty(-b; ab)_\infty(ab; ab)_\infty, \tag{2.1}$$

which is arguably the most useful and celebrated theorem in the theory of theta functions, and which was actually first discovered by Gauss. In Ramanujan's notation, perhaps the three most important special cases of (1.2) are defined by, together with their Jacobi triple product representations,

$$\varphi(q) := f(q, q) = \sum_{n=-\infty}^{\infty} q^{n^2} = (-q; q^2)_\infty^2(q^2; q^2)_\infty, \tag{2.2}$$

$$\psi(q) := f(q, q^3) = \sum_{n=0}^{\infty} q^{n(n+1)/2} = \frac{(q^2; q^2)_\infty}{(q; q^2)_\infty}, \tag{2.3}$$

$$f(-q) := f(-q, -q^2) = \sum_{n=-\infty}^{\infty} (-1)^n q^{n(3n-1)/2} = (q; q)_\infty. \tag{2.4}$$

The identity (2.4) is called Euler's pentagonal number theorem, which, in a less abbreviated notation, is given by

$$\sum_{n=-\infty}^{\infty} (-1)^n q^{n(3n-1)/2} = \prod_{n=1}^{\infty}(1 - q^n). \tag{2.5}$$

To discern its combinatorial interpretation, we first define the partition function $p(n)$ to be the number of ways the positive integer n can be written as a sum of positive integers, with the order of these positive integers immaterial. For example, $p(5) = 7$, because there are 7 ways to write 5 as a sum of positive integers, namely, 5, $4 + 1$, $3+2, 3+1+1, 2+2+1, 2+1+1+1, 1+1+1+1+1$. Observe that $\prod_{n=1}^{\infty}(1+q^n)$ generates the number of partitions of a positive integer into *distinct* parts. With this in mind, we offer a combinatorial interpretation of (2.5).

Corollary 2.1 (Combinatorial Version of Euler's Pentagonal Number Theorem). *Let $D_e(n)$ denote the number of partitions of n into an even number of distinct parts, and let $D_o(n)$ denote the number of partitions of n into an odd number of distinct parts. Then*

$$D_e(n) - D_o(n) = \begin{cases} (-1)^j, & \text{if } n = j(3j \pm 1)/2, \\ 0, & \text{otherwise}. \end{cases} \tag{2.6}$$

We next give perhaps the most important and useful theorem in which a q-product $(a; q)_n$ appears in the summands.

Theorem 2.2. *For* $|q|, |z| < 1$,

$$\sum_{n=0}^{\infty} \frac{(a;q)_n}{(q;q)_n} z^n = \frac{(az;q)_\infty}{(z;q)_\infty}. \tag{2.7}$$

Proof. Note that the product on the right side of (2.7) converges uniformly on compact subsets of $|z| < 1$ and so represents an analytic function on $|z| < 1$. Thus, we may write

$$F(z) := \frac{(az;q)_\infty}{(z;q)_\infty} = \sum_{n=0}^{\infty} A_n z^n, \qquad |z| < 1. \tag{2.8}$$

From the product representation in (2.8), we can readily verify that

$$(1-z)F(z) = (1-az)F(qz). \tag{2.9}$$

Equating coefficients of z^n, $n \geq 1$, on both sides of (2.9), we find that

$$A_n - A_{n-1} = q^n A_n - aq^{n-1} A_{n-1},$$

or

$$A_n = \frac{1-aq^{n-1}}{1-q^n} A_{n-1}, \qquad n \geq 1. \tag{2.10}$$

Iterating (2.10) and using the value $A_0 = 1$, which is readily apparent from (2.8), we deduce that

$$A_n = \frac{(a;q)_n}{(q;q)_n}, \qquad n \geq 0. \tag{2.11}$$

Using (2.11) in (2.8), we complete the proof of (2.7).

Corollary 2.3. *For* $|q| < 1$,

$$\sum_{n=0}^{\infty} \frac{z^n}{(q;q)_n} = \frac{1}{(z;q)_\infty}, \qquad |z| < 1, \tag{2.12}$$

and

$$\sum_{n=0}^{\infty} \frac{(-z)^n q^{n(n-1)/2}}{(q;q)_n} = (z;q)_\infty, \qquad |z| < \infty. \tag{2.13}$$

Theorem 2.2 is called the *q-analogue of the binomial theorem*, and in Section 4 we relate in detail why this name is given to this theorem. In its general form, Theorem 2.2 is due to A. L. Cauchy [30], while the special cases in Corollary 2.3 are due to Euler [36, Chapter 16]. The proof of (2.7) given above follows along the lines of that given by Euler and Cauchy and has been copied from the author's book [24, p. 8].

Although certain important theorems were found by Euler, Gauss, and Cauchy, the systematic development of the theory of q-series began with a paper by E. Heine [43] in 1847. In particular, the following theorem, called Heine's transformation, is perhaps his signature theorem.

Theorem 2.4. *For* $|t|$, $|b| < 1$,

$$ {}_2\phi_1(a, b; c; q, t) = \frac{(b; q)_\infty (at; q)_\infty}{(c; q)_\infty (t; q)_\infty} {}_2\phi_1(c/b, t; at; q, b). \tag{2.14} $$

In his lost notebook [58], Ramanujan, who independently discovered Heine's transformation [57, Chapter 16, Entry 6], [21, p. 15], found many applications of it [10, Chapter 1].

Two English mathematicians, F. H. Jackson and L. J. Rogers, at the end of the 19th and beginning of the 20th centuries devoted most of their mathematical careers to further developing the theory of q-series, but their efforts were not appreciated by their contemporary researchers. We first discuss Rogers.

L. J. Rogers, who was born on March 30, 1862 and died on September 12, 1933, was recognized during his lifetime for only a handful of his contributions to mathematics. As a mathematics student, he was elected to a Scholarship at Balliol College, Oxford in 1879, but he also earned a Bachelor of Music degree in 1884. In 1888, he became Professor of Mathematics at Yorkshire College, now called the University of Leeds. It appeared that music, in fact, took precedence over mathematics in his professional life. A. L. Dixon wrote one of the two obituaries that we have of Rogers; let us quote from that obituary [35].

> Rogers was a man of extraordinary gifts in many fields, and everything he did, he did well. Besides his mathematics and music he had many interests; he was a born linguist and phonetician, a wonderful mimic who delighted to talk broad Yorkshire, a first-class skater, and a maker of rock gardens.
>
> He did things, and did them well, because he liked doing them; but he had nothing of the professional outlook, and his knowledge of other people's work was vague. He had very little ambition or desire for recognition.

Nonetheless, the Rogers–Ramanujan identities (given below in (2.15) and (2.16)), the Rogers–Ramanujan continued fraction, and the Rogers–Szegö polynomials are important discoveries that are named after him. Moreover, as pointed out by G. H. Hardy, J. E. Littlewood, and G. Pólya [42, p. 25], Rogers [59] discovered Hölder's famous inequality one year before O. Hölder [45], and in a more symmetrical form.

During his lifetime, Rogers was singularly known for the two identities

$$ G(q) := \sum_{n=0}^{\infty} \frac{q^{n^2}}{(q; q)_n} = \frac{1}{(q; q^5)_\infty (q^4; q^5)_\infty}, \tag{2.15} $$

$$ H(q) := \sum_{n=0}^{\infty} \frac{q^{n(n+1)}}{(q; q)_n} = \frac{1}{(q^2; q^5)_\infty (q^3; q^5)_\infty}, \tag{2.16} $$

which he discovered in 1894 [60], but which were entirely ignored until Ramanujan rediscovered them about 20 years later. Let us quote from a second obituary [62] of Rogers. "Rogers' reputation as a mathematician rests almost entirely on a single incident." In other words, if it had not been for Ramanujan's rediscovery of these now famous identities, in the collective opinion of Rogers' contemporary mathematicians, his name would have been completely forgotten. For a history of these identities, see, for example, Ramanujan's *Collected Papers* [56, pp. 344–346], the monograph [4] by G. E. Andrews, Andrews's survey article [6], or the survey paper written by the author, Y.-S. Choi, and S.-Y. Kang [25].

The Rogers–Ramanujan continued fraction, also first studied by Rogers in [60], is defined by

$$R(q) := \frac{q^{1/5}}{1} \, \frac{}{+} \, \frac{q}{1} \, \frac{}{+} \, \frac{q^2}{1} \, \frac{}{+} \, \frac{q^3}{1} \, \frac{}{+} \cdots, \qquad |q| < 1. \tag{2.17}$$

The continued fraction (2.17) is intimately connected with the Rogers–Ramanujan functions by the beautiful theorem

$$R(q) = q^{1/5}\frac{H(q)}{G(q)} = q^{1/5}\frac{(q;q^5)_\infty(q^4;q^5)_\infty}{(q^2;q^5)_\infty(q^3;q^5)_\infty}, \tag{2.18}$$

by (2.15) and (2.16), which was also first proved by Rogers [60] and rediscovered by Ramanujan. Although Rogers established several of its properties, Ramanujan proved many more theorems about $R(q)$, most of which are found in his lost notebook [58]. For an account of a sizeable portion of Ramanujan's discoveries about $R(q)$, see the book by Andrews and the author [9].

We quote again from the obituary [62]; more extensive quotations may be found in Andrews's absorbing monograph [4].

> Of course the rest of Rogers' work was carefully studied; but, though everything he wrote was marked by a certain distinction, nothing else of first-rate importance was discovered. He was elected a fellow of the Royal Society in 1924, and relapsed into his former obscurity.
>
> To all who knew him but slightly, that incident must appear typical of Rogers. Fine abilities, they would say, wasted by freakishness and inconstancy! ... Rogers' major abilities can seldom have been excelled in extent or variety; and his minor accomplishments ranged from knitting to skating. ... Certainly he might have won great fame in music, either as scholar or executant; and surely no position in diplomacy would have been unattainable to one endowed with his easy mastery of languages, his quick intelligence, his sparkling wit, his fine presence, his athletic grace, his courtly charm that no woman could resist. Yet of what the world counts success he achieved practically nothing.

Of course, the first sentence of the quote above is complete nonsense. However, in defense of those in his time who might have attempted to understand Rogers' papers, it is to be admitted that Rogers wrote in a manner that makes his work very difficult to understand. We have discussed only the most famous of Rogers' discoveries, but in recent years, due to the efforts of Askey, Andrews, and others, considerably more of Rogers' mathematics has been discovered and appreciated than heretofore. However, some of Rogers' papers have yet to be thoroughly examined.

Jackson was contemporaneous with Rogers and even less appreciated for his work. The *Bible* in the theory of basic hypergeometric series is the text [39] by G. Gasper and M. Rahman. Readers of this book will find many results due to Jackson and rightly conclude that he, indeed, is one of the founders of the subject. For example, using Heine's transformation (2.14), one can derive Jackson's transformation formula [39, p. 10]

$$ {}_2\phi_1(a, b; c; q, z) = \frac{(az; q)_\infty}{(z; q)_\infty} {}_2\phi_2(a, c/b; c, az; q, bz). $$

As another example, Jackson proved that for each positive integer n [39, p. 17],

$$ {}_3\phi_2(a, b, q^{-n}; c, abc^{-1}q^{1-n}; q, q) = \frac{(c/a; q)_n (c/b; q)_n}{(c; q)_n (c/(ab); q)_n}. \tag{2.19} $$

Observe that the series on the left-hand side above terminates.

Jackson, who was born on August 16, 1870 and died on April 27, 1960 at the age of 89, authored 49 papers, most of them on q-series. Although he took the Tripos and was a Wrangler at the early age of 19, he chose the ministry as his profession. As an ordained minister, Jackson served for ten years as a chaplain and instructor in the Royal Navy, and held various positions in the Church of England throughout his career. Let us now quote from T. W. Chaundy's obituary [33] of this "amateur" mathematician.

> After a few practice pieces (as was usual in those days) titivating known results, his whole mathematical output was devoted to basic analogues or q-analogues (as they are obscurely called). .... The problem is very much that of unscrambling an egg; we have to reverse the substitution $q = 1$, and this involves us, especially with series, in inserting, variously, various unexpected q^N (often with N quadratic in n). Perhaps these were not unexpected to Jackson, for he seemed to have an especial flair for these extensions. .... He had a ceaseless and vivid imagination which (to this writer) seemed often to tax his powers of exposition. ... In all he was the enthusiastic amateur with gifts that, given the opportunity, might have led him to a professional status.

As further evidence of the contempt in which his work was regarded, he once read a paper before the London Mathematical Society, to which someone remarked, "Surely, Mr. President, we have heard all this before." Jackson quickly strode from the room and never darkened the pages of the Society again [33].

Ramanujan undoubtedly contributed more to q-series than anyone either before or after his time. His most famous theorem is his ${}_1\psi_1$ summation theorem, a theorem about bilateral hypergeometric series, which, in importance and usefulness, plays the same role in the theory of q-bilateral series as the q-binomial theorem plays in unilateral q-series. Our first task then is to extend the definition of $(a; q)_n$ to negative integers. To that end, define, for every integer n,

$$ (a; q)_n = \frac{(a; q)_\infty}{(aq^n; q)_\infty}. \tag{2.20} $$

Note that (2.20) is equivalent to (1.1) when $n \geq 0$.

Theorem 2.5 (Ramanujan's ${}_1\psi_1$ Summation). *For $|b/a| < |z| < 1$ and $|q| < 1$,*

$$\sum_{n=-\infty}^{\infty} \frac{(a;q)_n}{(b;q)_n} z^n = \frac{(az;q)_\infty (q/(az);q)_\infty (q;q)_\infty (b/a;q)_\infty}{(z;q)_\infty (b/(az);q)_\infty (b;q)_\infty (q/a;q)_\infty}. \tag{2.21}$$

Ramanujan's ${}_1\psi_1$ summation theorem was first stated by Ramanujan in his notebooks [57, Chapter 16, Entry 17]. It was found there by Hardy, who called it, "a remarkable formula with many parameters" and intimated that it could be established by employing the q-binomial theorem [41, pp. 222–223]. There now exist many proofs of Theorem 2.5, and a list of all known proofs (22 up to the printing of [10]), with a few brief descriptions of proofs, can be found in [10, Chapter 3, pp. 54–56]. In this source, one can also find several identities from the lost notebook that can be proved with the use of the ${}_1\psi_1$ summation theorem. The q-binomial theorem and the Jacobi triple product identity are both corollaries of Ramanujan's ${}_1\psi_1$ summation theorem. Connections of the q-binomial theorem and the ${}_1\psi_1$ theorem with q-analogues of the gamma and beta functions have been eloquently discussed by Askey [13], [14].

In closing our historical account, we remark that we do not know when it became standard to designate q as the variable. Heine used the letter x in place of q. In the theory of elliptic functions, q has been in use at least since the time of C. G. J. Jacobi, who used it in all of his papers, in particular, in his famous *Fundamenta Nova* [46]. Ramanujan used x as his variable in his earlier notebooks [57] but used q in his published papers [56] and lost notebook [58]. Rogers employed the notation q in all of his papers, and it is possible that the use of q in the theory of q-series and, more precisely, in the theory of basic hypergeometric series, was therefore instigated by Rogers.

3. The tools used to prove theorems about q-series

1. The proof of Theorem 2.2 illustrates the most common approach to proving identities for q-series. In that proof we established a functional equation relating $F(z)$ and $F(qz)$. This yielded a functional equation for the coefficients, which we iterated to complete the proof. Employing q-difference equations, i.e., equations relating functions at different arguments involving q, is the most potent method that we have at our disposal.

2. In the 1940's, W. N. Bailey [15], [16] introduced a powerful method for producing q-identities. One begins by defining a sequence $\{\beta_n\}$ in terms of any given arbitrary sequence $\{\alpha_n\}$. From these two sequences, one can then derive another pair of sequences $\{\alpha'_n\}$ and $\{\beta'_n\}$, each containing the same two free parameters. Of course, one needs to be clever in constructing the two original sequences in order that one can obtain two new meaningful and useful sequences. In recent years, several authors have derived a large number of beautiful identities by ingeniously employing versions of Bailey's Lemma. We now state Bailey's Lemma, as it is given in the text of Andrews, Askey, and Roy [8, p. 584].

Theorem 3.1. *Let* $\{\alpha_n\}$, $n \geq 0$, *be a sequence of complex numbers, and define*

$$\beta_n = \sum_{r=0}^{n} \frac{\alpha_r}{(q;q)_{n-r}(aq;q)_{n+r}}.$$

Define another sequence $\{\alpha'_n\}$ *by*

$$\alpha'_n := \frac{(\rho_1;q)_n(\rho_2;q)_n(aq/(\rho_1\rho_2);q)_n\alpha_n}{(aq/\rho_1;q)_n(aq/\rho_2;q)_n},$$

where ρ_1 *and* ρ_2 *are arbitrary non-zero complex numbers. Then*

$$\beta'_n = \sum_{r=0}^{n} \frac{\alpha'_r}{(q;q)_{n-r}(aq;q)_{n+r}},$$

where

$$\beta'_n := \sum_{j=0}^{\infty} \frac{(\rho_1;q)_j(\rho_2;q)_j(aq/(\rho_1\rho_2);q)_{n-j}\beta_j}{(q;q)_{n-j}(aq/\rho_1;q)_n(aq/\rho_2;q)_n}\left(\frac{aq}{\rho_1\rho_2}\right)^j.$$

Among those who have fruitfully developed the theory of Bailey pairs are Lucy Slater, Andrews, David Bressoud, S. Ole Warnaar, and Andrew Sills.

3. Using partial fractions is another effective technique for producing q-series identities. In the past few years, Andrews and this author have concluded that Ramanujan likely employed partial fractions far more than was formerly believed. See Chapter 12 in the first book [9] that Andrews and Berndt wrote on the lost notebook, the papers [7] and [23] by Andrews and Berndt, respectively, and S. H. Chan's thesis [31] and paper [32].

4. Combinatorial reasoning is an extremely compelling method for proving q-series identities. Although, arguably the combinatorics of q-series began with Euler's pentagonal number theorem given by (2.5) or (2.6), it lay dormant until it received a tremendous impetus with the bijections of J. J. Sylvester [66] and F. Franklin [38] in the 1880's, followed by P. A. MacMahon [51] early in the twentieth century. In the last few decades, the combinatorial efforts of several authors, led especially by Andrews, Bressoud, Frank Garvan, Krishnaswami Alladi, Sylvie Corteel, and Ae Ja Yee, have provided not only insightful proofs but have also led to new theorems as well. We briefly discuss some of these contributions.

Recall that prior to Corollary 2.1, we defined the ordinary partition function $p(n)$. Euler first established the generating function for $p(n)$,

$$\sum_{n=0}^{\infty} p(n)q^n = \frac{1}{(q;q)_\infty},$$

and so it is perhaps not surprising that partitions arise ubiquitously in combinatorially proving q-series identities. Euler found two further identities for $1/(q;q)_\infty$. First, a trivial consequence of Euler's theorem (2.12) is

$$\sum_{n=0}^{\infty} \frac{q^n}{(q;q)_n} = \frac{1}{(q;q)_\infty}. \tag{3.1}$$

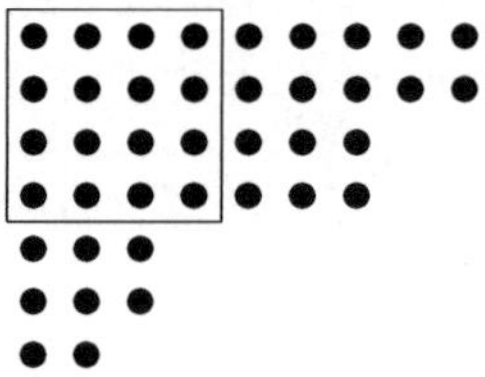

Figure 1. Durfee square of side $n = 4$

The identity (3.1) can easily be proved combinatorially. Observe that $1/(q;q)_n$ generates partitions into parts with each part less than or equal to n. The numerator q^n generates the part n. Thus, $q^n/(q;q)_n$ generates all partitions with largest part equal to n. The left side of (3.1) thus generates all partitions while sorting them out according to the largest part n. The partition $9+9+7+7+3+3+2$ in the Ferrers diagram depicted above is therefore one of the partitions counted by the summand with $n = 9$ on the left-hand side of (3.1). Euler also established the beautiful identity

$$\sum_{n=0}^{\infty} \frac{q^{n^2}}{(q;q)_n^2} = \frac{1}{(q;q)_\infty}. \tag{3.2}$$

The identity (3.2) can be proved combinatorially as follows. Consider the Ferrers graph of an arbitrary partition, and extract the largest *Durfee square* of side n. These nodes are then generated by q^{n^2}. The nodes to the right of the square form a partition into parts not exceeding n, while the nodes below the Durfee square also form a partition into parts less than or equal to n. These two partitions are generated by $1/(q;q)_n^2$, and so the generating function on the left side of (3.2) generates all partitions and separates them according to their largest Durfee square. Thus in Figure 1, the largest Durfee square is of size 4^2 with the partition at the right being $4+4+4+2+2$ and the partition below the square being $3+3+2$. See also Andrews's book [3, pp. 27–28].

Probably the simplest partition-theoretic q-identity is Euler's theorem

$$\frac{1}{(q;q^2)_\infty} = (-q;q)_\infty, \tag{3.3}$$

which is equivalent to the assertion that the number of partitions of a positive integer n into odd parts is equal to the number of partitions of n into distinct parts. Sylvester [66] developed a beautiful bijection between partitions with only odd parts and partitions with distinct parts in order to prove (3.3). This bijection is described by Andrews [5] in his interesting paper on Sylvester's work on partitions.

We remarked above that Euler's pentagonal number theorem has an elegant partition-theoretic interpretation given by Corollary 2.1. One might ask if there is a bijective proof of it, and indeed in 1881, Franklin [38] devised what is now widely known as Franklin's bijection to prove Corollary 2.1; see also [3, pp. 10–11]. Moreover, Franklin's bijection, like Sylvester's bijection, has been used on many occasions to give combinatorial proofs of further q-series identities. For example, see papers by Berndt and Yee [28] and Berndt, B. Kim, and Yee [26].

It was not until 1965 that a bijective proof of the Jacobi triple product identity (2.1) was developed by E. M. Wright [69], and since then it has had numerous applications; see, for example, the papers by Warnaar [68] and S. Kim [49].

Leading the modern era in providing combinatorial proofs of q-series identities is Andrews. We cite just one of his many papers on the subject, namely, [2] where a combinatorial proof of the useful Rogers–Fine identity [61], [37, p. 15]

$$\sum_{n=0}^{\infty} \frac{(\alpha; q)_n}{(\beta; q)_n} \tau^n = \sum_{n=0}^{\infty} \frac{(\alpha; q)_n (\alpha\tau q/\beta; q)_n \beta^n \tau^n q^{n^2-n} (1 - \alpha\tau q^{2n})}{(\beta; q)_n (\tau; q)_{n+1}}$$

is given.

The first bijective proof of the q-binomial theorem, Theorem 2.2, was given in 1987 by J. T. Joichi and D. Stanton [47]; see also I. Pak's survey, which provides several combinatorial proofs of partition and q-series identities. A variation of Joichi and Stanton's proof has been devised by Yee [70] to provide another bijective proof of the q-binomial theorem.

Lastly, in 2004, Yee [70] devised a remarkable bijective proof of Ramanujan's ${}_1\psi_1$ summation theorem, Theorem 2.5.

We conclude our discussion of combinatorial proofs by offering a few remarks about an identity

$$\sum_{n=0}^{\infty} a^n q^{n^2} = \sum_{n=0}^{\infty} \frac{(-q; q)_{n-1} a^n q^{n(n+1)/2}}{(-aq^2; q^2)_n} \tag{3.4}$$

found on page 28 in Ramanujan's lost notebook [58]. (The claims that we make below about partitions are not easy to establish [26].) Let D_n be the set of partitions into n distinct parts less than $2n$ such that the smallest part of each partition is 1, and if $2k - 1$ is the largest odd part, then all odd positive integers less than $2k - 1$ occur as parts. Let E_n be the set of partitions into even parts less than or equal to $2n$. In giving a bijective proof of (3.4), Berndt, Kim, and Yee [26] show that the right-hand side of (3.4) generates pairs of partitions (π, σ) with $\pi \in D_n$ and $\sigma \in E_n$, where the exponent of a denotes the number of parts of π minus the number of parts of σ. When we examine the left-hand side of (3.4), we see that the the partitions π and σ "cancel each other out," except on a very thin set, the squares, where only one single partition survives the cancellation. This example, which has also been studied combinatorially by Yee [71] and Alladi [1], and many other such identities illustrate why the combinatorial study of q-series identities is fascinating.

5. To complete our list of tools, we briefly mention two theories that have not been successful in establishing q-series identities. For example, although there are many proofs of the Rogers–Ramanujan identities [6], many involving theta functions, proofs have never been constructed using either the theory of elliptic functions or of modular forms. Theta functions play a prominent role in the theory of elliptic functions, which reached its peak in the late nineteenth and early twentieth centuries. However, in the past decade or so, Zhi-Guo Liu has written several papers demonstrating the power of elliptic functions in establishing many of Ramanujan's results on theta functions as

well as new theorems in the subject. Theta functions also play a leading role in the theory of modular forms. Although, as we have seen, theta functions are inextricably intertwined with the theory of basic hypergeometric series, the development of the theory of basic hypergeometric series has kept the theories of elliptic functions and modular forms in abeyance standing at the door. Will these theories ever be able to make substantial inroads into basic hypergeometric series?

4. The symbiosis of hypergeometric series and basic hypergeometric series

If $p \geq 0$, let $a_1, a_2, \ldots, a_{p+1}$ and $b_1, b_2, \ldots, b_p$ be arbitrary complex numbers, except that b_j, $1 \leq j \leq p$, cannot be a non-positive integer. The generalized hypergeometric function ${}_{p+1}F_p$ is defined by

$$ {}_{p+1}F_p\left(\begin{matrix} a_1, a_2, \ldots, a_{p+1} \\ b_1, b_2, \ldots, b_p \end{matrix}; z\right) := \sum_{n=0}^{\infty} \frac{(a_1)_n (a_2)_n \cdots (a_{p+1})_n}{(b_1)_n (b_2)_n \cdots (b_p)_n n!} z^n, \qquad |z| < 1, \tag{4.1} $$

where

$$ (a)_0 := 1, \qquad (a)_n := a(a+1)\cdots(a+n-1), \qquad n \geq 1, \tag{4.2} $$

is called the *rising* or *shifted factorial*. If the number of parameters is "small," then in place of the left-hand side of (4.1), we write ${}_{p+1}F_p(a_1, a_2, \ldots, a_{p+1}; b_1, b_2, \ldots, b_p; z)$. Observe that the quotient of successive terms in (4.1) is a rational function of n, and conversely every such series with this property is a hypergeometric series.

Replacing a by q^a in (1.1), where $a > 0$, and using L'Hôspital's rule while letting q tend to 1, we find that

$$ \begin{aligned} \lim_{q \to 1} \frac{(q^a; q)_n}{(1-q)^n} &= \lim_{q \to 1} \frac{1-q^a}{1-q} \frac{1-q^{a+1}}{1-q} \cdots \frac{1-q^{a+n-1}}{1-q} \\ &= a(a+1)\cdots(a+n-1) = (a)_n. \end{aligned} \tag{4.3} $$

The equality (4.3) demonstrates an intimate relation between basic hypergeometric series defined in (1.3) and hypergeometric series defined in (4.1). In particular, formally,

$$ \lim_{q \to 1^-} {}_{p+1}\phi_p\left(\begin{matrix} q^{a_1}, q^{a_2}, \ldots, q^{a_{p+1}} \\ q^{b_1}, q^{b_2}, \ldots, q^{b_p} \end{matrix}; q, z\right) = {}_{p+1}F_p\left(\begin{matrix} a_1, a_2, \ldots, a_{p+1} \\ b_1, b_2, \ldots, b_p \end{matrix}; z\right). $$

One could define many other q-series, such that when we let $q \to 1$, they tend to the same hypergeometric series. Of course, most of these q-series would not play meaningful roles in any theory of q-series. The series and theorems in q-series that have counterparts in the theory of hypergeometric series are called *q-analogues*. Does every theorem about hypergeometric series have a natural, meaningful q-analogue? Is it possible for a theorem about ordinary hypergeometric series to have more than one q-analogue?

Let us begin with a q-analogue of the binomial coefficient $\binom{n}{m}$. Let n and m denote integers. Then the Gaussian polynomial or q-binomial coefficient is defined by

$$\begin{bmatrix} n \\ m \end{bmatrix} := \begin{cases} \frac{(q;q)_n}{(q;q)_m(q;q)_{n-m}}, & \text{if } 0 \le m \le n, \\ 0, & \text{otherwise.} \end{cases} \tag{4.4}$$

If we let q tend to 1 in (4.4) and employ (4.3), we easily see that

$$\lim_{q \to 1} \begin{bmatrix} n \\ m \end{bmatrix} = \binom{n}{m}.$$

We are now able to answer the last question at the close of the penultimate paragraph. It is easy to show by induction on n that, for $n \ge 1$,

$$\begin{bmatrix} n \\ m \end{bmatrix} = \begin{bmatrix} n-1 \\ m-1 \end{bmatrix} + q^m \begin{bmatrix} n-1 \\ m \end{bmatrix}, \tag{4.5}$$

$$\begin{bmatrix} n \\ m \end{bmatrix} = q^{n-m} \begin{bmatrix} n-1 \\ m-1 \end{bmatrix} + \begin{bmatrix} n-1 \\ m \end{bmatrix}. \tag{4.6}$$

Both (4.5) and (4.6) are q-analogues of Pascal's familiar recurrence formula for binomial coefficients, which is obtained by letting $q \to 1$ in either (4.5) or (4.6). Thus, more than one q-analogue may exist!

Next we show that Theorem 2.2 is justified in being christened the *q-binomial theorem*. In (2.7), replace a by q^a, where we now assume that a is an integer. After simplifying the right-hand side of (2.7) and then letting q tend to 1, we deduce the ordinary binomial theorem in the form

$$\sum_{n=0}^{\infty} \frac{(a)_n}{n!} z^n = (1-z)^{-a}. \tag{4.7}$$

If a is not an integer, letting $q \to 1$ in (2.7) still yields (4.7), but it is considerably more difficult to obtain the right-hand side of (4.7) [8, p. 491, Theorem 10.2.4].

We provide another example in which a famous theorem about ordinary hypergeometric series has an analogue in the theory of basic hypergeometric series. In 1812, Gauss proved that if $\mathrm{Re}(c-a-b) > 0$ [8, p. 66, Theorem 2.2.2],

$${}_2F_1(a, b; c; 1) = \frac{\Gamma(c)\Gamma(c-a-b)}{\Gamma(c-a)\Gamma(c-b)}. \tag{4.8}$$

The special case

$${}_2F_1(-n, a; c; 1) = \frac{(c-a)_n}{(c)_n}, \tag{4.9}$$

where n is a non-negative integer, is called the Chu–Vandermonde Theorem, because S.-C. Chu in 1303 [34] and A. T. Vandermonde in 1772 [67] had previously discovered (4.9). The q-analogue of (4.8) is due to Heine. For $|c/(ab)| < 1$,

$${}_2\phi_1(a, b; c; q, c/(ab)) = \frac{(c/a; q)_\infty (c/b; q)_\infty}{(c; q)_\infty (c/(ab); q)_\infty}. \tag{4.10}$$

Note that Euler's identity (3.2) is the special case $a = b = 0, c = q$ of (4.10). To show that (4.10) is a q-analogue of (4.8), it is perhaps best to introduce a q-analogue of the classical gamma function. Observe that

$$[n]_q! := \frac{(1-q)(1-q^2)\cdots(1-q^n)}{(1-q)^n}, \qquad n \geq 0, \tag{4.11}$$

is clearly an analogue of $n!$. Observe also that we may rewrite (4.11) in the form

$$[n]_q! = \frac{(q;q)_\infty}{(q^{n+1};q)_\infty}(1-q)^{-n}, \qquad |q| < 1. \tag{4.12}$$

But (4.12) can be used to define $[n]_q!$ even when n is *not* a non-negative integer. Hence, in analogy with the classical gamma function, we are led to define its q-analogue by

$$\Gamma_q(x) := \frac{(q;q)_\infty}{(q^x;q)_\infty}(1-q)^{1-x}, \qquad |q| < 1.$$

Then it can be shown that [8, p. 495, Corollary 10.3.4]

$$\lim_{q \to 1^-} \Gamma_q(x) = \Gamma(x). \tag{4.13}$$

If we now replace a, b, and c in (4.10) by q^a, q^b, and q^c, respectively, let q tend to 1, and employ (4.13), we immediately deduce (4.8). Setting $b = q^{-n}$ in (4.10) and reversing the order of summation, we can obtain a q-analogue of the Chu–Vandermonde Theorem (4.9) in the form

$${}_2\phi_1(q^{-n}, a; c; q, q) = \frac{(c/a;q)_n}{(c;q)_n}a^n. \tag{4.14}$$

Observe that Jackson's theorem (2.19) is a generalization of (4.14), which can be seen by letting $b \to \infty$ in (2.19). If we had not inverted the order of summation above, we would have obtained a second analogue of the Chu–Vandermonde theorem [39, p. 14].

Ramanujan derived a multitude of beautiful continued fractions for quotients of gamma functions that are found in his notebooks [57]. Proofs of these continued fractions can be found in the author's books [20, Chapter 12], [22, Chapter 32]. Methods for proving these continued fraction formulas are varied and at times *ad hoc*. Ramanujan evidently had a systematic procedure for proving these continued fractions, but we do not know what it is. As an illustration, we cite Entry 25 from Chapter 12 of Ramanujan's second notebook [57], [20, p. 140]. Suppose that either n is an odd integer and x is any complex number, or that n is any complex number and $\operatorname{Re} x > 0$. Then

$$\frac{\Gamma\left(\frac{1}{4}(x+n+1)\right)\Gamma\left(\frac{1}{4}(x-n+1)\right)}{\Gamma\left(\frac{1}{4}(x+n+3)\right)\Gamma\left(\frac{1}{4}(x-n+3)\right)} = \frac{4}{x-}\,\frac{n^2-1^2}{2x-}\,\frac{n^2-3^2}{2x-}\,\frac{n^2-5^2}{2x-}\cdots. \tag{4.15}$$

On the other hand, consider another continued fraction, Entry 12 in Chapter 16 of Ramanujan's second notebook [57], [21, p. 24]. Suppose that a, b, and q are complex numbers such that $|ab| < 1$ and $|q| < 1$, or that $a = b^{2m+1}$ for some integer m. Then

$$\frac{(a^2q^3;q^4)_\infty(b^2q^3;q^4)_\infty}{(a^2q;q^4)_\infty(b^2q;q^4)_\infty} = \frac{1}{1-ab} + \frac{(a-bq)(b-aq)}{(1-ab)(1+q^2)} + \frac{(a-bq^3)(b-aq^3)}{(1-ab)(1+q^4)} + \cdots. \tag{4.16}$$

At a first glance, it does not appear that (4.16) is a q-analogue of (4.15). However, if we replace a and b by q^a and q^b, respectively, then replace q^4 by q, set $2a = x+n$ and $2b = x-n$, multiply both sides of (4.16) by $(1-q)$, and lastly let $q \to 1$, we formally obtain (4.15).

We had earlier asked if there were theorems about ordinary hypergeometric series that do not have q-analogues. We provide one example for which q-analogues indeed do exist, but they are not *faithful* analogues, in that they do not have the shapes that one would suspect. Consider Clausen's identity [8, p. 116]

$${}_2F_1^2\left(a, b; a+b+\frac{1}{2}; z\right) = {}_3F_2\left(2a, 2b, a+b; a+b+\frac{1}{2}, 2a+2b; z\right). \tag{4.17}$$

This identity was used by Ramanujan in deriving his 17 famous series for $1/\pi$ [55], which were not entirely proved in print until 1987 when Jonathan and Peter Borwein [29, Chapter 5] used (4.17), modular equations, and special values of theta functions to prove all of them. As an example, we offer

$$\frac{4}{\pi} = \sum_{n=0}^{\infty}(6n+1)\frac{\left(\frac{1}{2}\right)_n^3}{n!^3}\frac{1}{4^n}. \tag{4.18}$$

For an expository discussion of proofs of these 17 series representations and many more such series formulas for $1/\pi$, see the survey article by N. D. Baruah, Berndt, and H. H. Chan [17].

Gasper and Rahman [39, p. 232, Eq. (8.8.3); p. 234, Eq. (8.8.12)] have derived two analogues of (4.17). The first, which is a consequence of the second, is given by

$$\left\{{}_4\phi_3\left(\begin{matrix} a, b, abz, ab/z \\ ab\sqrt{q}, -ab\sqrt{q}, -ab \end{matrix}; q, q\right)\right\}^2 = {}_5\phi_4\left(\begin{matrix} a^2, b^2, ab, abz, ab/z \\ a^2b^2, ab\sqrt{q}, -ab\sqrt{q}, -ab \end{matrix}; q, q\right),$$

under the assumption that the series terminate. The more general q-analogue of Gasper and Rahman expresses

$$\{{}_2\phi_1(a, b; aq/b; q, zq/b)\}^2$$

as a sum of two ${}_5\phi_4$'s.

We have observed that the q-binomial theorem is an analogue of the binomial theorem. Readers may ask if the corresponding bilateral sum theorem, Ramanujan's

${}_1\psi_1$ summation theorem (2.21), has an analogous forerunner in the theory of hypergeometric series. To answer this question, we should define a bilateral hypergeometric series. For *every* integer n, define

$$(a)_n := \frac{\Gamma(a+n)}{\Gamma(a)},$$

which agrees with our previous definition (4.2) of $(a)_n$ if n is a nonnegative integer. The bilateral hypergeometric series ${}_pH_p$ is defined for complex parameters $a_1, a_2, \dots, a_p$ and $b_1, b_2, \dots, b_p$ by

$${}_pH_p\left(\begin{matrix} a_1, a_2, \dots, a_p; \\ b_1, b_2, \dots, b_p; \end{matrix} z\right) := \sum_{n=-\infty}^{\infty} \frac{(a_1)_n (a_2)_n \cdots (a_p)_n}{(b_1)_n (b_2)_n \cdots (b_p)_n} z^n, \tag{4.19}$$

provided that the parameters are chosen so that poles do not arise. There is only a meager theory of bilateral hypergeometric series, which is mostly covered in Slater's book [65, Chapter 6], primarily because, if they converge, they converge only on $|z| = 1$ under the condition

$$\operatorname{Re}(b_1 + b_2 + \cdots + b_p - a_1 - a_2 - \cdots - a_p) > 1. \tag{4.20}$$

Despite the limitations necessarily imposed on ${}_pH_p$, Ramanujan's ${}_1\psi_1$ summation and the q-binomial theorem are analogues of the evaluation [50]

$${}_1H_1\left(\begin{matrix} a; \\ c; \end{matrix} e^{i\theta}\right) = (1 - e^{i\theta})^{c-a-1} e^{i(1-c)(\theta-\pi)} \frac{\Gamma(1-a)\Gamma(c)}{\Gamma(c-a)}, \qquad 0 < \theta < 2\pi. \tag{4.21}$$

5. Theta functions, q-series, and mock theta functions in Ramanujan's lost notebook

In Section 3, we discussed the combinatorics of (3.4). The series on the left-hand side of (3.4) is a *partial theta function.* We offer now some q-series representations for certain *complete* theta functions that are found in Ramanujan's lost notebook [58, p. 35], [10, pp. 32–35, Entries 1.7.7, 1.7.9, 1.7.6, 1.7.8]. If $f(a, b)$ is defined by (1.2) and $\varphi(q)$ is defined by (2.2), then

$$\sum_{n=0}^{\infty} \frac{(-1)^n q^{(n+1)(n+2)/2}}{(q;q)_n (1 - q^{2n+1})} = q f(q, q^7),$$

$$\sum_{n=0}^{\infty} \frac{(-1)^n q^{n(n+1)/2}}{(q;q)_n (1 - q^{2n+1})} = f(q^3, q^5),$$

$$\varphi(-q) \sum_{n=0}^{\infty} \frac{(-q;q^2)_n q^{(n+1)(n+2)/2}}{(q;q)_n (q;q^2)_{n+1}} = q f(-q, -q^7),$$

$$\varphi(-q) \sum_{n=0}^{\infty} \frac{(-q;q^2)_n q^{n(n+1)/2}}{(q;q)_n (q;q^2)_{n+1}} = f(-q^3, -q^5).$$

As in (3.4), the partitions generated on the left-hand sides above mostly "cancel" each other, as we see from the right-hand sides. The first two are proved combinatorially in [26].

As another example, we offer [58, p. 36], [9, p. 235, Entry 9.4.7]

$$\sum_{n=0}^{\infty} \frac{(-1)^n q^n}{(-q^2;q^2)_n} = \sum_{n=0}^{\infty} q^{n(3n+1)/2}(1-q^{2n+1}). \tag{5.1}$$

The function on the right-hand side is a *false theta function.* Note that the latter series is the same as Euler's pentagonal number series (2.5), except that the signs of infinitely many terms have been changed. The identity (5.1) has a combinatorial proof [26], [52].

We conclude this section with the two identities

$$\sum_{n=0}^{\infty} \frac{q^{n^2}}{(-q;q)_n} = \frac{(q^5;q^5)_\infty (q^5;q^{10})_\infty}{(q;q^5)_\infty (q^4;q^5)_\infty} + 2 - 2\sum_{n=0}^{\infty} \frac{q^{5n^2}}{(q;q^5)_{n+1}(q^4;q^5)_n}, \tag{5.2}$$

$$\sum_{n=0}^{\infty} \frac{q^{n^2+n}}{(-q;q)_n} = \frac{(q^5;q^5)_\infty (q^5;q^{10})_\infty}{(q^2;q^5)_\infty (q^3;q^5)_\infty} + \frac{2}{q} - \frac{2}{q}\sum_{n=0}^{\infty} \frac{q^{5n^2}}{(q^2;q^5)_{n+1}(q^3;q^5)_n}, \tag{5.3}$$

from Ramanujan's lost notebook [58]. The functions on the left-hand sides of (5.2) and (5.3) are fifth order mock theta functions. The first quotients on the right-hand sides of (5.2) and (5.3) can be expressed in terms of theta functions. On pages 18–20 in his lost notebook [58], Ramanujan stated two classes of five identities of which (5.2) and (5.3) are representatives of the two classes, respectively. Andrews and Garvan [11] showed that if one identity in each group of five could be proved, then the remaining four identities would follow in each case. These became known as the *mock theta conjectures*, and they were proved by D. Hickerson [44] in 1988.

The identities (5.2) and (5.3) have interesting combinatorial interpretations. We state, but not prove, the combinatorial equivalent of (5.2). No combinatorial proof of either (5.2) or (5.3) has ever been given, i.e., there do not currently exist combinatorial proofs of the mock theta conjectures. To explicate the first combinatorial interpretation, it is necessary to make a couple of definitions and introduce some notation.

Definition 5.1. *The rank of a partition equals the largest part minus the number of parts.*

For example, the rank of the partition $4+1$ is 2. Let $N(a,b,n)$ denote the number of partitions of n with rank congruent to a modulo b.

Definition 5.2. *We define $\rho_0(n)$ to be the number of partitions of n with unique smallest part and all other parts $\leq$ the double of the smallest part.*

For example, $\rho_0(5) = 3$, with the relevant partitions being 5, $2+3$, and $1+2+2$.

The first mock theta conjecture.

$$N(1,5,5n) = N(0,5,5n) + \rho_0(n).$$

For example, if $n = 5$, then $N(1, 5, 25) = 393$, $N(0, 5, 25) = 390$, and, as observed above, $\rho_0(5) = 3$.

6. Concluding remarks

We have seen that initially the theory of basic hypergeometric series was developed more or less in isolation with few connections with other parts of mathematics. For this reason, the subject's development was slow and scarcely appreciated. The Rogers–Ramanujan identities (2.15) and (2.16) perhaps provided the first examples to break these chains of confinement and link q-series with partitions at a deeper level. Since that time, q-series have not only played a leading role in the theory of partitions, but they have appeared prominently in other areas of number theory and combinatorics, e.g., in sums of squares. See [24] for some of these applications.

Today, as a subject in analysis, the work of Gasper, Rahman, Askey, Mourad Ismail, and others has continued the work of the early pioneers, Heine, Rogers, Jackson, and Ramanujan, in giving us a beautiful coherent theory still in the process of active development.

Many other bonds with q-series now exist. For example, q-series have appeared in various ways in physics. In particular, see the work of R. J. Baxter in statistical mechanics, e.g., [19], [18], and papers (either solely or coauthored) by A. Schilling [12], [53], [63], [64].

In this introduction to q-series, we have ignored several topics. First, we have not discussed bibasic hypergeometric series, which possess two independent bases, say q and p [39, pp. 80–88]. Second, we have neglected to mention multiple q-series, a subject to which Steve Milne has made numerous contributions. Third, we have not discussed q-integration, the exclusive topic of the monograph [48]. Fourth, one of the current active topics in q-series are elliptic and theta hypergeometric series [39, Chapter 11]. Very briefly and oversimplified, a theta hypergeometric series has in its summands products of theta functions. Hjalmar Rosengren, Michael Schlosser, Vyacheslav Spiridonov, and Warnaar are four of the leading researchers in the subject.

Where should readers begin to learn the subject of q-series? If one wants to learn how q-series interact with the theory of partitions, readers should read Andrews's book [3, esp. Chapter 2] and N. J. Fine's book [37]. If one wishes to learn how q-series articulate with theta functions and number theory, in particular from the point of view of Ramanujan, read Berndt's book [24]. For a highly motivated introduction to the subject of q-series, one should read Chapter 10 in the book of Andrews, Askey, and Roy [8]. Lastly, if you want to be an expert in q-series, or if you really just want to learn the subject very well, then you must read Gasper and Rahman's text [39].

7. Acknowledgments

The author is grateful to Richard Askey, Tom Koornwinder, Rob Rhoades, Michael Schlosser, Ae Ja Yee, and the referees for helpful comments.

References

[1] K. Alladi, A partial theta function identity of Ramanujan and its number theoretic interpretation, *Ramanujan J.*, to appear.

[2] G. E. Andrews, Two theorems of Gauss and allied identities proved arithmetically, *Pacific J. Math.*, **41** (1972) 563–578.

[3] G. E. Andrews, The Theory of Partitions, Addison–Wesley, Reading, MA (1976); reissued: Cambridge University Press, Cambridge (1998).

[4] G. E. Andrews, Partitions: Yesterday and Today, New Zealand Mathematical Society, Wellington (1979).

[5] G. E. Andrews, J. J. Sylvester, Johns Hopkins and partitions, in A Century of Mathematics in America, Part I (P. Duren, ed.), American Mathematical Society, Providence, RI (1988) 21–40.

[6] G. E. Andrews, On the proofs of the Rogers–Ramanujan identities, in q-Series and Partitions (D. Stanton, ed.), Springer-Verlag, New York (1989) 1–14.

[7] G. E. Andrews, Ramanujan and partial fractions, in Contributions to the History of Indian Mathematics, Vol. 3, Culture and History of Mathematics, G. Emch, R. Sridharan, and M. Srinivas, eds., Hindustan Book Agency, New Delhi (2005) 251–260.

[8] G. E. Andrews, R. A. Askey and R. Roy, Special Functions, Cambridge University Press, Cambridge (1999).

[9] G. E. Andrews and B. C. Berndt, Ramanujan's Lost Notebook, Part I, Springer, New York (2005).

[10] G. E. Andrews and B. C. Berndt, Ramanujan's Lost Notebook, Part II, Springer, New York (2009).

[11] G. E. Andrews and F. G. Garvan, Ramanujan's "lost" notebook VI: The mock theta conjectures, *Adv. Math.*, **89** (1991) 242–255.

[12] G. E. Andrews, A. Schilling and S. O. Warnaar, An A_2 Bailey lemma and Rogers-Ramanujan-type identities, *J. Amer. Math. Soc.*, **12** (1999) 677–702.

[13] R. Askey, Ramanujan's extensions of the gamma and beta functions, *Amer. Math. Monthly*, **87** (1980) 346–359.

[14] R. A. Askey, Ramanujan and hypergeometric series and basic hypergeometric series, in Ramanujan International Symposium on Analysis (Pune, 1987), N. K. Thakare, ed., Macmillan of India, New Delhi, (1989) 1–83; reprinted in [27, pp. 277–324].

[15] W. N. Bailey, A note on certain q-identities, *Quart. J. Math. (Oxford)*, **19** (1941) 157–166.

[16] W. N. Bailey, Identities of Rogers–Ramanujan type, *Proc. London Math. Soc.*, **50** no. 2, 1–10.

[17] N. D. Baruah, B. C. Berndt and H. H. Chan, Ramanujan's series for $1/\pi$: A survey, Math. Student, (2007) (special centenary volume issued in March, 2008), 1–24; *Amer. Math. Monthly*, **116** (2009) 567–587.

[18] R. J. Baxter, Exactly Solved Models in Statistical Mechanics, Academic Press, London (1982).

[19] R. J. Baxter, Ramanujan's identities in statistical mechanics, in Ramanujan Revisited, G. E. Andrews, R. A. Askey, B. C. Berndt, K. G. Ramanathan and R. A. Rankin, eds., Academic Press, Boston (1988) 69–84.

[20] B. C. Berndt, Ramanujan's Notebooks, Part II, Springer-Verlag, New York (1989).

[21] B. C. Berndt, Ramanujan's Notebooks, Part III, Springer-Verlag, New York (1991).

[22] B. C. Berndt, Ramanujan's Notebooks, Part V, Springer-Verlag, New York (1998).

[23] B. C. Berndt, An unpublished manuscript of Ramanujan on infinite series identities, *J. Ramanujan Math. Soc.*, **19** (2004) 57–74.

[24] B. C. Berndt, Number Theory in the Spirit of Ramanujan, American Mathematical Society, Providence, RI (2006).

[25] B. C. Berndt, Y.-S. Choi and S.-Y. Kang, The problems submitted by Ramanujan to the Journal of the Indian Mathematical Society, in Continued Fractions: From Analytic

Number Theory to Constructive Approximation, B. C. Berndt and F. Gesztesy, eds., *Contem. Math.*, **236**, American Mathematical Society, Providence, RI, (1999) 15–56; reprinted and updated in [27, pp. 215–258].

[26] B. C. Berndt, B. Kim and A. J. Yee, Ramanujan's lost notebook: Combinatorial proofs of identities associated with Heine's transformation or partial theta functions, *J. Comb. Thy. Ser. A*, **117** (2010) 957–973.

[27] B. C. Berndt and R. A. Rankin, Ramanujan: Essays and Surveys, American Mathematical Society, Providence (2001) London Mathematical Society, London (2001).

[28] B. C. Berndt and A. J. Yee, Combinatorial proofs of identities in Ramanujan's lost notebook associated with the Rogers–Fine identity and false theta functions, *Ann. of Combinatorics*, 7 (2003) 409–423.

[29] J. M. Borwein and P. B. Borwein, Pi and the AGM; A Study in Analytic Number Theory and Computational Complexity, Wiley, New York (1987).

[30] A. L. Cauchy, Oeuvres, Ser. 1, Vol. 8, Gauthier-Villars, Paris (1893).

[31] S. H. Chan, On Cranks of Partitions, Generalized Lambert Series, and Basic Hypergeometric Series, Ph.D. thesis, University of Illinois at Urbana–Champaign, Urbana (2005).

[32] S. H. Chan, Generalized Lambert series, *Proc. London Math. Soc.*, **91** (2005) no. 3, 598–622.

[33] T. W. Chaundy, Frank Hilton Jackson, *J. London Math. Soc.*, **37** (1962) 126–128.

[34] S.-C. Chu, Precious Mirror of the Four Elements, in Chinese, 1303; English trans., J. Hoe, The Jade Mirror of the Four Unknowns, Mingming Bookroom, New Zealand (2007); referenced in J. Needham, Science and Civilisation in China, Vol. III: Mathematics and the Sciences of the Heavens and the Earth, Cambridge University Press, Cambridge (1959).

[35] A. L. Dixon, Leonard James Rogers, *J. London Math. Soc.*, **9** (1934) 237–240.

[36] L. Euler, Introductio in Analysin Infinitorum, Marcum-Michaelem Bousquet, Lausannae (1748).

[37] N. J. Fine, Basic Hypergeometric Series and Applications, American Mathematical Society, Providence, RI (1988).

[38] F. Franklin, Sur le dévelopement du produit infini $(1-x)(1-x^2)(1-x^3)\cdots$, *Comptes Rendus*, **82** (1881) 448–450.

[39] G. Gasper and M. Rahman, Basic Hypergeometric Series, 2nd. ed., Cambridge University Press, Cambridge (2004).

[40] C. F. Gauss, Werke, Vol. 3, K. Gesell. Wiss., Göttingen (1866).

[41] G. H. Hardy, Ramanujan, Cambridge University Press, Cambridge, 1940; reprinted by Chelsea, New York, 1960; reprinted by the American Mathematical Society, Providence, RI (1999).

[42] G. H. Hardy, J. E. Littlewood and G. Pólya, Inequalities, University Press, Cambridge (1934).

[43] E. Heine, Untersuchungen über die Reihe

$$1+\frac{(1-q^{\alpha})(1-q^{\beta})}{(1-q)(1-q^{\gamma})}\cdot x+\frac{(1-q^{\alpha})(1-q^{\alpha+1})(1-q^{\beta})(1-q^{\beta+1})}{(1-q)(1-q^{2})(1-q^{\gamma})(1-q^{\gamma+1})}\cdot x^{2}+\cdots,$$

J. Reine Angew. Math., **34** (1847) 285–328.

[44] D. Hickerson, A proof of the mock theta conjectures, *Invent. Math.*, **94** (1988) 639–660.

[45] O. Hölder, Über einen Mittelwertsatz, Göttinger Nachrichten (1889) 38–47.

[46] C. G. J. Jacobi, Fundamenta Nova Theoriae Functionum Ellipticarum, Sumptibus Fratrum Bornträger, Regiomonti (1829).

[47] J. T. Joichi and D. Stanton, Bijective proofs of basic hypergeometric series identities, *Pacific J. Math.*, **127** (1987) 103–120.

[48] V. Kac and P. Cheung, Quantum Calculus, Springer, New York (2002).

[49] S. Kim, Bijective proofs of partition identities arising from modular equations, *J. Combin. Thy. Ser. A.*, **116** (2009) 699–712.

[50] T. Koornwinder, Quantum groups and q-special functions, in Representations of Lie groups and quantum groups, V. Baldoni and M. A. Picardello, eds., Longman House, Burnt Mill, England (1994) 46–128.

[51] P. A. MacMahon, Combinatory Analysis, Vol. II, Cambridge University Press, Cambridge, 1916; reprinted by AMS-Chelsea, Providence, RI (1960).

[52] M. Monks, A combinatorial proof of a mock theta function identity of Ramanujan, preprint.

[53] M. Okado, A. Schilling and M. Shimozono, Crystal bases and q-identities, *Contemp. Math.*, **291** (2001) 29–53.

[54] I. Pak, Partition bijections, a survey, *Ramanujan J.*, **12** (2006) 5–75.

[55] S. Ramanujan, Modular equations and approximations to π, *Quart. J. Math. (Oxford)*, **45** (1914) 350–372.

[56] S. Ramanujan, Collected Papers, Cambridge University Press, Cambridge, 1927; reprinted by Chelsea, New York, 1962; reprinted by the American Mathematical Society, Providence, RI (2000).

[57] S. Ramanujan, Notebooks (2 volumes), Tata Institute of Fundamental Research, Bombay (1957).

[58] S. Ramanujan, The Lost Notebook and Other Unpublished Papers, Narosa, New Delhi (1988).

[59] L. J. Rogers, An extension of a certain theorem in inequalities, *Mess. Math.*, **17** (1888) 145–150.

[60] L. J. Rogers, Second memoir on the expansion of certain infinite products, *Proc. London Math. Soc.*, **25** (1894) 318–343.

[61] L. J. Rogers, On two theorems of combinatory analysis and some allied identities, *Proc. London Math. Soc.*, **16** (1917) 315–336.

[62] L. J. Rogers, Obituary (signed N.R.C.), *Nature*, **132** (1933) 701–702.

[63] A. Schilling, $X = M$ Theorem: Fermionic formulas and rigged configurations, *Mathematical Society of Japan Memoirs*, **17** (2007) 75–104.

[64] A. Schilling and S. O. Warnaar, Conjugate Bailey pairs. From configuration sums and fractional-level string functions to Bailey's lemma, *Contemp. Math.*, **297** (2002) 227–255.

[65] L. J. Slater, Generalized Hypergeometric Functions, Cambridge University Press, Cambridge (1966).

[66] J. J. Sylvester, A constructive theory of partitions arranged in three acts, an interact, and an exodion, *Amer. J. Math.*, **5** (1884) 231–330; **6** (1886) 334–336; reprinted in Collected Papers of J. J. Sylvester, Vol. 4, Cambridge University Press, London (1912) 1–83; reprinted by Chelsea, New York (1974).

[67] A. T. Vandermonde, Mémoire sur des irrationnelles de différens ordres avec une application au cercle, *Mém. Acad. Roy. Sci. Paris*, (1772) 489–498.

[68] S. O. Warnaar, A generaliztion of the Farkas and Kra partition theorem for modulus 7, *J. Combin. Thy. Ser. A.*, **110** (2005) 43–52.

[69] E. M. Wright, An enumerative proof of an identity of Jacobi, *J. London Math. Soc.*, **40** (1965) 55–57.

[70] A. J. Yee, Combinatorial proofs of Ramanujan's ${}_1\psi_1$ summation and the q-Gauss summation, *J. Combin. Thy. Ser. A.*, **105** (2004) 63–77.

[71] A. J. Yee, Ramanujan's partial theta series and parity in partitions, *Ramanujan J.*, to appear.

Ramanujan Rediscovered No. 14, 2010, pp. 53–62.

A Proof of the General Theta Transformation Formula

Bruce C. Berndt[1], Chadwick Gugg[2], Sarachai Kongsiriwong[3] and Johann Thiel[4]

[1,4]Department of Mathematics, University of Illinois at Urbana-Champaign, 1409 West Green Street, Urbana, IL 61801, USA
[2]Department of Mathematics, University of Toledo, 2801 West Bancroft Street, Toledo, Ohio 43606, USA
[3]Department of Mathematics, Prince of Songkla University, Hat Yai, Songkhla, Thailand 90112
e-mail: berndt@illinois.edu; Chadwick.Gugg@utoledo.edu; sarachai.k@psu.ac.th; jthiel2@math.uiuc.edu

Abstract. A new proof, based on an idea of K. Venkatachaliengar, of the transformation formula for Ramanujan's general theta function is given.

Keywords. Theta function, transformation formula, Ramanujan's general theta function.

In Memory of K. Venkatachaliengar

1. Introduction

In Ramanujan's notation for theta functions, set

$$f(a,b) := \sum_{n=-\infty}^{\infty} a^{n(n+1)/2} b^{n(n-1)/2}, \qquad |ab| < 1.$$

Arguably, the two most important properties of $f(a,b)$ are the Jacobi triple product identity and the modular transformation formula, which can be written in two different ways [3, p. 36, Entry 20], [2, pp. 252–253, Entry 7].

Theorem 1.1. *Let $\alpha, \beta > 0$, with $\alpha\beta = \pi$. Let z be any complex number. Then*

$$e^{z^2/4}\sqrt{\alpha}\, f(e^{-\alpha^2+iz\alpha}, e^{-\alpha^2-iz\alpha}) = \sqrt{\beta}\, f(e^{-\beta^2+z\beta}, e^{-\beta^2-z\beta}) \tag{1.1}$$

and

$$e^{z^2/4}\sqrt{\alpha}\left\{\frac{1}{2} + \sum_{n=1}^{\infty} e^{-\alpha^2 n^2}\cos(\alpha z n)\right\} = \sqrt{\beta}\left\{\frac{1}{2} + \sum_{n=1}^{\infty} e^{-\beta^2 n^2}\cosh(\beta z n)\right\}. \tag{1.2}$$

The most familiar proof of Theorem 1.1, especially in the form (1.2), is via Poisson's summation formula. That proof is briefly sketched in [2, p. 253]. It should be emphasized that only in special cases is $f(a,b)$ a modular form. In these special cases, especially in the case $a = b = q$, more proofs are known. In particular, there are three important special cases of Theorem 1.1. For two of them, following Ramanujan, set

$$\varphi(q) := f(q,q) = \sum_{n=-\infty}^{\infty} q^{n^2} \quad \text{and} \quad \psi(q) := f(q,q^3) = \sum_{n=0}^{\infty} q^{n(n+1)/2}, \qquad |q| < 1.$$

Then [3, p. 43, Entry 27 (i), (ii)] if $\alpha\beta = \pi$ with $\operatorname{Re} \alpha^2, \beta^2 > 0$,

$$\sqrt{\alpha}\varphi(e^{-\alpha^2}) = \sqrt{\beta}\varphi(e^{-\beta^2}) \tag{1.3}$$

and

$$2\sqrt{\alpha}\psi(e^{-2\alpha^2}) = \sqrt{\beta}e^{\alpha^2/4}\varphi(-e^{-\beta^2}). \tag{1.4}$$

For the third major special instance, in [4], the first author and K. Venkatachaliengar gave a proof of Theorem 1.1 for

$$\begin{aligned} f(-q) &:= f(-q,-q^2) \\ &= \sum_{n=-\infty}^{\infty} (-1)^n q^{n(3n-1)/2} = e^{-2\pi i\tau/24}\eta(\tau), \quad q = e^{2\pi i\tau}, \ \operatorname{Im}\tau > 0, \end{aligned} \tag{1.5}$$

where $\eta(\tau)$ is the Dedekind eta-function. If $\operatorname{Re} \alpha, \beta > 0$, this transformation formula is given by

$$e^{-\alpha/12}\sqrt[4]{\alpha}f(-e^{-2\alpha}) = e^{-\beta/12}\sqrt[4]{\beta}f(-e^{-2\beta}). \tag{1.6}$$

There are several known proofs of (1.6). A virtue of the proof given in [4] is that it is quite elementary.

It is natural to ask if the ideas from [4] can be used to prove the more general theta transformation formula given in Theorem 1.1. In fact, in their doctoral theses [7], [8], the second and third authors of this paper used the methods from [4] to establish (1.4) and (1.3), respectively. The purpose of this paper is to give a proof of the general transformation formula in the form (1.1) along the lines of those in [4]. This more general proof encounters serious difficulties that are not found in the proofs of (1.6), (1.3), and (1.4). In particular, in order to complete the last step in the proof, we are required to prove that we can invert the order of summation in a certain conditionally convergent double series. We think that this challenging problem is of interest in itself. In Section 2, we give the first part of our proof of Theorem 1.1, and in Section 3, we give the second portion of our proof, i.e., justifying that we can indeed invert the order of summation in the double series in question.

In an earlier paper [9], Venkatachaliengar sketched short proofs of the transformation formulas (1.3) and (1.6) for, respectively, the classical theta function $\varphi(q)$ and the Dedekind eta-function $\eta(\tau)$. The proof for $\varphi(q)$ is similar to that given in Whittaker and Watson's text [11, pp. 475–476], but Venkatachaliengar's determination of the constant is simpler. In his monograph [10, pp. 32–35], Venkatachaliengar sketched another proof of the transformation formula for $\varphi(q)$, with the details more fully worked out later by S. Cooper [5].

2. First part of our proof

It will be slightly convenient to reformulate (1.1). Let $a, b > 0$ with $ab = \pi^2$. Then

$$e^{z^2/4} a^{1/4} f(e^{-a+iz\sqrt{a}}, e^{-a-iz\sqrt{a}}) = b^{1/4} f(e^{-b+z\sqrt{b}}, e^{-b-z\sqrt{b}}). \tag{2.1}$$

We now prove (2.1).

Using the Jacobi triple product identity,

$$f(a,b) = \prod_{n=0}^{\infty} (1 + a(ab)^n)(1 + b(ab)^n)(1 - (ab)^{n+1}), \quad |ab| < 1,$$

we can rewrite (2.1) in the form

$$e^{z^2/4} a^{1/4} \prod_{n=0}^{\infty} (1 + e^{-(2n+1)a+iz\sqrt{a}})(1 + e^{-(2n+1)a-iz\sqrt{a}}) \prod_{n=1}^{\infty} (1 - e^{-2na})$$
$$= b^{1/4} \prod_{n=0}^{\infty} (1 + e^{-(2n+1)b+z\sqrt{b}})(1 + e^{-(2n+1)b-z\sqrt{b}}) \prod_{n=1}^{\infty} (1 - e^{-2nb}). \tag{2.2}$$

Taking logarithms on both sides of (2.2), we find that

$$\frac{z^2}{4} + \frac{1}{4}\log a + \sum_{n=0}^{\infty} \log(1 + e^{-(2n+1)a+iz\sqrt{a}})$$
$$+ \sum_{n=0}^{\infty} \log(1 + e^{-(2n+1)a-iz\sqrt{a}}) + \sum_{n=1}^{\infty} \log(1 - e^{-2na})$$
$$= \frac{1}{4}\log b + \sum_{n=0}^{\infty} \log(1 + e^{-(2n+1)b+z\sqrt{b}})$$
$$+ \sum_{n=0}^{\infty} \log(1 + e^{-(2n+1)b-z\sqrt{b}}) + \sum_{n=1}^{\infty} \log(1 - e^{-2nb}). \tag{2.3}$$

Using the Maclaurin series for $\log(1 - w)$, inverting the order of summation, and then summing the resulting geometric series, we easily find that

$$\sum_{n=0}^{\infty} \log(1 + e^{-(2n+1)a+iz\sqrt{a}}) = -\sum_{m=1}^{\infty} \frac{(-1)^m}{m} \frac{e^{ma+izm\sqrt{a}}}{e^{2ma} - 1}.$$

Using this and similar expansions for the five remaining logarithmic series in (2.3), we find that, if $|z|$ is sufficiently small,

$$-2\sum_{m=1}^{\infty} \frac{(-1)^m}{m} \frac{e^{ma}\cos(zm\sqrt{a})}{e^{2ma} - 1} + 2\sum_{m=1}^{\infty} \frac{(-1)^m}{m} \frac{e^{mb}\cosh(zm\sqrt{b})}{e^{2mb} - 1}$$
$$-\sum_{m=1}^{\infty} \frac{1}{m} \left\{ \frac{1}{e^{2ma} - 1} - \frac{1}{e^{2mb} - 1} \right\} = \frac{1}{4}\log\frac{b}{a} - \frac{z^2}{4}. \tag{2.4}$$

Now, from [4, Eq. (2)],

$$\sum_{m=1}^{\infty}\frac{1}{m}\left\{\frac{1}{e^{2ma}-1}-\frac{1}{e^{2mb}-1}\right\}=\frac{1}{4}\log\frac{a}{b}+\frac{b-a}{12}. \tag{2.5}$$

Thus, using (2.5) in (2.4), we see that we must show that

$$-2\sum_{m=1}^{\infty}\frac{(-1)^m}{m}\frac{e^{ma}\cos(zm\sqrt{a})}{e^{2ma}-1}+2\sum_{m=1}^{\infty}\frac{(-1)^m}{m}\frac{e^{mb}\cosh(zm\sqrt{b})}{e^{2mb}-1}$$
$$=\frac{b-a}{12}-\frac{z^2}{4}. \tag{2.6}$$

Now from [2, p. 268], we know the partial fraction decomposition

$$\frac{\cosh(\phi my)}{\sinh(\pi my)}=\frac{1}{\pi my}-\frac{my}{\pi}\sum_{\substack{k=-\infty\\k\neq 0}}^{\infty}\frac{(-1)^k\cos(k\phi)}{k(imy-k)}. \tag{2.7}$$

Set $\pi y = a$ and $\phi = \pi iz/\sqrt{a}$ in (2.7). For the remainder of the proof, we will need to assume that z is purely imaginary and, as mentioned above, sufficiently small in modulus. At the end of our proof, we appeal to analytic continuation. Using the relation $ab=\pi^2$ and simplifying, we find that

$$\frac{\cos(zm\sqrt{a})}{\sinh(ma)}=\frac{1}{ma}-\frac{m}{b}\sum_{\substack{k=-\infty\\k\neq 0}}^{\infty}\frac{(-1)^k\cosh(kz\sqrt{b})}{k(ima/\pi-k)}$$
$$=\frac{1}{ma}+\frac{2m}{b}\sum_{k=1}^{\infty}\frac{(-1)^k\cosh(kz\sqrt{b})}{k^2+m^2a/b}. \tag{2.8}$$

If we now use (2.8) in the first expression in (2.6), we find that

$$-2\sum_{m=1}^{\infty}\frac{(-1)^m}{m}\frac{e^{ma}\cos(zm\sqrt{a})}{e^{2ma}-1}$$
$$=-\sum_{m=1}^{\infty}\frac{(-1)^m}{m}\left\{\frac{1}{ma}+2m\sum_{k=1}^{\infty}\frac{(-1)^k\cosh(zk\sqrt{b})}{bk^2+am^2}\right\}$$
$$=\frac{b}{12}-2\sum_{m=1}^{\infty}\sum_{k=1}^{\infty}\frac{(-1)^{m+k}\cosh(zk\sqrt{b})}{bk^2+am^2}. \tag{2.9}$$

Next, recall that [6, p. 44, Formula 1.422, no. 3]

$$\frac{1}{\sinh y}=\frac{1}{y}+\frac{2y}{\pi^2}\sum_{k=1}^{\infty}\frac{(-1)^k}{y^2/\pi^2+k^2}. \tag{2.10}$$

Using (2.10) and the relation $ab = \pi^2$, we find that

$$
\begin{aligned}
2\sum_{m=1}^{\infty} & \frac{(-1)^m}{m}\frac{e^{mb}\cosh(zm\sqrt{b})}{e^{2mb}-1} \\
&= \sum_{m=1}^{\infty}\frac{(-1)^m}{m}\frac{\cosh(zm\sqrt{b})}{\sinh(mb)} \\
&= \sum_{m=1}^{\infty}\frac{(-1)^m}{m}\cosh(zm\sqrt{b})\left\{\frac{1}{mb}+2mb\sum_{k=1}^{\infty}\frac{(-1)^k}{b^2m^2+\pi^2k^2}\right\} \\
&= \frac{1}{b}\sum_{m=1}^{\infty}\frac{(-1)^m}{m^2}\cosh(zm\sqrt{b}) + 2\sum_{m=1}^{\infty}\sum_{k=1}^{\infty}\frac{(-1)^{m+k}\cosh(zm\sqrt{b})}{bm^2+ak^2}. \qquad (2.11)
\end{aligned}
$$

We examine the first sum on the far right-hand side of (2.11). Set $z = ix$ and recall [1, p. 805, Eq. 23.1.18] the Fourier expansion of the second Bernoulli polynomial $B_2(t)$

$$t^2 - t + \frac{1}{6} =: B_2(t) = \frac{1}{\pi^2}\sum_{k=1}^{\infty}\frac{\cos(2\pi kt)}{k^2}, \qquad 0 \le t \le 1. \qquad (2.12)$$

Thus, from (2.12), if $|x\sqrt{b}/(2\pi)| < 1$,

$$
\begin{aligned}
\frac{1}{b}\sum_{m=1}^{\infty} & \frac{(-1)^m}{m^2}\cos(xm\sqrt{b}) \\
&= \frac{1}{b}\left\{\sum_{m=1}^{\infty}\frac{\cos(2xm\sqrt{b})}{4m^2} - \sum_{m=0}^{\infty}\frac{\cos(x(2m+1)\sqrt{b})}{(2m+1)^2}\right\} \\
&= \frac{1}{b}\left\{\frac{1}{2}\sum_{m=1}^{\infty}\frac{\cos(2xm\sqrt{b})}{m^2} - \sum_{m=1}^{\infty}\frac{\cos(xm\sqrt{b})}{m^2}\right\} \\
&= \frac{1}{b}\left\{\frac{\pi^2}{2}\left(\left(\frac{x\sqrt{b}}{\pi}\right)^2 - \left(\frac{x\sqrt{b}}{\pi}\right) + \frac{1}{6}\right) - \pi^2\left(\left(\frac{x\sqrt{b}}{2\pi}\right)^2 - \left(\frac{x\sqrt{b}}{2\pi}\right) + \frac{1}{6}\right)\right\} \\
&= \frac{x^2}{4} - \frac{a}{12} \\
&= -\frac{z^2}{4} - \frac{a}{12}. \qquad (2.13)
\end{aligned}
$$

We now employ (2.13) in (2.11), then use (2.9) and (2.11) in (2.6), and find that the latter equality now reduces to the new equality

$$-\sum_{m=1}^{\infty}\sum_{k=1}^{\infty}\frac{(-1)^{m+k}\cosh(zk\sqrt{b})}{bk^2+am^2} + \sum_{m=1}^{\infty}\sum_{k=1}^{\infty}\frac{(-1)^{m+k}\cosh(zm\sqrt{b})}{bm^2+ak^2} = 0.$$

If we invert the roles of the summation variables m and k in the first sum above, we see that we are now required to show that, for $x \in \mathbb{R}$ and $ab = \pi^2$,

$$\sum_{k=1}^{\infty}\sum_{m=1}^{\infty}\frac{(-1)^{m+k}\cos(xm\sqrt{b})}{bm^2+ak^2}=\sum_{m=1}^{\infty}\sum_{k=1}^{\infty}\frac{(-1)^{m+k}\cos(xm\sqrt{b})}{bm^2+ak^2}. \tag{2.14}$$

In other words, our task is to show that we can invert the order of summation in the double sums above.

3. Justifying the inversion in order of summation

Denote the right-hand side of (2.14) by A. Since A converges, for every $N \in \mathbb{N}$ there is an integer $M_N \geq N$ such that if $\ell \geq M_N$, then

$$\left|\sum_{m=\ell}^{\infty}\sum_{k=1}^{\infty}\frac{(-1)^{m+k}\cos(xm\sqrt{b})}{bm^2+ak^2}\right|<\frac{1}{N}.$$

So,

$$\begin{aligned}
A &= \sum_{m=1}^{M_N}\sum_{k=1}^{\infty}\frac{(-1)^{m+k}\cos(xm\sqrt{b})}{bm^2+ak^2}+O\left(\frac{1}{N}\right)\\
&= \sum_{m=1}^{M_N}\sum_{k=1}^{M_Nm^2}\frac{(-1)^{m+k}\cos(xm\sqrt{b})}{bm^2+ak^2}\\
&\quad +\sum_{m=1}^{M_N}\sum_{k=M_Nm^2+1}^{\infty}\frac{(-1)^{m+k}\cos(xm\sqrt{b})}{bm^2+ak^2}+O\left(\frac{1}{N}\right)\\
&= \sum_{m=1}^{M_N}\sum_{k=1}^{M_Nm^2}\frac{(-1)^{m+k}\cos(xm\sqrt{b})}{bm^2+ak^2}\\
&\quad +\sum_{m=1}^{M_N}\frac{(-1)^{m}\cos(xm\sqrt{b})}{bm^2}\sum_{k=M_Nm^2+1}^{\infty}\frac{(-1)^k}{1+(a/b)(k/m)^2}+O\left(\frac{1}{N}\right).
\end{aligned}$$

The inner sum of the second double sum is a convergent alternating series, and so we can estimate it crudely by

$$\left|\sum_{k=M_Nm^2+1}^{\infty}\frac{(-1)^k}{1+(a/b)(k/m)^2}\right|\leq\frac{b}{a(M_Nm+1/m)^2}=O_{a,b}\left(\frac{1}{(M_Nm)^2}\right).$$

Combining this estimate with the work above yields

$$A = \sum_{m=1}^{M_N} \sum_{k=1}^{M_N m^2} \frac{(-1)^{m+k} \cos(xm\sqrt{b})}{bm^2 + ak^2} + O_{a,b}\left(\frac{1}{M_N^2} \sum_{m=1}^{\infty} \frac{1}{m^4}\right) + O\left(\frac{1}{N}\right)$$

$$= \sum_{m=1}^{M_N} \sum_{k=1}^{M_N m^2} \frac{(-1)^{m+k} \cos(xm\sqrt{b})}{bm^2 + ak^2} + O_{a,b}\left(\frac{1}{N}\right). \tag{3.1}$$

Note that the constant implied by the error term of (3.1) is independent of N.

Now let B denote the left-hand side of (2.14). Since B converges, for every $N \in \mathbb{N}$ there is an integer $M'_N \geq N$ such that if $\ell \geq M'_N$, then

$$\left| \sum_{k=\ell}^{\infty} \sum_{m=1}^{\infty} \frac{(-1)^{m+k} \cos(xm\sqrt{b})}{bm^2 + ak^2} \right| < \frac{1}{N}.$$

So,

$$B = \sum_{k=1}^{M'_N} \sum_{m=1}^{\infty} \frac{(-1)^{m+k} \cos(xm\sqrt{b})}{bm^2 + ak^2} + O\left(\frac{1}{N}\right)$$

$$= \sum_{k=1}^{M'_N} \sum_{m=1}^{M'_N k^2} \frac{(-1)^{m+k} \cos(xm\sqrt{b})}{bm^2 + ak^2}$$

$$+ \sum_{k=1}^{M'_N} \sum_{m=M'_N k^2+1}^{\infty} \frac{(-1)^{m+k} \cos(xm\sqrt{b})}{bm^2 + ak^2} + O\left(\frac{1}{N}\right)$$

$$= \sum_{k=1}^{M'_N} \sum_{m=1}^{M'_N k^2} \frac{(-1)^{m+k} \cos(xm\sqrt{b})}{bm^2 + ak^2}$$

$$+ \sum_{k=1}^{M'_N} \frac{(-1)^k}{ak^2} \sum_{m=M'_N k^2+1}^{\infty} \frac{(-1)^m \cos(xm\sqrt{b})}{1 + (b/a)(m/k)^2} + O\left(\frac{1}{N}\right).$$

The inner sum of the second double sum is estimated by

$$\left| \sum_{m=M'_N k^2+1}^{\infty} \frac{(-1)^m \cos(xm\sqrt{b})}{1 + (b/a)(m/k)^2} \right| \leq \sum_{m=M'_N k^2+1}^{\infty} \frac{1}{1 + (b/a)(m/k)^2}$$

$$\leq \frac{ak^2}{b} \sum_{m=M'_N k^2+1}^{\infty} \frac{1}{m^2}$$

$$= O_{a,b}\left(\frac{1}{M'_N}\right).$$

Putting this estimate back into the work above yields

$$B = \sum_{k=1}^{M'_N} \sum_{m=1}^{M'_N k^2} \frac{(-1)^{m+k} \cos(xm\sqrt{b})}{bm^2 + ak^2} + O_{a,b}\left(\frac{1}{M'_N} \sum_{m=1}^{\infty} \frac{1}{k^2} \right) + O\left(\frac{1}{N}\right)$$

$$= \sum_{k=1}^{M'_N} \sum_{m=1}^{M'_N k^2} \frac{(-1)^{m+k} \cos(xm\sqrt{b})}{bm^2 + ak^2} + O_{a,b}\left(\frac{1}{N}\right). \tag{3.2}$$

As in (3.1), the constant implied by the O-symbol in (3.2) does not depend on N. Note that (3.1) and (3.2) also hold for $M''_N = \max\{M_N, M'_N\}$.

We now switch the roles of m and k in (3.2) and take differences to find that

$$A - B = \sum_{m=1}^{M''_N} \sum_{k=1}^{M''_N m^2} \frac{(-1)^{m+k} \cos(xm\sqrt{b})}{bm^2 + ak^2}$$

$$- \sum_{m=1}^{M''_N} \sum_{k=1}^{M''_N m^2} \frac{(-1)^{m+k} \cos(xk\sqrt{b})}{bk^2 + am^2} + O_{a,b}\left(\frac{1}{N}\right)$$

$$= \sum_{m=1}^{M''_N} \sum_{k=1}^{M''_N m^2} \left\{ \frac{(-1)^{m+k} \cos(xm\sqrt{b})}{bm^2 + ak^2} - \frac{(-1)^{m+k} \cos(xk\sqrt{b})}{bk^2 + am^2} \right\} + O_{a,b}\left(\frac{1}{N}\right)$$

$$= \sum_{m=2}^{M''_N} \sum_{k=M''_N+1}^{M''_N m^2} \left\{ \frac{(-1)^{m+k} \cos(xm\sqrt{b})}{bm^2 + ak^2} - \frac{(-1)^{m+k} \cos(xk\sqrt{b})}{bk^2 + am^2} \right\}$$

$$+ O_{a,b}\left(\frac{1}{N}\right)$$

$$= \sum_{m=2}^{M''_N} \sum_{k=M''_N+1}^{M''_N m^2} \frac{(-1)^{m+k} \cos(xm\sqrt{b})}{bm^2 + ak^2}$$

$$- \sum_{k=M''_N+1}^{M''^3_N} \sum_{m=\lfloor\sqrt{k/M''_N}\rfloor}^{M''_N} \frac{(-1)^{m+k} \cos(xk\sqrt{b})}{bk^2 + am^2} + O_{a,b}\left(\frac{1}{N}\right). \tag{3.3}$$

We need to estimate the two sums above. First,

$$\left| \sum_{m=2}^{M''_N} \sum_{k=M''_N+1}^{M''_N m^2} \frac{(-1)^{m+k} \cos(xm\sqrt{b})}{bm^2 + ak^2} \right| \le \sum_{m=2}^{M''_N} \left| \sum_{k=M''_N+1}^{M''_N m^2} \frac{(-1)^{m+k} \cos(xm\sqrt{b})}{bm^2 + ak^2} \right|$$

$$\le \sum_{m=2}^{M''_N} \frac{1}{|b|m^2} \left| \sum_{k=M''_N+1}^{M''_N m^2} \frac{(-1)^k}{1 + (a/b)(k/m)^2} \right|$$

$$\leq \sum_{m=2}^{M''_N} \frac{1}{|b|m^2}\left(\frac{1}{1+(a/b)(M''_N/m)^2}\right)$$

$$\leq \sum_{m=2}^{M''_N} \frac{1}{|a|M''^2_N}$$

$$= O_{a,b}\left(\frac{1}{N}\right).$$

Second,

$$\left|\sum_{k=M''_N+1}^{M''^3_N} \sum_{m=\lfloor\sqrt{k/M''_N}\rfloor}^{M''_N} \frac{(-1)^{m+k}\cos(xk\sqrt{b})}{bk^2+am^2}\right|$$

$$\leq \sum_{k=M''_N+1}^{M''^3_N} \left|\sum_{m=\lfloor\sqrt{k/M''_N}\rfloor}^{M''_N} \frac{(-1)^{m+k}\cos(xk\sqrt{b})}{bk^2+am^2}\right|$$

$$\leq \sum_{k=M''_N+1}^{M''^3_N} \frac{1}{|b|k^2}\left|\sum_{m=\lfloor\sqrt{k/M''_N}\rfloor}^{M''_N} \frac{(-1)^m}{1+(a/b)(m/k)^2}\right|$$

$$\leq \sum_{k=M''_N+1}^{M''^3_N} \frac{1}{|b|k^2}\left(\frac{1}{1+(a/b)k^{-2}}\right)$$

$$= O_{a,b}\left(\frac{1}{N}\right).$$

Using these last two estimates in (3.3), we conclude that

$$A - B = O_{a,b}\left(\frac{1}{N}\right),$$

and so the proof of (2.14) is complete.

The authors are grateful to Shaun Cooper for helpful comments. In particular, he informed us of Venkatachaliengar's proofs of theta function transformation formulas delineated at the end of Section 1.

References

[1] M. Abramowitz and I. A. Stegun, eds., Handbook of Mathematical Functions, Dover, New York (1965).

[2] B. C. Berndt, Ramanujan's Notebooks, Part II, Springer-Verlag, New York (1989).

[3] B. C. Berndt, Ramanujan's Notebooks, Part III, Springer-Verlag, New York (1991).
[4] B. C. Berndt and K. Venkatachaliengar, On the transformation formula for the Dedekind eta-function, in Symbolic Computation, Number Theory, Special Functions, Physics and Combinatorics, F. G. Garvan and M. E. H. Ismail, eds., Kluwer, Dordrecht (2001) pp. 73–77.
[5] S. Cooper, Cubic theta functions, *J. Comp. Appl. Math.*, **160** (2003) 77–94.
[6] I. S. Gradshteyn and I. M. Ryzhik, eds., Table of Integrals, Series, and Products, 5th ed., Academic Press, San Diego (1994).
[7] C. Gugg, Modular Identities for the Rogers-Ramanujan Functions and Analogues, Ph.D. Thesis, University of Illinois at Urbana–Champaign (2010).
[8] S. Kongsiriwong, Theta Functions and Related Infinite Series, Ph.D. Thesis, University of Illinois at Urbana–Champaign (2003).
[9] K. Venkatachaliengar, Elliptic modular functions and Picard's theorem, in Hyperbolic Complex Analysis (Proc. All India Sem., Ramanujan Inst., Univ. Madras, Madras, 1977), pp. 21–41; Publ. Ramanujan Inst., 4, Univ. Madras, Madras (1979).
[10] K. Venkatachaliengar, Development of Elliptic Functions According to Ramanujan, Technical Report, No. 2, Department of Mathematics, Madurai Kamaraj University, Madurai (1988).
[11] E. T. Whittaker and G. N. Watson, A Course of Modern Analysis, 4th ed., Cambridge University Press, Cambridge (1966).

Ramanujan Rediscovered No. 14, 2010, pp. 63–72.

Some Applications of Ramanujan's Remarkable Summation Formula

S. Bhargava

Department of Studies in Mathematics, University of Mysore, Manasagangothri, Mysore 570 006, India
e-mail: sribhargava@hotmail.com

Abstract. K. Venkatachaliengar had an enduring interest in formulas from Ramanujan's Notebooks. One of Venkatachaliengar's favorite formulas was 'Ramanujan's remarkable summation formula' for which he found several applications and gave elegant, self contained proofs. Many applications for what is more commonly referred to now as Ramanujan's ${}_1\psi_1$ summation formula have been discovered. Also, as is well known, the remarkable formula has, as its special cases, many classical formulas of seminal importance in the theory of q-series and number theory. This paper presents more seemingly new and interesting applications of the remarkable formula.

Keywords. Theta functions, modular equations, Ramanujan's remarkable summation formula.

In memory of Prof. K. Venkatachaliengar

1. Introduction

Two of the identities that are widely used in the classical theory of theta and elliptic functions, theory of q-series and number theory are the Euler-Cauchy q-binomial theorem [29,20],

$$\sum_{n=0}^{\infty} \frac{(a)_n}{(q)_n} z^n = \frac{(az)_\infty}{(z)_\infty}, \quad |q| < 1, \quad |z| < 1, \tag{1.1}$$

and the Jacobi triple product identity [37],

$$\sum_{n=-\infty}^{\infty} q^{n^2} z^n = (-qz; q^2)_\infty (-qz^{-1}; q^2)_\infty (q^2; q^2)_\infty, \tag{1.2}$$

with $z \neq 0$ and $|q| < 1$. Here and in what follows, as usual,

$$(a)_0 := (a; q)_0 := 1, (a)_n := (a; q)_n = \prod_{k=0}^{n-1} (1 - aq^k),$$

and, for $|q| < 1$,

$$(a)_\infty := (a; q)_\infty = \prod_{k=0}^{\infty} (1 - aq^k).$$

Also of importance in these theories is the quintuple product identity [27],

$$\sum_{n=-\infty}^{\infty} q^{3n^2+n}(q^{-3n}z^{3n} - q^{3n+1}z^{-3n-1})$$
$$= (qz;q^2)_\infty(q/z;q^2)_\infty(z^2;q^4)_\infty(q^4/z^2;q^4)_\infty(q^2;q^2)_\infty, \tag{1.3}$$

with $z \neq 0$ and $|q| < 1$.

In Entry 17 of Chapter 16 of his second notebook, Ramanujan [40] states the formula,

$$1+\sum_{n=1}^{\infty}\frac{(\alpha^{-1};q^2)_n(-\alpha qz)^n}{(\beta q^2;q^2)_n}+\sum_{n=1}^{\infty}\frac{(\beta^{-1};q^2)_n(-\beta q/z)^n}{(\alpha q^2;q^2)_n}$$
$$=\frac{(-qz;q^2)_\infty(-q/z;q^2)_\infty(q^2;q^2)_\infty(\alpha\beta q^2;q^2)_\infty}{(-\alpha qz;q^2)_\infty(-\beta q/z;q^2)_\infty(\alpha q^2;q^2)_\infty(\beta q^2;q^2)_\infty}, \tag{1.4}$$

with $|q| < 1$ and $|\beta q| < |z| < 1/|\alpha q|$ and mentions that the classical triple product identity (1.2) can be proved from his formula (1.4). (Indeed, put $\alpha = \beta = 0$.)

It was G. H. Hardy [32, pp. 222–223] who unearthed (1.4) and called it "a remarkable formula with many parameters" (now also known by the names "Ramanujan's remarkable summation formula", "Ramanujan's ${}_1\psi_1$ sum" and by similar names) and indicated that it could be proved from the classical q-binomial theorem (1.1). On the other hand, more interestingly, it turns out that (1.1) is also a special case of the remarkable formula (1.4). The classical triple product identity (1.2) can also be proven using (1.4). An easy proof of (1.1) follows from (1.4) by setting $b = q$ in the following equivalent form of (1.4),

$$\sum_{n=-\infty}^{\infty}\frac{(a)_n}{(b)_n}z^n=\frac{(az)_\infty(qa^{-1}z^{-1})_\infty(q)_\infty(ba^{-1})_\infty}{(z)_\infty(ba^{-1}z^{-1})_\infty(b)_\infty(qa^{-1})_\infty}, \tag{1.5}$$

with $|b/a| < |z| < 1$ and $|q| < 1$, and

$$(a)_{-n} := (a;q)_{-n} := \frac{(a)_\infty}{(aq^{-n})_\infty}, \quad n = 1, 2, 3, \ldots.$$

The form (1.5) also tells us that the remarkable sum (1.4) can be seen as a bilateral analogue of the q-binomial theorem (1.1). The remarkable formula (1.4) can also be viewed as a generalization of the quintuple product identity (1.3). See, for example, [18,27]. Thus we have the beautiful fact that Ramanujan's remarkable formula (1.4) is a common generalization of all the three important classical identities (1.1), (1.2), and (1.3).

Ever since Hardy brought (1.4) to light through his book [32, pp. 222–223], where he has also applied it to establish some modular equations of Ramanujan, there have been many applications and proofs of (1.4) – most of the proofs depending on its special case (1.1) as per Hardy's suggestion. However, it was K. Venkatachaliengar [43] who, one could infer, gave (1.4) its pride of place in literature above the celebrated identities of Euler and Cauchy (1.1) and of Jacobi (1.2), by publishing elegant and self contained proofs in the spirit of Ramanujan. He [43,23–25] also demonstrated the important and powerful role of (1.4) in Ramanujan's development of theta and elliptic functions. For instance, he [43] established from the remarkable formula the product representations for Jacobi's elliptic functions (which Ramanujan defined via their series representations). References [1–19,21–28,30,31,36,38–43] make up but a partial list of works containing numerous proofs, applications and ramifications of the remarkable sum (1.4). A glance at the titles of the articles cited under these references reveals the spread of the remarkable identity into various areas such as, theta functions, elliptic functions, cubic elliptic functions, modular equations, eta function identities, q-integrals, representations of numbers as sums of powers involving divisor functions and partition functions, and other areas. Alternative proofs are also available in the literature for some of the consequences of the remarkable sum. See [33–35], for instance.

The main objective of the present paper is to demonstrate some seemingly new applications of (1.4). Our results complement those in [19]. Our proofs utilize the results found in Chapter 16 of Ramanujan's second notebook [40], which are well known and follow by simple algebraic manipulations of series and products. See [1,6], for instance.The fundamental technique employed in the paper is to express products of theta functions as Lambert series via Ramanujan's ${}_1\psi_1$ summation formula. This is a standard means of showing many of the theta function identities in Ramanujan's notebooks, including some of the modular equations due to Ramanujan. The results of the paper show that the technique can be utilized to obtain altogether new results. Moreover, the strategy may be worthwhile in providing elementary proofs in the spirit of Ramanujan for entries in [40] (such as those in Chapter 21) that have heretofore only been proven using the theory of modular forms.

2. Main results

The following definitions of various theta functions as well as relations among them, namely (2.1)–(2.4) and (2.6)–(2.10) are found in Chapter 16 of Ramanujan's notebook [1,6,40]. For the role the function $b(q)$ defined in (2.5) plays in the cubic theory of theta and elliptic functions, one may see, for example [24,25]. Identities (2.13) and (2.14) are respectively analogous to Entries 10(iv) and 10(v) of Chapter 19 of Ramanujan's second notebook [40, 6 p. 262], namely

$$\varphi^2(q) - \varphi^2(q^5) = 4qf(q,q^5)f(q^3,q^7)$$

and

$$\psi^2(q) - q\psi^2(q^5) = f(q,q^4)f(q^2,q^3).$$

Although (2.13) and (2.14) are similar in structure to special cases of Entry 29 of Chapter 16 in Ramanujan's second notebook [40], identities (2.13)–(2.15) do not seem to be direct consequences solely of Entry 29. We must pair Entry 29 with Ramanujan's remarkable formula. In contrast to identities (2.13) and (2.14), identity (2.15) does not seem to be analogous to any entry in [40].

$$f(a,b) := \sum_{n=-\infty}^{\infty} a^{n(n+1)/2} b^{n(n-1)/2} = (-a;ab)_\infty(-b;ab)_\infty(ab;ab)_\infty \tag{2.1}$$

$$\varphi(q) := f(q,q) = \sum_{n=-\infty}^{\infty} q^{n^2} = \frac{(-q;-q)_\infty}{(q;-q)_\infty} \tag{2.2}$$

$$\psi(q) := f(q,q^3) = \frac{1}{2}f(1,q) = \sum_{n=0}^{\infty} q^{n(n+1)/2} = \frac{(q^2;q^2)_\infty}{(q;q^2)_\infty} \tag{2.3}$$

$$f(-q) := f(-q,-q^2) = \sum_{n=-\infty}^{\infty} (-1)^n q^{n(3n-1)/2} = (q;q)_\infty \tag{2.4}$$

$$b(q) := \frac{f^3(-q)}{f(-q^3)} \tag{2.5}$$

$$f(a,b)f(c,d) + f(-a,-b)f(-c,-d) = 2f(ac,bd)f(ad,bc) \tag{2.6}$$

$$f(ac,bd)f(ad,bc) + af\left(\frac{b}{c},\frac{c}{b}abcd\right)f\left(\frac{b}{d},\frac{d}{b}abcd\right) = f(a,b)f(c,d) \tag{2.7}$$

$$f(a,b)f(-a,-b) = f(-a^2,-b^2)\varphi(-ab) \tag{2.8}$$

$$\varphi(-q)\varphi(q) = \varphi^2(-q^2) \tag{2.9}$$

and

$$\varphi(q)\psi(q^2) = \psi^2(q), \tag{2.10}$$

for $|q| < 1$ and $ab = cd$ with $|ab| < 1$.

As one can find from [1,6, pp. 34–36, 40, 45–46], the identities contained in (2.1)–(2.4) and (2.6)–(2.10) follow by simple algebraic manipulations of series and products.

Our main results are contained in the following theorem and the corollary that follows.

Theorem 2.1. *Given $|q| < |z| < |1/q|$ and $|q| < 1$, we have*

$$\varphi^2(-q^2)\frac{f(qz,q/z)}{f(-qz,-q/z)} + \varphi^2(-q^6)\frac{f(q^3z,q^3/z)}{f(-q^3z,-q^3/z)} = 2\varphi^2(-q^6)\frac{f(-q^2,-q^{10})f(-q^6z^2,-q^6/z^2)}{f(-qz,-q^5/z)f(-q^5z,-q/z)}, \tag{2.11}$$

and

$$\psi^2(q)\frac{f(z,q^2/z)}{f(qz,q/z)} - q\psi^2(q^3)\frac{f(z,q^6/z)}{f(q^3z,q^3/z)} = \psi^2(q^3)\frac{f(q,q^2)f(z,q^3/z)}{f(qz,q^5/z)f(q^5z,q/z)}. \tag{2.12}$$

Corollary 2.2. *Given* $|q| < 1$, *we have*

$$\varphi^2(q) + \varphi^2(q^3) = 2\varphi^2(-q^6)\frac{f(q,q^5)}{f(-q,-q^5)}, \tag{2.13}$$

$$\psi^2(q^2) - q\psi^2(q^6) = \psi^3(q^3)\frac{f(q,q^2)}{f^2(q,q^5)}, \tag{2.14}$$

and

$$\varphi^2(q)\frac{b(q)}{b(-q)} + \varphi^2(q^3)\frac{b(q^3)}{b(-q^3)} = 2\varphi^3(-q^6)\frac{b(q^{12})}{b^2(q^6)}\frac{f(-q,-q^5)f(-q^2,-q^{10})}{f(-q^3,-q^{15})}. \tag{2.15}$$

Proof of Theorem 2.1.

Proof of (2.11). Putting $\alpha = -1 = \beta$ in the remarkable formula (1.4), we have, on using (2.1) together with (2.2)

$$\varphi^2(-q^2)\frac{f(qz,q/z)}{f(-qz,-q/z)} = 1 + 2\sum_{n=1}^{\infty}\frac{q^n z^n}{1+q^{2n}} + 2\sum_{n=1}^{\infty}\frac{q^n z^{\ n}}{1+q^{2n}}. \tag{2.16}$$

On expanding each term of each series of (2.16) in terms of geometric series, then interchanging the order of summation and finally, summing each of the inner geometric series we have,

$$\varphi^2(-q^2)\frac{f(qz,q/z)}{f(-qz,-q/z)} = 1 + 2\sum_{n=0}^{\infty}\frac{(-1)^n q^{2n+1}z}{1-q^{2n+1}z} + 2\sum_{n=0}^{\infty}\frac{(-1)^n q^{2n+1}/z}{1-q^{2n+1}/z}. \tag{2.17}$$

Trisecting each of the series in (2.17) we have,

$$\varphi^2(-q^2)\frac{f(qz,q/z)}{f(-qz,-q/z)}$$
$$= 1 + 2\sum_{k=0}^{2}\left[\sum_{n=0}^{\infty}\frac{(-1)^{n+k}q^{6n+2k+1}z}{1-q^{6n+2k+1}z} + \sum_{n=0}^{\infty}\frac{(-1)^{n+k}q^{6n+2k+1}z^{-1}}{1-q^{6n+2k+1}z^{-1}}\right].$$

Adding this to the equation obtained from (2.17) on changing q to q^3 we have

$$\varphi^2(-q^2)\frac{f(qz,q/z)}{f(-qz,-q/z)} + \varphi^2(-q^6)\frac{f(q^3z,q^3/z)}{f(-q^3z,-q^3/z)}$$
$$= 2 + 2\sum_{k=0}^{1}\left[\sum_{n=0}^{\infty}\frac{(-1)^n q^{6n+4k+1}z}{1-q^{6n+4k+1}z} + \sum_{n=0}^{\infty}\frac{(-1)^n q^{6n+4k+1}z^{-1}}{1-q^{6n+4k+1}z^{-1}}\right].$$

On expanding each term of each series in the last expression in terms of geometric series, then interchanging the order of summation and summing each of the inner geometric series, we have

$$\varphi^2(-q^2)\frac{f(qz,q/z)}{f(-qz,-q/z)}+\varphi^2(-q^6)\frac{f(q^3z,q^3/z)}{f(-q^3z,-q^3/z)}$$
$$=2+2\sum_{n=1}^{\infty}\frac{(q^n+q^{5n})(z^n+z^{-n})}{1+q^{6n}}. \tag{2.18}$$

Now, changing q to q^3 and then z to zq^{-2} in (2.16), we obtain

$$\varphi^2(-q^6)\frac{f(qz,q^5/z)}{f(-qz,-q^5/z)}=1+2\sum_{n=1}^{\infty}\frac{q^nz^n}{1+q^{6n}}+2\sum_{n=1}^{\infty}\frac{q^{5n}z^{-n}}{1+q^{6n}}. \tag{2.19}$$

Changing z to z^{-1} in (2.19) and adding the resulting equation to (2.19) and combining the result with that in (2.18), we have

$$\varphi^2(-q^2)\frac{f(qz,q/z)}{f(-qz,-q/z)}+\varphi^2(-q^6)\frac{f(q^3z,q^3/z)}{f(-q^3z,-q^3/z)}$$
$$=\varphi^2(-q^6)\left[\frac{f(qz,q^5/z)}{f(-qz,-q^5/z)}+\frac{f(q^5z,q/z)}{f(-q^5z,-q/z)}\right]$$
$$=2\varphi^2(-q^6)\frac{f(-q^2,-q^{10})f(-q^6z^2,-q^6z^{-2})}{f(-qz,-q^5/z)f(-q^5z,-q/z)},$$

on employing (2.6) with $a=qz$, $b=q^5/z$, $c=-q/z$ and $d=-q^5z$. This proves (2.11).

Proof of (2.12). Putting $\alpha=q=\beta^{-1}$ and changing z to z/q in (1.4) we have, for $|q|<1$ and $|q|<|z|<|q^{-1}|$,

$$\frac{(-z;q^2)(-q^2/z;q^2)_\infty(q^2;q^2)^2_\infty}{(-qz;q^2)_\infty(-q/z;q^2)_\infty(q^3;q^2)_\infty(q;q^2)_\infty}$$
$$=1+\sum_{n=1}^{\infty}\frac{(q^{-1};q^2)_n(-qz)^n}{(q;q^2)_n}+\sum_{n=1}^{\infty}\frac{(q;q^2)_n(-q/z)^n}{(q^3;q^2)_n}.$$

Dividing the above identity throughout by $1/(1-q)$, we have, on using (2.1) and (2.3),

$$\psi^2(q)\frac{f(z,q^2/z)}{f(qz,q/z)}=\sum_{n=0}^{\infty}\frac{(-1)^nq^nz^{n+1}}{1-q^{2n+1}}+\sum_{n=0}^{\infty}\frac{(-1)^nq^nz^{-n}}{1-q^{2n+1}} \tag{2.20}$$
$$=\sum_{n=0}^{\infty}\frac{q^nz}{1+zq^{2n+1}}+\sum_{n=0}^{\infty}\frac{q^n}{1+z^{-1}q^{2n+1}}. \tag{2.21}$$

The equality displayed on line (2.21) follows by expanding each term of each series in (2.20) in terms of geometric series, then interchanging the order of summation and finally, summing each of the inner geometric series. Trisecting each of the series in (2.20) we get,

$$\psi^2(q)\frac{f(z,q^2/z)}{f(qz,q/z)} = \sum_{k=0}^{2}\left[\sum_{n=0}^{\infty}\frac{q^{3n+k}z}{1+zq^{6n+2k+1}} + \sum_{n=0}^{\infty}\frac{q^{3n+k}}{1+z^{-1}q^{6n+2k+1}}\right]. \quad (2.22)$$

Changing q to q^3 in (2.21) and subtracting q times the resulting identity from (2.22) we have,

$$\psi^2(q)\frac{f(z,q^2/z)}{f(qz,q/z)} - q\psi^2(q^3)\frac{f(z,q^6/z)}{f(q^3z,q^3/z)}$$

$$= \sum_{k=0}^{1}\left[\sum_{n=0}^{\infty}\frac{q^{3n+2k}z}{1+zq^{6n+4k+1}} + \sum_{n=0}^{\infty}\frac{q^{3n+2k}}{1+z^{-1}q^{6n+4k+1}}\right] \quad (2.23)$$

$$= \sum_{n=0}^{\infty}\frac{(-1)^n(q^n+q^{5n+2})(z^{n+1}+z^{-n})}{(1-q^{6n+3})}. \quad (2.24)$$

The equality displayed on line (2.24) follows by expanding each term of each series in (2.23) in terms of geometric series, then interchanging the order of summation and finally, summing each of the inner geometric series.

Now, changing q to q^3 in (2.20) and z to zq^{-2}, and multiplying the resulting identity by q^2 we have

$$q^2\psi^2(q^3)\frac{f(q^{-2}z,q^8z^{-1})}{f(qz,q^5z^{-1})} = \sum_{n=0}^{\infty}\frac{(-1)^n(q^nz^{n+1}+q^{5n+2}z^{-n})}{1-q^{6n+3}}. \quad (2.25)$$

Changing z to z^{-1} in (2.25) and multiplying the resulting identity by z, we have

$$zq^2\psi^2(q^3)\frac{f(q^{-2}z^{-1},q^8z)}{f(qz^{-1},q^5z)} = \sum_{n=0}^{\infty}\frac{(-1)^n(q^nz^{-n}+q^{5n+2}z^{n+1})}{1-q^{6n+3}}.$$

Adding this to (2.25) and combining the result with that in (2.24), we may rewrite the left side of (2.12) as

$$q^2\psi^2(q^3)\frac{f(q^{-2}z,q^8z^{-1})}{f(qz,q^5z^{-1})} + zq^2\psi(q^3)\frac{f(q^8z,q^{-2}z^{-1})}{f(q^5z,qz^{-1})}$$

$$= \psi^2(q^3)\frac{f(q,q^2)f(z,q^3z^{-1})}{f(qz,q^5z^{-1})f(q^5z,qz^{-1})}.$$

The last equation follows by employing (2.7) with $a = z$, $b = q^3/z$, $c = q^{-2}$ and $d = q^5$. This proves (2.12) and the proof of the theorem is complete. □

Proof of Corollary 2.2.

Proof of (2.13). By putting $z = 1$ in (2.11) and using identity (2.9) twice, (2.8) and the definition in (2.2), we have

$$\begin{aligned}\varphi^2(q) + \varphi^2(q^3) &= 2\varphi^3(-q^6)\frac{f(-q^2, -q^{10})}{f^2(-q, -q^5)}\\ &= 2\varphi^3(-q^6)\frac{f(q, q^5)f(-q, -q^5)}{\varphi(-q^6)f^2(-q, -q^5)}\\ &= 2\varphi^2(-q^6)\frac{f(q, q^5)}{f(-q, -q^5)}.\end{aligned}$$

This proves (2.13).

Proof of (2.14). By putting $z = 1$ in (2.12) and using the identity (2.10) and the definition in (2.3), we get

$$\frac{\psi^2(q)\psi(q^2)}{\varphi(q)} - \frac{q\psi^2(q^3)\psi(q^6)}{\varphi(q^3)} = \psi^3(q^3)\frac{f(q, q^2)}{f^2(q, q^5)},$$

or, equivalently

$$\psi^2(q^2) - q\psi^2(q^6) = \psi^3(q^3)\frac{f(q, q^2)}{f^2(q, q^5)}.$$

This proves (2.14).

Proof of (2.15). We can deduce (2.15) from (2.11) by putting $z = \omega$, the cube root of unity, and using the relevant identities and definitions in (2.1)–(2.10). We omit the details of simplifications as they are similar to the ones above.

S. Cooper has communicated to the author that the expressions on the right sides of (2.13) and (2.14) can, by trivial transformations, be put in the form

$$2\frac{\varphi(-q^2)\varphi(-q^3)\varphi(-q^6)}{\varphi(-q)}$$

and

$$\frac{\psi(q^2)\psi(q^3)\varphi(q^3)}{\psi(q)}$$

respectively. □

Acknowledgement

The author is thankful to the referee for his/her valuable suggestions, Professor S. Cooper for motivation and personal discussions and Dr. K. R. Vasuki and Mr. G. Sharath for their help. The author is also thankful to the organizers of the conference 'Ramanujan Rediscovered 2009' held at Bangalore during June 1–5, 2009. The present paper is based on the invited talk given by the author at the conference.

References

[1] C. Adiga, B. C. Berndt, S. Bhargava and G. N. Watson, Chapter 16 of Ramanujan's Second Notebook: The Theta functions and q-series, *Mem. Am. Math. Soc.*, No. 315, Vol. 53, *Amer. Math. Soc.*, Providence, RI (1985).

[2] R. P. Agarwal, Resonance of Ramanujan's Mathematics, Vol. III, New Age International Publishers, New Delhi (1999).

[3] G. E. Andrews, On Ramanujan's summation of ${}_1\psi_1$, *Proc. Amer. Math. Soc.*, **22** (1969) 552–553.

[4] G. E. Andrews, On a transformation of bilateral series with applications, *Proc. Amer. Math. Soc.*, **25** (1970) 554–558.

[5] R. Askey, Ramanujan's ${}_1\psi_1$ and formal Laurent series, *Indian J. Math.*, **29** (1987) 101–105.

[6] B. C. Berndt, Ramanujan's Notebooks, Part III, Springer-Verlag, New York (1991).

[7] B. C. Berndt, Number Theory in the Spirit of Ramanujan, Vol. 34, Student Mathematical Library, American Mathematical Society, Providence, RI (2000).

[8] B. C. Berndt, S. H. Chan, B. P. Yeap and Ae Ja Yee, A reciprocity theorem for certain q-series found in Ramanujan's lost notebook, *The Ramanujan J.*, **13** (2007) 27–37.

[9] S. Bhargava, On Ramanujan's remarkable summation formula, *The Math. Student*, **63** (1996) 181–192.

[10] S. Bhargava, A series for the quintuple product in the context of Ramanujan's remarkable formula, *The Math. Student*, **70** (2001) 121–124.

[11] S. Bhargava and Chandrashekar Adiga, Simple Proofs of Jacobi's two and four square theorems, *Inter. J. Math. Ed. Sci. Tech.*, **19** (1988) 779–782.

[12] S. Bhargava and D. D. Somashekara, Some eta-function identities deducible from Ramanujan's ${}_1\psi_1$ summation, *J. Math. Anal. Appl.*, **176** (1993) 554–560.

[13] S. Bhargava, Chandrashekar Adiga and D. D. Somashekara, Ramanujan's ${}_1\psi_1$ summation and an interesting generalization of an identity of Jacobi, *Indian J. Math.*, **41** (1999) 39–42.

[14] S. Bhargava, Chandrashekar Adiga and D. D. Somashekara, Representation of an integer as the sum of a square and thrice a square, *Inter. J. Math. Ed. Sci. Tech.*, **24** (1993) 493–496.

[15] S. Bhargava, Chandrashekar Adiga and D. D. Somashekara, Three square theorem as an application of Andrews' identity, *Fibonacci Quart.*, **31** (1993) 129–133.

[16] S. Bhargava, Chandrashekar Adiga and D. D. Somashekara, Ramanujan's remarkable summation formula and an interesting convolution identity, *Bull. Austral. Math. Soc.*, **47** (1993) 155–162.

[17] S. Bhargava, Chandrashekar Adiga and D. D. Somashekara, Number of representations of an integer as the sum of a square and five times a square, *J. Ind. Math. Soc.*, **65** (1998) 249–251.

[18] S. Bhargava, Chandrashekar Adiga and M. S. Mahadeva Naika, Ramanujan's remarkable identity as a 2-parameter generalization of the quintuple product identity, *Tamkang J. Math.*, **33** (2002) 285–288.

[19] S. Bhargava, K. R. Vasuki and K. R. Rajanna, On certain identities of Ramanujan for ratios of eta-functions, Preprint.

[20] A. Cauchy, Oeuvres, *Ser. 1*, **8** Gauthier-Villars, Paris (1893).

[21] S. H. Chan, A short proof of Ramanujan's ${}_1\psi_1$ summation formula, *J. Approx. Thy.*, **132** (2005) 149–153.

[22] Chandrashekar Adiga, A study of some identities stated by Srinivasa Ramanujan in his 'lost' notebook and earlier works, Ph.D. Thesis, University of Mysore, Mysore (1983).

[23] S. Cooper, The development of elliptic functions according to Ramanujan and Venkatachaliengar, http://iims.messey.ac.nz/research/letters/see, Vol. 1 (2000).

[24] S. Cooper, Cubic theta functions, *J. Comp. App. Math.*, **160** (2003) 77–94.

[25] S. Cooper, Cubic elliptic functions, *The Ramanujan J.*, **11** (2006) 355–397.
[26] S. Cooper, On sums of an even number of squares, and an even number of triangular numbers: an elementary approach based on Ramanujan's ${}_1\psi_1$ summation formula, *Contemporary Math.*, **291** (2001).
[27] S. Cooper, The quintuple product identity, *Inter. J. Number Thy.*, **2** (2006) 115–161.
[28] S. Cooper and H. Y. Lam, Sums of two, four, six and eight squares and triangular numbers: An elementary approach, *Ind. J. Math.*, **44** (2002) 21–40.
[29] L. Euler, Introductio in Analysin Infinitorum, Marcum-Michaelem Bousquet, Lausanne (1748).
[30] N. J. Fine, Basic Hypergeometric Series and Applications, Mathematical Surveys and Monographs 27, American Mathematical Society, Providence, RI (1988).
[31] W. Hahn, Beiträge zur Theorie der Heineschen Reihen, *Math. Nachr.*, **2** (1949) 340–379.
[32] G. H. Hardy, Ramanujan: Twelve lectures on subjects suggested by his life and work, Cambridge University Press, Cambridge 1940; reprinted by the American mathematical society, Providence, RI (1999).
[33] M. D. Hirschhorn, A simple proof of Jacobi's four-square theorem, *J. Austral. Math. Soc. Ser. A*, **32** (1982) 61–67.
[34] M. D. Hirschhorn, A simple proof of Jacobi's two-square theorem, *Amer. Math. Monthly*, **92** (1985) 579–580.
[35] M. D. Hirschhorn, A simple proof of Jacobi's four-square theorem, *Proc. Amer. Math. Soc.*, **101** (1987) 436–438.
[36] M. Jackson, On Lerch's transcendent and the basic bilateral hypergeometric series ${}_3\psi_2$, *J. London Math. Soc.*, **25** (1950) 189–196.
[37] C. G. J. Jacobi, Fundamenta Nova Theoriae Functionum Ellipticarum, Bornträger, Regiomonti (1829).
[38] W. P. Johnson, How Cauchy missed Ramanujan's ${}_1\psi_1$ summation, *Amer. Math. Monthly*, **111** (2004) 791–800.
[39] V. Ramamani, Some Identities Conjectured by Ramanujan Found in His Lithographed Notes Connected with Partition Theory and Elliptic Modular Functions-Their Proofs-Interconnection with Various Other Topics in the Theory of Numbers and Some Generalizations Thereon, Ph.D. Thesis, University of Mysore (1971).
[40] S. Ramanujan, Notebooks (2 volumes), Tata Institute of Fundamental Research, Mumbai (1957).
[41] S. Ramanujan, The Lost Notebook and other unpublished papers, Narosa, New Delhi (1988).
[42] D. D. Somashekara, Contributions to the Theory of Special Functions and Number Theory Motivated by Works of Srinivasa Ramanujan, Ph.D. Thesis, University of Mysore, Mysore (1992).
[43] K. Venkatachaliengar, Development of Elliptic Functions According to Ramanujan, Tech. Report 2, Madurai Kamaraj University, Madurai (1988).

Ramanujan Rediscovered No. 14, 2010, pp. 73–78.

A New Proof of Winquist's Identity

Song Heng Chan

Division of Mathematical Sciences, School of Physical and Mathematical Sciences, Nanyang Technological University, Singapore 637371, Singapore
e-mail: chansh@ntu.edu.sg

Abstract. We present a new proof of Winquist's identity via Halphen's identity. We also derive an identity due to Weierstrass via Halphen's identity.

Keywords. Winquist's identity, Halphen's identity.

2000 Subject Classification: 33D15, 33E05, 11P83

1. Introduction

Let $p(n)$ denote the number of unrestricted partitions of n, and let $|q| < 1$ throughout this article. In [18], L. Winquist gave the first elementary proof of the Ramanujan congruence

$$p(11n+6) \equiv 0 \pmod{11}.$$

Winquist's proof was based on the identity

$$\begin{aligned}
&[a, b, ab, a/b; q]_\infty (q;q)_\infty^2 \\
&= \sum_{m=-\infty}^{\infty} \sum_{n=-\infty}^{\infty} (-1)^{m+n} q^{(3m^2+3n^2+3m+n)/2} \\
&\quad \times (a^{-3m}b^{-3n} - a^{-3m}b^{3n+1} - a^{-3n+1}b^{-3m-1} + a^{3n+2}b^{-3m-1}), \qquad (1.1)
\end{aligned}$$

where we use the notations

$$(x;q)_\infty := \prod_{n=0}^{\infty}(1-xq^n),$$

$$[x;q]_\infty := (x;q)_\infty (q/x;q)_\infty,$$

$$[x_1, \ldots, x_k; q]_\infty := [x_1;q]_\infty \cdots [x_k;q]_\infty.$$

Identity (1.1) was later labeled Winquist's identity. Since Winquist's original proof of (1.1) in [18], many new proofs were discovered, and they include, H. H. Chan,

This work was partially supported by Nanyang Technological University Academic Research Fund, project number RG57/07.

S. Cooper, and P. C. Toh [5], Chan, Z.-G. Liu, and S. T. Ng [6], S.-Y. Kang [11], S. Konsiriwong and Liu [12], Liu [13]. Proofs and generalizations of (1.3) also appear in L. Carlitz and M. V. Subbarao [4], M. D. Hirschhorn [10] and W. Chu [8].

Applying the Jacobi triple product identity [3, pp. 33–36],

$$\sum_{n=-\infty}^{\infty} a^{n(n+1)/2} b^{n(n-1)/2} = (-a; ab)_\infty (-b; ab)_\infty (ab; ab)_\infty \tag{1.2}$$

on the right side of (1.1), we restate (1.1) in the equivalent form,

$$\begin{aligned}
&[a, b, ab, a/b; q]_\infty (q; q)_\infty^2 \\
&\quad = [a^3; q^3]_\infty (q^3; q^3)_\infty^2 ([b^3 q; q^3]_\infty - b[b^3 q^2; q^3]_\infty) \\
&\quad\quad - \frac{a}{b} [b^3; q^3]_\infty (q^3; q^3)_\infty^2 ([a^3 q; q^3]_\infty - a[a^3 q^2; q^3]_\infty).
\end{aligned} \tag{1.3}$$

The main purpose of this article is to present a new proof of Winquist's identity via the following identity, which we refer to as Halphen's identity.

$$\begin{aligned}
H(a, b, c, q) &:= \frac{[ab, bc, ca]_\infty (q)_\infty^2}{[a, b, c, abc]_\infty} \\
&= 1 + F(a, q) + F(b, q) + F(c, q) - F(abc, q),
\end{aligned} \tag{1.4}$$

where

$$F(x, q) := \sum_{k=0}^{\infty} \frac{xq^k}{1 - xq^k} - \sum_{k=1}^{\infty} \frac{q^k/x}{1 - q^k/x}, \quad (|q| < 1). \tag{1.5}$$

Halphen's identity in the form above was first discovered and presented in [1]. However, equivalent versions of (1.4) in terms of Weierstrassian elliptic functions appeared in the literature much earlier, as exercises in the classical text by Whittaker and Watson [17, Examples 19 and 20, p. 458] and also in G. H. Halphen [9, p. 187]. Halphen's identity is closely related to the fundamental multiplicative identity of K. Venkatachaliengar [16, p. 37].

Most of proofs of Winquist's identity involve either multiplying several product-series type identities together and then simplifying, or via the method of elliptic functions [6], [13]. The proof that we present in Section 2 involves essentially specializing the variables in [10] and then applying the identity [7, Corollary 3.1]

$$\frac{[xy]_\infty (q)_\infty^2}{[x, y]_\infty} = \sum_{k=-\infty}^{\infty} \frac{x^k}{1 - yq^k}, \quad (|q| < |x| < 1), \tag{1.6}$$

which is another classical identity, perhaps better known as a corollary of Ramanujan's ${}_1\psi_1$ summation formula. This suggests possibly that (1.3) was known before Winquist's paper [18] but was never published, as it was not thought to be of importance. Proofs of both (1.4) and (1.6) via partial fractions were discovered by the author in [7], and so one can obtain (1.3) via the partial fraction decomposition of an appropriate infinite product and simplifications.

Recently, in [8, Theorem 1.1], Chu rediscovered an identity of Weierstrass (3.1), from which he derives many identities, including Winquist's identity. In the final section of this article, we show that Weierstrass's identity (3.1) also follows from Halphen's identity (1.4).

2. A new proof of Winquist's identity

We first state and prove a simple lemma.

Lemma 2.1. *For $|q| < 1$, we have*

$$1 + F(a,q) = -F(1/a,q). \tag{2.1}$$

Proof. Since both series are absolutely convergent for $|q| < 1$, we find that

$$\begin{aligned}
F(a,q) &= \sum_{k=0}^{\infty} \frac{aq^k}{1-aq^k} - \sum_{k=1}^{\infty} \frac{q^k/a}{1-q^k/a} \\
&= \frac{a}{1-a} + \frac{\frac{1}{a}}{1-\frac{1}{a}} + \sum_{k=1}^{\infty} \frac{q^k/\frac{1}{a}}{1-q^k/\frac{1}{a}} - \sum_{k=0}^{\infty} \frac{\frac{1}{a}q^k}{1-\frac{1}{a}q^k} \\
&= -1 - F(1/a,q).
\end{aligned}$$

□

Proof of Winquist's identity. Let ω be a cubic root of unity. From Halphen's identity (1.4), we have

$$H(a\omega, b\omega^2, \omega^2/b, q) = 1 + F(a\omega) + F(b\omega^2) + F(\omega^2/b) - F(a\omega^2). \tag{2.2}$$

Simplifying the left side of (2.2), we obtain

$$\begin{aligned}
H(a\omega, b\omega^2, \omega^2/b, q) &= \frac{[ab, a/b, \omega; q]_\infty (q;q)_\infty^2}{[a\omega, b\omega^2, \omega^2/b, a\omega^2; q]_\infty} \\
&= -b\omega(1-\omega)\frac{[a, b, ab, a/b; q]_\infty (q;q)_\infty^2}{[a^3, b^3, q; q^3]_\infty}.
\end{aligned} \tag{2.3}$$

Note that

$$\begin{aligned}
&F(a\omega) - F(a\omega^2) \\
&= \sum_{k=0}^{\infty} \frac{a\omega q^k}{1-a\omega q^k} - \sum_{k=1}^{\infty} \frac{\omega^2 q^k/a}{1-\omega^2 q^k/a} - \sum_{k=0}^{\infty} \frac{a\omega^2 q^k}{1-a\omega^2 q^k} + \sum_{k=1}^{\infty} \frac{\omega q^k/a}{1-\omega q^k/a} \\
&= (\omega-\omega^2)\sum_{k=0}^{\infty} \frac{(1-aq^k)aq^k}{1-a^3q^{3k}} + (\omega-\omega^2)\sum_{k=1}^{\infty} \frac{(1-q^k/a)q^k/a}{1-q^{3k}/a^3} \\
&= (\omega-\omega^2)\sum_{k=0}^{\infty} \frac{(1-aq^k)aq^k}{1-a^3q^{3k}} + (\omega-\omega^2)\sum_{k=1}^{\infty} \frac{(1-aq^{-k})aq^{-k}}{1-a^3q^{-3k}}
\end{aligned}$$

$$= a(\omega - \omega^2) \sum_{k=-\infty}^{\infty} \frac{q^k}{1 - a^3 q^{3k}} - a^2(\omega - \omega^2) \sum_{k=-\infty}^{\infty} \frac{q^{2k}}{1 - a^3 q^{3k}}$$

$$= a(\omega - \omega^2) \frac{[a^3 q; q^3]_\infty (q^3; q^3)_\infty^2}{[a^3, q; q^3]_\infty} - a^2(\omega - \omega^2) \frac{[a^3 q^2; q^3]_\infty (q^3; q^3)_\infty^2}{[a^3, q^2; q^3]_\infty}, \tag{2.4}$$

where we applied (1.6) twice in the last equality. By invoking (2.1), we obtain similarly,

$$1 + F(b\omega^2) + F(\omega^2/b)$$
$$= F(b\omega^2) - F(b\omega)$$
$$= b^2(\omega - \omega^2) \frac{[b^3 q^2; q^3]_\infty (q^3; q^3)_\infty^2}{[b^3, q^2; q^3]_\infty} - b(\omega - \omega^2) \frac{[b^3 q; q^3]_\infty (q^3; q^3)_\infty^2}{[b^3, q; q^3]_\infty}. \tag{2.5}$$

Substituting (2.3) on the left side of (2.2), and (2.4) and (2.5) on the right side of (2.2) and dividing by $-b\omega(1 - \omega)$, we obtain

$$\frac{[a, b, ab, a/b; q]_\infty (q; q)_\infty^2}{[a^3, b^3, q; q^3]_\infty}$$
$$= \frac{[b^3 q; q^3]_\infty (q^3; q^3)_\infty^2}{[b^3, q; q^3]_\infty} - b \frac{[b^3 q^2; q^3]_\infty (q^3; q^3)_\infty^2}{[b^3, q^2; q^3]_\infty}$$
$$- \frac{a}{b} \frac{[a^3 q; q^3]_\infty (q^3; q^3)_\infty^2}{[a^3, q; q^3]_\infty} + \frac{a^2}{b} \frac{[a^3 q^2; q^3]_\infty (q^3; q^3)_\infty^2}{[a^3, q^2; q^3]_\infty}.$$

Upon multiplying both sides by $[a^3, b^3, q; q^3]_\infty$, we obtain Winquist's identity in the form (1.3). □

3. Proof of an identity of Weierstrass

In Example 5 on page 451 of Whittaker and Watson's book [17], a three term relation sigma function identity is given and credited to Weierstrass, with a reference to p. 47 of the edition of his lectures by Schwarz [15]. That very same sigma function identity was also stated by B. Riemann in Equation (10) on p. 139 of [14], with full credit given to Weierstrass. In q notations, Weierstrass's identity was first discovered by W. N. Bailey [2]. This identity was recently rediscovered by Chu [8, Theorem 1.1], from which he derives many identities, including Winquist's identity. We state Weierstrass's identity in the form given by Chu in [8, Theorem 1.1].

Theorem 3.1 (Weierstrass's identity). *Let A, b, c, d, e be five complex parameters satisfying $A^2 = bcde$. Then the following theta function identity is true:*

$$[A/b, A/c, A/d, A/e; q]_\infty - [b, c, d, e; q]_\infty = b[A, A/bc, A/bd, A/be; q]_\infty. \tag{3.1}$$

In this section, we give a proof of (3.1) via Halphen's identity (1.4).

Proof. Applying Halpen's identity (1.4) to the difference

$$\begin{aligned} H(a,b,c,q) - H(a,b,d,q) &= 1 + F(a) + F(b) + F(c) - F(abc) \\ &\quad - 1 - F(a) - F(b) - F(d) + F(abd) \\ &= F(c) - F(d) + F(abd) - F(abc) \\ &= 1 + F(c) + F(1/d) + F(abd) - F(abc) \\ &= H(c, 1/d, abd), \end{aligned}$$

where in the penultimate identity, we invoked (2.1) to $F(d)$. Rewriting this identity in terms of the infinite products, we obtain

$$\frac{[ab, bc, ca; q]_\infty (q;q)_\infty^2}{[a,b,c,abc;q]_\infty} - \frac{[ab, bd, da; q]_\infty (q;q)_\infty^2}{[a,b,d,abd;q]_\infty}$$
$$= \frac{[c/d, abcd, ab; q]_\infty (q;q)_\infty^2}{[c, 1/d, abc, abd; q]_\infty}.$$

Multiplying throughout by $\frac{[a,b,c,d,abc,abd;q]_\infty}{[ab;q]_\infty (q;q)_\infty^2}$, we have

$$[ac, bc, d, abd; q]_\infty - [ad, bd, c, abc; q]_\infty = [c/d, abcd, a, b; q]_\infty \frac{[d;q]_\infty}{[1/d;q]_\infty}.$$

Upon simplifying and rearranging, we arrive at

$$[ad, bd, c, abc; q]_\infty - [ac, bc, d, abd; q]_\infty = d[c/d, abcd, a, b; q]_\infty.$$

Replacing a, b, c, and d by ce/A, cd/A, A/c, and A^2/cde, respectively, we recover (3.1) and this completes the proof. □

Acknowledgement

Part of this work was done during the author's visit at the University of Illinois. The author thanks Professor Bruce Berndt for his hospitality. The author also thanks Professor Berndt and Sun Kim for their helpful discussions, Shaun Cooper who uncovered several misprints in an earlier version of this article, and Michael Schlosser for references, information, and suggestions on how to introduce (3.1) in the article.

References

[1] G. E. Andrews, R. Lewis and Z.–G. Liu, An identity relating a theta function to a sum of Lambert series, *Bull. London Math. Soc.*, **33** (2001) 25–31.

[2] W. N. Bailey, Series of hypergeometric type which are infinite in both directions, *Quart. J. Math. (Oxford)*, **7** (1936) 105–115.

[3] B. C. Berndt, Ramanujan's Notebooks, Part III, Springer–Verlag, New York (1991).

[4] L. Carlitz and M. V. Subbarao, On a combinatorial identity of Winquist and its generalization, *Duke Math. J.*, **39** (1972) 165–172.

[5] H. H. Chan, S. Cooper and P. C. Toh, Ramanujan's Eisenstein series and powers of Dedekind's eta-function, *J. Lond. Math. Soc. (2)*, **75(1)** (2007) 225–242.

[6] H. H. Chan, Z.-G. Liu and S. T. Ng, Elliptic functions and the quintuple, Hirschhorn and Winquist product identities, *Int. J. Number Theory*, **1(1)** (2005) 33–43.

[7] S. H. Chan, Generalized Lambert series identities, *Proc. London Math. Soc. (3)*, **91** (2005) 598–622.

[8] W. Chu, Theta function identities and Ramanujan's congruences on the partition function, *Q. J. Math.*, **56(4)** (2005) 491–506.

[9] G. H. Halphen, Traité des fonctions elliptiques et de leurs applications vol. 1, Gauthier-Villars, Paris (1888).

[10] M. D. Hirschhorn, A generalisation of Winquist's identity and a conjecture of Ramanujan, *J. Indian Math. Soc.*, **51** (1987) 49–55.

[11] S.-Y. Kang, A new proof of Winquist's identity, *J. Combin. Theory Ser. A*, **78** (1997) 313–318.

[12] S. Kongsiriwong and Z.-G. Liu, Uniform proofs of q-series-product identities, *Results Math.*, **44(3–4)** (2003) 312–339.

[13] Z.-G. Liu, An addition formula for the Jacobian theta function and its applications, *Adv. Math.*, **212(1)** (2007) 389–406.

[14] G. F. B. Riemann, Elliptische Functionen, Vorlesungen von Bernhard Riemann. Mit Zusätzen herausgegeben von Hermann Strahl, Teubner-Verlag, Leipzig (1899).

[15] H. Schwarz and K. Weierstrass, Formeln und Lehrsätze zum Gebrauche der elliptischen Functionen, Göttingen (1885).

[16] K. Venkatachaliengar, Development of elliptic functions according to Ramanujan, Department of Mathematics, Madurai Kamaraj University, Technical Report 2 (1988).

[17] E. T. Whittaker and G. N. Watson, A Course of Modern Analysis, fourth edition, Cambridge University Press (1952).

[18] L. Winquist, An elementary proof of $p(11m + 6) \equiv 0 \pmod{11}$, *J. Comb. Thy.*, **6** (1969) 56–59.

Ramanujan Rediscovered No. 14, 2010, pp. 79–86.

Theta Function Identities for Level 15

Shaun Cooper

Institute of Information and Mathematical Sciences, Massey University-Albany,
Private Bag 102904, North Shore Mail Centre, Auckland, New Zealand
e-mail: s.cooper@massey.ac.nz

Abstract. Six infinite products are investigated. Their series expansions are determined and these imply several linear relations among the products. As applications, new and simple proofs can be given for theorems on the number of representations of a positive integer by the quadratic forms $j^2+jk+4k^2$, $2j^2+jk+2k^2$, j^2+15k^2 and $3j^2+5k^2$, and by the quadratic polynomials $j(j+1)/2+15k(k+1)/2$ and $3j(j+1)/2+5k(k+1)/2$.

Keywords. Eta-product, quadratic form, Ramanujan, theta function.

2000 Subject Classification: Primary: 11F11; Secondary: 33E05

Dedicated to the memory of Professor K. Venkatachaliengar

1. Introduction and definitions

Let τ be a complex number with positive imaginary part, and let $q = \exp(2\pi i\tau)$, so that $|q| < 1$. For any positive integer n, define η_n by

$$\eta_n = q^{n/24}\prod_{j=1}^{\infty}(1-q^{nj}).$$

Let

$$Z_1 = Z_1(q) = \frac{\eta_3^2\eta_5^2}{\eta_1\eta_{15}}, \quad Z_2 = Z_2(q) = \frac{\eta_1^2\eta_{15}^2}{\eta_3\eta_5},$$

$$Y_1 = Y_1(q) = \frac{\eta_1\eta_6\eta_{10}\eta_{15}}{\eta_2\eta_{30}}, \quad Y_2 = Y_2(q) = \frac{\eta_2\eta_3\eta_5\eta_{30}}{\eta_1\eta_{15}},$$

$$Y_3 = Y_3(q) = \frac{\eta_1\eta_6\eta_{10}\eta_{15}}{\eta_3\eta_5}, \quad Y_4 = Y_4(q) = \frac{\eta_2\eta_3\eta_5\eta_{30}}{\eta_6\eta_{10}}.$$

In this work, series expansions for all six of these eta-quotients will be derived by series manipulations. As consequences, new and simple proofs can be given for theorems on the number of representations of a positive integer by each of the quadratic forms

$$j^2+jk+4k^2, \quad 2j^2+jk+2k^2, \quad j^2+15k^2, \quad 3j^2+5k^2, \tag{1}$$

and by the quadratic polynomials

$$\frac{j(j+1)}{2}+\frac{15k(k+1)}{2}, \quad \frac{3j(j+1)}{2}+\frac{5k(k+1)}{2}. \tag{2}$$

Some brief historical details about the number of representations of integers by the forms in (1) and the polynomials in (2) will be given in Section 4.

The standard notation for infinite products will be used throughout:

$$(a_1, a_2, \ldots, a_k; q)_\infty = \prod_{j=0}^{\infty}(1-a_1q^j)(1-a_2q^j)\cdots(1-a_kq^j).$$

2. Main results

The main result of this work is

Theorem 2.1. *Let $\left(\frac{a}{b}\right)$ denote the Jacobi symbol. Then:*

$$Z_1 = 1+\sum_{j=1}^{\infty}\sum_{k=1}^{\infty}\left(\frac{j}{15}\right)q^{jk}, \quad Z_2 = \sum_{j=1}^{\infty}\sum_{k=1}^{\infty}\left(\frac{j}{3}\right)\left(\frac{k}{5}\right)q^{jk},$$

$$Y_1 = 1+\sum_{j=1}^{\infty}\sum_{k=1}^{\infty}\left(\frac{j}{15}\right)(-1)^j q^{jk}, \quad Y_2 = \sum_{\substack{j=1 \\ j \text{ odd}}}^{\infty}\sum_{k=1}^{\infty}\left(\frac{j}{15}\right)q^{jk},$$

$$Y_3 = \sum_{\substack{j=1 \\ j \text{ odd}}}^{\infty}\sum_{k=1}^{\infty}\left(\frac{j}{3}\right)\left(\frac{k}{5}\right)q^{jk}, \quad Y_4 = \sum_{j=1}^{\infty}\sum_{k=1}^{\infty}\left(\frac{j}{3}\right)\left(\frac{k}{5}\right)(-1)^{j+1}q^{jk}.$$

Proof. We shall use the identity

$$\frac{(xy, xz, yz, qx^{-1}y^{-1}, qx^{-1}z^{-1}, qy^{-1}z^{-1}, q, q; q)_\infty}{(x, y, z, xyz, qx^{-1}, qy^{-1}, qz^{-1}, qx^{-1}y^{-1}z^{-1}; q)_\infty}$$
$$= \frac{1}{2}\left(\frac{1+x}{1-x}+\frac{1+y}{1-y}+\frac{1+z}{1-z}-\frac{1+xyz}{1-xyz}\right)$$
$$+\sum_{k=1}^{\infty}\frac{q^j}{1-q^j}(x^j+y^j+z^j-(xyz)^j-x^{-j}-y^{-j}-z^{-j}+(xyz)^{-j}), \tag{3}$$

which holds for $|q| < |x|, |y|, |z|, |xyz| < |q|^{-1}$. This result was known in the 19th century, e.g., see [9, p. 187]. The identity (3) is discussed, and references to other proofs are given, in [5].

For any positive integer n, let $\omega_n = \exp(2\pi i/n)$. If we replace (x, y, z) with $(\omega_{15}, \omega_{15}^2, \omega_{15}^4)$ in (3) and simplify the infinite product, the result is

$$\frac{\eta_3^2\eta_5^2}{\eta_1\eta_{15}} = 1+\sum_{j=1}^{\infty}h(j)\frac{q^j}{1-q^j}, \tag{4}$$

where $h(j)$ is a function with period 15. If we compare coefficients of $q, q^2, \ldots, q^{15}$ on both sides of (4), we find that

$$h(j) = \begin{cases} 1 & \text{if } j \equiv 1,\ 2,\ 4 \text{ or } 8 \pmod{15}, \\ -1 & \text{if } j \equiv 7,\ 11,\ 13 \text{ or } 14 \pmod{15}, \\ 0 & \text{otherwise} \end{cases}$$
$$= \left(\frac{j}{15}\right).$$

The result for Z_1 now follows. The results for Y_1 and Y_2 may be proved similarly, by replacing (x, y, z, q) with $(\omega_{30}^7, \omega_{30}^{11}, \omega_{30}^{13}, q)$ and $(q^{-1}, q^{11}, q^{13}, q^{30})$, respectively.

To prove the result for Z_2, we manipulate the series as follows:

$$\begin{aligned}\sum_{j=1}^{\infty}\sum_{k=1}^{\infty}\left(\frac{j}{3}\right)\left(\frac{k}{5}\right)q^{jk} &= \sum_{r=1}^{4}\left(\frac{r}{5}\right)\sum_{j=0}^{\infty}\sum_{k=0}^{\infty}\left(q^{(3j+1)(5k+r)} - q^{(3j+2)(5k+5-r)}\right) \\ &= \sum_{r=1}^{4}\left(\frac{r}{5}\right)\sum_{k=0}^{\infty}\left(\frac{q^{5k+r}}{1-q^{15k+3r}} - \frac{q^{10k+10-2r}}{1-q^{15k+15-3r}}\right) \\ &= \sum_{r=1}^{4}\left(\frac{r}{5}\right)\sum_{k=-\infty}^{\infty}\frac{q^{5k+r}}{1-q^{15k+3r}} \\ &= \frac{(q^{15};q^{15})_\infty^2}{(q^5,q^{10};q^{15})_\infty}\sum_{r=1}^{4}\left(\frac{r}{5}\right)q^r\frac{(q^{5+3r},q^{10-3r};q^{15})_\infty}{(q^{3r},q^{15-3r};q^{15})_\infty}, \qquad (5)\end{aligned}$$

where Ramanujan's ${}_1\psi_1$ summation formula [2, pp. 31–33] has been used in the last step. Now M. D. Hirschhorn [10] has proved that

$$\sum_{r=1}^{4}\left(\frac{r}{5}\right)a^r\frac{(q^r a^5, q^{5-r}a^{-5}; q^5)_\infty}{(q^r, q^{5-r}; q^5)_\infty} = a\frac{(a, qa^{-1}, a^2, qa^{-2}, q; q)_\infty}{(q^5;q^5)_\infty}.$$

Replacing q with q^3 and then setting $a = q$ in Hirschhorn's identity, we find that (5) simplifies to

$$\sum_{j=1}^{\infty}\sum_{k=1}^{\infty}\left(\frac{j}{3}\right)\left(\frac{k}{5}\right)q^{jk} = q\frac{(q^{15};q^{15})_\infty^2}{(q^5,q^{10};q^{15})_\infty}\frac{(q,q^2,q^2,q,q^3;q^3)_\infty}{(q^{15};q^{15})_\infty} = \frac{\eta_1^2\eta_{15}^2}{\eta_3\eta_5}.$$

This proves the result for Z_2. The result for Y_3 may be proved by the same procedure (replacing q with q^6 and then taking $a = q$ in Hirschhorn's identity) and we omit the details. It remains to prove the result for Y_4.

From the results in Theorem 2.1 already proved, it is clear that

$$Z_1(q) + Y_1(q) = 2Z_1(q^2).$$

Multiplying by $\eta_1\eta_2\eta_{15}\eta_{30}/\eta_3\eta_5\eta_6\eta_{10}$ gives

$$\frac{\eta_2\eta_3\eta_5\eta_{30}}{\eta_6\eta_{10}} + \frac{\eta_1^2\eta_{15}^2}{\eta_3\eta_5} = 2\frac{\eta_1\eta_6\eta_{10}\eta_{15}}{\eta_3\eta_5}.$$

It follows that

$$Y_4 = 2Y_3 - Z_2 = \sum_{j=1}^{\infty}\sum_{k=1}^{\infty}\left(\frac{j}{3}\right)\left(\frac{k}{5}\right)(-1)^{j+1}q^{jk}. \qquad \square$$

The right hand sides in Theorem 2.1 have all been expressed as double sums. Each may be expressed as a single sum by summing over either j or k using geometric series. We have used double series because the coefficient of q^n can easily be extracted, and also to achieve consistency among the formulas.

Let

$$\varphi(q) = \sum_{j=-\infty}^{\infty} q^{j^2} \quad \text{and} \quad \psi(q) = \sum_{j=0}^{\infty} q^{j(j+1)/2},$$

so that $\varphi(q)$ and $\psi(q)$ are the generating functions for the squares and triangular numbers, respectively. By Jacobi's triple product identity [2, pp. 35–36], it is well-known that

$$\varphi(-q) = \frac{\eta_1^2}{\eta_2} \quad \text{and} \quad q^{1/8}\psi(q) = \frac{\eta_2^2}{\eta_1}.$$

The next two theorems involve generating functions for the quadratic forms in (1) and the quadratic polynomials in (2).

Theorem 2.2. *The following identities hold:*

$$\varphi(-q^3)\varphi(-q^5) = Y_1 + Y_4,$$
$$\varphi(-q)\varphi(-q^{15}) = Y_1 - Y_4,$$
$$2q\psi(q^3)\psi(q^5) = Y_2 + Y_3,$$
$$2q^2\psi(q)\psi(q^{15}) = Y_2 - Y_3.$$

Proof. From the series expansions in Theorem 2.1, it can easily be shown that

$$Z_1(q) = Z_1(q^2) + Y_2(q) \quad \text{and} \quad Y_1(q) = Z_1(q^2) - Y_2(q).$$

Multiplying by $\eta_1\eta_{15}/\eta_6\eta_{10}$ gives

$$\frac{\eta_3^2\eta_5^2}{\eta_6\eta_{10}} = \frac{\eta_1\eta_6\eta_{10}\eta_{15}}{\eta_2\eta_{30}} + \frac{\eta_2\eta_3\eta_5\eta_{30}}{\eta_6\eta_{10}} \quad \text{and} \quad \frac{\eta_1^2\eta_{15}^2}{\eta_2\eta_{30}} = \frac{\eta_1\eta_6\eta_{10}\eta_{15}}{\eta_2\eta_{30}} - \frac{\eta_2\eta_3\eta_5\eta_{30}}{\eta_6\eta_{10}},$$

and this proves the first two results. The third and fourth results can be proved in a similar way, by multiplying the identities

$$2Z_1(q^2) = Z_1(q) + Y_1(q) \quad \text{and} \quad 2Y_2(q) = Z_1(q) - Y_1(q)$$

by $\eta_2\eta_{30}/\eta_3\eta_5$. $\square$

Theorem 2.3. *The following identities hold:*

$$\sum_{j=-\infty}^{\infty}\sum_{k=-\infty}^{\infty} q^{j^2+jk+4k^2} = Z_1 + Z_2,$$

$$\sum_{j=-\infty}^{\infty}\sum_{k=-\infty}^{\infty} q^{2j^2+jk+2k^2} = Z_1 - Z_2.$$

Proof. Breaking the sum into the cases k even and k odd gives

$$\sum_{j=-\infty}^{\infty}\sum_{k=-\infty}^{\infty} q^{j^2+jk+4k^2}$$

$$= \sum_{j=-\infty}^{\infty}\sum_{k=-\infty}^{\infty} q^{(j+k)^2+15k^2} + \sum_{j=-\infty}^{\infty}\sum_{k=-\infty}^{\infty} q^{\left(j+k+\frac{1}{2}\right)^2+15\left(k+\frac{1}{2}\right)^2}$$

$$= \varphi(q)\varphi(q^{15}) + 4q^4\psi(q^2)\psi(q^{30}).$$

By Theorem 2.2, this is

$$\sum_{j=-\infty}^{\infty}\sum_{k=-\infty}^{\infty} q^{j^2+jk+4k^2} = Y_1(-q) - Y_4(-q) + 2Y_2(q^2) - 2Y_3(q^2). \tag{6}$$

It is easy to check, using Theorem 2.1, that

$$Y_1(-q) + 2Y_2(q^2) = Z_1(q) \quad \text{and} \quad -Y_4(-q) - 2Y_3(q^2) = Z_2(q). \tag{7}$$

The first result now follows on combining (6) and (7).

To prove the second result, use the series rearrangement

$$\sum_{j=-\infty}^{\infty}\sum_{k=-\infty}^{\infty} c(j,k) = \sum_{j=-\infty}^{\infty}\sum_{k=-\infty}^{\infty} c(j+k, j-k) + \sum_{j=-\infty}^{\infty}\sum_{k=-\infty}^{\infty} c(j+k+1, j-k)$$

to get

$$\sum_{j=-\infty}^{\infty}\sum_{k=-\infty}^{\infty} q^{2j^2+jk+2k^2} = \varphi(q^3)\varphi(q^5) + 4q^2\psi(q^6)\psi(q^{10}).$$

The remainder of the proof is similar to the proof of the first result. We omit the details. □

3. Representation by quadratic forms and by quadratic polynomials

The results in Theorems 2.2 and 2.3, together with the series expansions in Theorem 2.1, yield formulas for the number of representations of a positive integer by the quadratic forms in (1) and the quadratic polynomials in (2). We illustrate by giving the details for the forms $j^2 + jk + 4k^2$ and $2j^2 + jk + 2k^2$. To describe the results, let p denote any prime number and define sets S^0, S^+ and S^- by

$$S^0 = \left\{p : \left(\frac{p}{15}\right) = 0\right\} = \{3, 5\},$$

$$S^+ = \left\{p : \left(\frac{p}{15}\right) = 1\right\} = \{p : p \equiv 1, 2, 4, \text{ or } 8 \pmod{15}\},$$

$$S^- = \left\{p : \left(\frac{p}{15}\right) = -1\right\} = \{p : p \equiv 7, 11, 13, \text{ or } 14 \pmod{15}\}.$$

Let n be a positive integer and write its prime factorization as

$$n = \prod_{p \text{ prime}} p^{\alpha_p} = \prod_{p \in S^0} p^{\alpha_p} \prod_{p \in S^+} p^{\alpha_p} \prod_{p \in S^-} p^{\alpha_p}.$$

Let

$$m = \prod_{p \in S^+} p^{\alpha_p} \prod_{p \in S^-} p^{\alpha_p}$$

so that $n = 3^{\alpha_3} 5^{\alpha_5} m$, where $(3, m) = (5, m) = 1$.

If we denote the coefficients of q^n in $Z_1(q)$ and $Z_2(q)$ by $z_1(n)$ and $z_2(n)$, respectively, then the first two results in Theorem 2.1 imply

$$z_1(n) = \prod_{p \in S^+} (1 + \alpha_p) \prod_{p \in S^-} \left(\frac{1 + (-1)^{\alpha_p}}{2}\right) \tag{8}$$

and

$$z_2(n) = (-1)^{\alpha_3 + \alpha_5} \left(\frac{m}{5}\right) z_1(n). \tag{9}$$

It follows that for any prime p,

$$z_1(p) = 1 + \left(\frac{p}{15}\right) \quad \text{and} \quad z_2(p) = \left(\frac{p}{3}\right) + \left(\frac{p}{5}\right).$$

Using these in Theorem 2.3 we recover the classical result that a prime p is represented by the quadratic form $j^2 + jk + 4k^2$ if and only if

$$\left(1 + \left(\frac{p}{3}\right)\right)\left(1 + \left(\frac{p}{5}\right)\right) > 0,$$

that is, when $p \equiv 1$ or $4 \pmod{15}$. Similarly, a prime p is represented by the quadratic form $2j^2 + jk + 2k^2$ if and only if

$$\left(1 - \left(\frac{p}{3}\right)\right)\left(1 - \left(\frac{p}{5}\right)\right) > 0,$$

that is, when $p = 3$, $p = 5$, or $p \equiv 2$ or $8 \pmod{15}$.

Analogous results for the other quadratic forms in (1), and the quadratic polynomials in (2) can be deduced by the same procedure. For $k = 1, 2, 3$ or 4, let $y_k(n)$ denote the coefficient of q^n in $Y_k(q)$. Then the results of Theorem 2.1 imply

$$y_1(n) = \left(\frac{\alpha_2 - 1}{\alpha_2 + 1}\right) z_1(n), \quad y_2(n) = \frac{1}{(\alpha_2 + 1)} z_1(n),$$

$$y_3(n) = \frac{1}{(\alpha_2 + 1)} z_2(n), \quad y_4(n) = -\left(\frac{\alpha_2 - 1}{\alpha_2 + 1}\right) z_2(n)$$

where $z_1(n)$ and $z_2(n)$ are given by (8) and (9). The substitution of these results in Theorem 2.2 yields the number of representations of a positive integer by each of $j^2 + 15k^2$, $3j^2 + 5k^2$, $j(j+1)/2 + 15k(k+1)/2$ and $3j(j+1)/2 + 5k(k+1)/2$.

4. Discussion

The question of which primes are represented by the quadratic form $j^2 + mk^2$ for a given positive integer m is very old. The book [6] by D. A. Cox gives a comprehensive answer to this question, starting from work of Fermat and progressing to describe modern techniques.

Series expansions for $q\psi(q^3)\psi(q^5) + q^2\psi(q)\psi(q^{15})$ and $\varphi(q^3)\varphi(q^5) + \varphi(q)\varphi(q^{15})$ were stated without proof by S. Ramanujan in [13, Chapter 20, Entry 10 (v), (vi)]. These results have been proved by B. C. Berndt [2, pp. 379–383], while different proofs have been given by K. S. Williams [18]. These results of Ramanujan, as well as Parts (i)–(vii) of Entry 9 of Chapter 20 in Ramanujan's second notebook [13], follow immediately from our Theorems 2.1 and 2.2.

An analysis of the number of representations of an integer coprime to 2, 3 and 5 by each of the quadratic forms $j^2 + jk + 4k^2$ and $2j^2 + jk + 2k^2$ has been given by L. E. Dickson [7, p. 87]. The determination of the number of representations by the forms $j^2 + 15k^2$ and $3j^2 + 5k^2$ was set as an exercise [7, p. 86, Ex. 8].

Dickson's results have been extended to all integers by N. A. Hall [8].

The number of representations of a positive integer n by the equivalent, positive-definite, primitive, integral binary quadratic forms in a genus of discriminant $d < 0$ has been given by J. G. Huard, P. Kaplan and K. S. Williams [11, Theorem 8.1]. Their result with $d = -15$ can be shown to be equivalent to our formulas (8) and (9). (See also [15, p. 159].) The number of representations of a positive integer by the forms $j^2 + 15k^2$ and $3j^2 + 5k^2$ is given by Z.-H. Sun and K. S. Williams [15, p. 163]. Formulas for the number of representations of a positive integer by the quadratic polynomials in (2) are originally due to Sun and Williams [16, p. 369]. These results have been greatly extended by Sun [14].

The results in Theorem 2.1 for the series Y_1 and Y_4 were proved by A. Berkovich and H. Yesilyurt [1]. Our proof is the same, except we use Hirschhorn's identity for the last step, whereas the proof in [1] makes use of series manipulations and one of Ramanujan's "forty identities" for the Rogers-Ramanujan functions, e.g., see [3]. Berkovich and Yesilyurt used their results to deduce explicit formulas for the number of representations by the forms $j^2 + 15k^2$ and $3j^2 + 5k^2$.

A proof of Theorem 2.1 for the series Z_2 has been given by H. H. Chan, S. H. Chan and Z.-G. Liu [4]. It relies on a theta function identity proved by Liu [12, Th. 1].

The results in Thm. 2.3 have been greatly generalized by P. C. Toh [17].

Acknowledgement

The author is grateful to the anonymous referee for helpful comments

References

[1] A. Berkovich and H. Yesilyurt, Ramanujan's identities and representations of integers by certain binary and quaternary quadratic forms, *Ramanujan J.*, **20** (2009) 375–408.

[2] B. C. Berndt, Ramanujan's Notebooks, Part III, Springer-Verlag, New York (1991).

[3] B. C. Berndt, G. Choi, Y.-S. Choi, H. Hahn, B. P. Yeap, A. J. Yee, H. Yesilyurt and J. Yi, Ramanujan's forty identities for the Rogers-Ramanujan functions, *Mem. Amer. Math. Soc.*, **188** (2007), no. 880.

[4] H. H. Chan, S. H. Chan and Z.-G. Liu, The Rogers-Ramanujan continued fraction and a new Eisenstein series identity, *J. Number Theory*, **129** (2009) 1786–1797.

[5] S. Cooper, On the number of representations of integers by certain quadratic forms, II, *Journal of Combinatorics and Number Theory*, **1**, Issue 2 (2009) 153–182.

[6] D. A. Cox, Primes of the form $x^2 + ny^2$, Wiley, New York (1989).

[7] L. E. Dickson, Introduction to the theory of numbers, University of Chicago Press, 1929. Republished by Dover, New York (1957).

[8] N. A. Hall, The number of representations function for binary quadratic forms, *Amer. J. Math.*, **62** (1940) 589–598.

[9] G.-H. Halphen, Traité des fonctions elliptiques et de leurs applications, vol. 1, Gauthier-Villars, Paris (1886).

[10] M. D. Hirschhorn, An identity of Ramanujan, and applications, q-series from a contemporary perspective (South Hadley, MA, 1998), 229–234, *Contemp. Math.*, **254** *Amer. Math. Soc.*, Providence, RI (2000).

[11] J. G. Huard, P. Kaplan and K. S. Williams, The Chowla-Selberg formula for genera, *Acta Arith.*, **73** (1995) 271–301.

[12] Z.-G. Liu, A theta function identity and the Eisenstein series on $\Gamma_0(5)$, *J. Ramanujan Math. Soc.*, **22** (2007) 283–298.

[13] S. Ramanujan, Notebooks (2 volumes), Tata Institute of Fundamental Research, Bombay (1957).

[14] Z.-H. Sun, On the number of representations of n by $ax(x-1)/2+by(y-1)/2$, *J. Number Theory*, **129** (2009) 971–989.

[15] Z.-H. Sun and K. S. Williams, On the number of representations of n by $ax^2+bxy+cy^2$, *Acta Arith.*, **122** (2006) 101–171.

[16] Z.-H. Sun and K. S. Williams, Ramanujan identities and Euler products for a type of Dirichlet series, *Acta Arith.*, **122** (2006) 349–393.

[17] P. C. Toh, Representations of certain binary quadratic forms as Lambert series, *Acta Arith.*, **143** (2010) 227–237.

[18] K. S. Williams, Some Lambert series expansions of products of theta functions, *Ramanujan J.*, **3** (1999) 367–384.

Ramanujan Rediscovered No. 14, 2010, pp. 87–111.

Biranks for Partitions into 2 Colors

F. G. Garvan

Department of Mathematics, University of Florida, Gainesville, Florida 32611-8105
e-mail: fgarvan@ufl.edu

Abstract. In 2003, Hammond and Lewis defined a statistic on partitions into 2 colors which combinatorially explains certain well known partition congruences mod 5. We give two analogs of Hammond and Lewis's birank statistic. One analog is in terms of Dyson's rank and the second uses the 5-core crank due to Garvan, Kim and Stanton. We discuss Andrews's bicrank statistic and how it may be extended. We also generalize the Hammond-Lewis birank to a multirank for multipartitions and the Andrews bicrank to a multicrank for extended multipartitions. Both of these give combinatorial interpretations for multipartition congruences modulo all primes $t > 3$.

Keywords. Partition congruences, Dyson's rank, crank, 5-cores, multipartitions.

2000 Subject Classification: Primary: 11P83; Secondary: 05A17, 05A19, 11F03, 11F20, 11F33, 11P81, 20C30

This paper is dedicated to the memory of my friend Richard Lewis (1942–2007)

1. Introduction

Hammond and Lewis [17] found some elementary results for 2-colored partitions mod 5. Let $E(q) = \prod_{n=1}^{\infty}(1 - q^n)$, and

$$\sum_{n=0}^{\infty} p_{-2}(n)q^n = \frac{1}{E(q)^2},$$

which is the generating function for pairs of partitions (π_1, π_2) (or 2-colored partitions). Throughout this paper we refer to such pairs of partitions as *bipartitions*. It is not hard to show that

$$p_{-2}(5n + 2) \equiv p_{-2}(5n + 3) \equiv p_{-2}(5n + 4) \equiv 0 \pmod 5. \tag{1.1}$$

These congruences are well-known [7,12,26]. Hammond and Lewis [17] found a crank for these congruences. By crank we mean a statistic that divides the relevant partitions into equinumerous classes. They define

$$\text{birank}(\pi_1, \pi_2) = \#(\pi_1) - \#(\pi_2), \tag{1.2}$$

The author was supported in part by NSA Grant H98230-09-1-0051. The main results of this paper were presented as part of a talk "The Rank and Crank of Partitions—Dedicated to the Memory of Richard P. Lewis" given on Friday, March 14, 2008 at the Gainesville Partitions, q-Series and Modular Forms Conference.

where $\#(\pi)$ denotes the number of parts in the partition π. They show that the residue of the birank mod 5 divides the bipartitions of n into 5 equal classes provided $n \equiv 2, 3$ or 4 (mod 5). The proof is elementary. It relies on Jacobi's triple product identity and the method of [13], which uses roots of unity. We have found two other analogs.

1.1 First analog – The Dyson-birank

Dyson [9] defined the *rank* of a partition as the largest part minus the number of parts. We define

$$\text{Dyson-birank}(\pi_1, \pi_2) = \text{rank}(\pi_1) + 2\,\text{rank}(\pi_2). \tag{1.3}$$

This statistic was independently suggested by P. R. Hammond [16, p. 68], who asked whether it could explain the congruences in (1.1). In Section 3, we show that the residue of the Dyson-birank mod 5 does indeed divide the bipartitions of n into 5 equal classes provided $n \equiv 2$, or 4 (mod 5). Unfortunately the Dyson-birank does not work if $n \equiv 3$ (mod 5). Nonetheless, for the other residue classes this is a surprising and deep result because of the nature of the rank generating function. The proof depends on known results for the rank mod 5 due to Atkin and Swinnerton-Dyer [4].

1.2 Second analog – The 5-core-birank

In [15] new statistics were defined in terms of t-cores which gave new combinatorial interpretations of Ramanujan's partition congruences mod 5, 7 and 11. For example, for a partition π the 5-*core-crank* is defined as

$$\text{5-core-crank}(\pi) = r_1 + 2r_2 - 2r_3 - r_4, \tag{1.4}$$

where r_k is the number of cells labelled k in the 5-residue diagram of π. In the t-residue diagram of π a cell in the i-th row and j-column is labelled by the least nonnegative residue of $j - i$ (mod t). Then in [15] we proved combinatorially that the residue of the 5-core-crank divides the partitions of $5n + 4$ into 5 equal classes. We define

$$\text{5-core-birank}(\pi_1, \pi_2) = \text{5-core-crank}(\pi_1) + 2\,(\text{5-core-crank}(\pi_2)). \tag{1.5}$$

In Section 4, we show that the 5-core-birank divides the bipartitions of n into 5 equal classes for $n \equiv 2, 3$ or 4 (mod 5). This is quite a surprising result. The proof relies on the 5-dissection of the 5-core-crank generating function for 5-cores.

The crank of a partition is defined to be the largest part if it contains no ones and otherwise it is the difference between number of parts larger than the number of ones, and the number of ones. The crank gives a combinatorial interpretation of Ramanujan's partition congruences mod 5, 7 and 11 and solves a problem of Dyson [9], [10, p. 52]. See [2]. This crank is different to the 5-core crank given in [15]. It is natural to ask whether there is a crank analog of the birank. This question has been answered in part by Andrews [1]. In Section 6, we consider Andrews' result and how it may be extended. In [1], Andrews also considered congruences for more general multipartitions. In Section 7, we give multipartition analogs of the Hammond-Lewis birank which explain these more general congruences. In Section 8, we extend Andrews bicrank to multicranks of what we call extended multipartitions, and give alternative explanations of our multipartition congruences. In Section 9, we close with some further problems.

Notation. For a partition π we denote the sum of parts by $|\pi|$. We will use the standard q-notation:

$$(z;q)_n = (z)_n = \begin{cases} \prod_{j=0}^{n-1}(1-zq^j), & n>0 \\ 1, & n=0, \end{cases}$$

and

$$(z;q)_\infty = (z)_\infty = \lim_{n\to\infty}(z;q)_n = \prod_{n=1}^{\infty}(1-zq^{(n-1)}),$$

where $|q|<1$. We will also use the following notation for Jacobi-type theta-products:

$$J_{a,m}(q) := (q^a;q^m)_\infty(q^{m-a};q^m)_\infty(q^m;q^m)_\infty.$$

2. The Hammond-Lewis birank

For completeness we include some details of the Hammond-Lewis birank. For a bipartition $\pi = (\pi_1,\pi_2)$ we denote the sum of parts by

$$|\pi| = |\pi_1| + |\pi_2|. \tag{2.1}$$

We denote the Hammond-Lewis birank by

$$\text{HL-birank}(\pi) = \#(\pi_1) - \#(\pi_2), \tag{2.2}$$

where $\#(\pi)$ denotes the number of parts in the partition π. The HL-birank generating function is

$$\sum_{\pi=(\pi_1,\pi_2)} z^{\text{HL-birank}(\pi)} q^{|\pi|} = \frac{1}{(zq;q)_\infty(z^{-1}q;q)_\infty}. \tag{2.3}$$

We let $N_{\rm HL}(m,t,n)$ denote the number of bipartitions $\pi = (\pi_1,\pi_2)$ of n with HL-birank congruent to $m \pmod t$. Suppose ζ is primitive 5th root of unity. By letting $z=\zeta$ in (2.3) and using Jacobi's triple product identity, Hammond and Lewis found that

$$\begin{aligned}\sum_{n=0}^{\infty}\sum_{k=0}^{4}\zeta^k N_{\rm HL}(k,5,n)q^n &= \frac{1}{(\zeta q;q)_\infty(\zeta^{-1}q;q)_\infty} \\ &= \frac{(\zeta^2 q;q)_\infty(\zeta^{-2}q;q)_\infty(q;q)_\infty}{(q^5;q^5)_\infty}, \\ &= (q^{25};q^{25})_\infty\left(\frac{1}{J_{1,5}(q^5)} + (\zeta+\zeta^{-1})\frac{q}{J_{2,5}(q^5)}\right).\end{aligned} \tag{2.4}$$

Since the coefficient of q^n on the right side of (2.4) is zero when $n \equiv 2, 3$ or $4 \pmod 5$, Hammond and Lewis's main result follows.

Theorem 2.1 ([17]). *The residue of the HL-birank mod 5 divides the bipartitions of n into 5 equal classes provided $n \equiv 2$, 3 or 4 (mod 5).*

We illustrate Theorem 2.1 for the case $n = 3$.

Bipartitions of 3	HL-birank (mod 5)
$(3, -)$	$1 - 0 \equiv 1$
$(2 + 1, -)$	$2 - 0 \equiv 2$
$(1 + 1 + 1, -)$	$3 - 0 \equiv 3$
$(2, 1)$	$1 - 1 \equiv 0$
$(1 + 1, 1)$	$2 - 1 \equiv 1$
$(1, 2)$	$1 - 1 \equiv 0$
$(1, 1 + 1)$	$1 - 2 \equiv 4$
$(-, 3)$	$0 - 1 \equiv 4$
$(-, 2 + 1)$	$0 - 2 \equiv 3$
$(-, 1 + 1 + 1)$	$0 - 3 \equiv 2$

Thus

$$N_{\mathrm{HL}}(0, 5, 3) = N_{\mathrm{HL}}(1, 5, 3) = N_{\mathrm{HL}}(2, 5, 3) = N_{\mathrm{HL}}(3, 5, 3) = N_{\mathrm{HL}}(4, 5, 3) = 2,$$

and we see that the residue of the HL-birank mod 5 divides the 10 bipartitions of 3 into 5 equal classes.

3. The Dyson-birank

Dyson [9], [10, p. 52] defined the rank of a partition as the largest part minus the number of parts. We define the Dyson-analog of the birank for bipartitions $\pi = (\pi_1, \pi_2)$ by

$$\text{Dyson-birank}(\pi) = \operatorname{rank}(\pi_1) + 2\operatorname{rank}(\pi_2). \tag{3.1}$$

In this section we prove

Theorem 3.1. *The residue of the Dyson-birank* mod 5 *divides the bipartitions of n into 5 equal classes provided $n \equiv 2$, or 4* (mod 5).

We let $N_{\mathrm{D}}(m, t, n)$ denote the number of bipartitions $\pi = (\pi_1, \pi_2)$ with Dyson-birank congruent to m (mod t). We illustrate Theorem 3.1 for the case $n = 2$.

Bipartitions of 2	Dyson-birank (mod 5)
$(2, -)$	$1 + 0 \equiv 1$
$(1 + 1, -)$	$-1 + 0 \equiv 4$
$(1, 1)$	$0 + 0 \equiv 0$
$(-, 2)$	$0 + 2 \equiv 2$
$(-, 1 + 1)$	$0 - 2 \equiv 3$

Thus

$$N_{\mathrm{D}}(0,5,2) = N_{\mathrm{D}}(1,5,2) = N_{\mathrm{D}}(2,5,2) = N_{\mathrm{D}}(3,5,2) = N_{\mathrm{D}}(4,5,2) = 1,$$

and we see that the residue of the Dyson-birank mod 5 divides the 5 bipartitions of 2 into 5 equal classes. We note that Theorem 3.1 does not hold for $n \equiv 3 \pmod 5$. The first counterexample occurs when $n = 13$. The Dyson-birank mod 5 fails to divide the 1770 bipartitions of 13 into 5 equal classes. We have

$$N_{\mathrm{D}}(0,5,13) = 358, \quad \text{but}$$
$$N_{\mathrm{D}}(1,5,13) = N_{\mathrm{D}}(2,5,13) = N_{\mathrm{D}}(3,5,13) = N_{\mathrm{D}}(4,5,13) = 353.$$

To prove Theorem 3.1 we need the 5-dissection of the rank generating function when $z = \zeta$. The Dyson-rank generating function is

$$f(z,q) = \sum_{\pi} z^{\mathrm{rank}(\pi)} q^{|\pi|} = 1 + \sum_{n=1}^{\infty} \frac{q^{n^2}}{(zq;q)_n (z^{-1}q;q)_n}. \tag{3.2}$$

We let $N(m,t,n)$ denote the number of ordinary partitions of n with rank congruent to m mod t. Then

$$\begin{aligned} f(\zeta,q) &= \sum_{n=0}^{\infty}\sum_{k=0}^{4} \zeta^k N(k,5,n) q^n = 1 + \sum_{n=1}^{\infty} \frac{q^{n^2}}{(\zeta q;q)_n (\zeta^{-1}q;q)_n} \\ &= (A(q^5) - (3+\zeta^2+\zeta^3)\,\phi(q^5)) + q\,B(q^5) + q^2(\zeta+\zeta^4)\,C(q^5) \\ &\quad + q^3((1+\zeta^2+\zeta^3)\,D(q^5) + (1+2\zeta^2+2\zeta^3)\,\psi(q^5)), \end{aligned} \tag{3.3}$$

where

$$A(q) = \frac{E^2(q^5)\,J_{2,5}(q)}{J_{1,5}^2(q)}, \tag{3.4}$$

$$B(q) = \frac{E^2(q^5)}{J_{1,5}(q)}, \tag{3.5}$$

$$C(q) = \frac{E^2(q^5)}{J_{2,5}(q)}, \tag{3.6}$$

$$D(q) = \frac{E^2(q^5)\,J_{1,5}(q)}{J_{2,5}^2(q)}, \tag{3.7}$$

$$\begin{aligned} \phi(q) &= -1 + \sum_{n=0}^{\infty} \frac{q^{5n^2}}{(q;q^5)_{n+1}(q^4;q^5)_n} \\ &= \frac{q}{E(q^5)} \sum_{m=-\infty}^{\infty} (-1)^m \frac{q^{\frac{15}{2}m(m+1)}}{1-q^{5m+1}}, \end{aligned} \tag{3.8}$$

and

$$\psi(q) = \frac{1}{q}\left\{-1 + \sum_{n=0}^{\infty} \frac{q^{5n^2}}{(q^2;q^5)_{n+1}(q^3;q^5)_n}\right\}$$

$$= \frac{q}{E(q^5)} \sum_{m=-\infty}^{\infty} (-1)^m \frac{q^{\frac{15}{2}m(m+1)}}{1-q^{5m+2}}. \tag{3.9}$$

Equation (3.3) has an unusual history. It is originally due to Ramanujan since it appears in the Lost Notebook. It predates, but is closely related to Dyson's conjectures on the rank [9], which were proved by Atkin and Swinnerton-Dyer [4]. As pointed out in [13] and [14], equation (3.3) is actually equivalent to one of Atkin and Swinnerton-Dyer's main results. Dyson, Atkin and Swinnerton-Dyer were unaware of Ramanujan's result.

The Dyson-birank generating function is

$$\sum_{\pi=(\pi_1,\pi_2)} z^{\text{Dyson-birank}(\pi)} q^{|\pi|} = f(z,q) f(z^2,q), \tag{3.10}$$

where $f(z,q)$ is the generating function for the Dyson rank of ordinary partitions given in (3.2). Thus we have

$$\sum_{n=0}^{\infty}\sum_{k=0}^{4} \zeta^k N_{\mathrm{D}}(k,5,n) q^n = f(\zeta,q)\, f(\zeta^2,q). \tag{3.11}$$

Using only (3.3) and the fact that

$$B^2(q) = A(q)\,C(q), \quad C^2(q) = B(q)\,D(q), \tag{3.12}$$

we find that the coefficient of q^n in the q-expansion of $f(\zeta,q)\, f(\zeta^2,q)$ is zero if $n \equiv 2$, or $4 \pmod 5$. Theorem 3.1 then follows from (3.11). Although Theorem 3.1 does not hold when $n \equiv 3 \pmod 5$, there is some simplification in the product $f(\zeta,q)\, f(\zeta^2,q)$. We find that

$$\sum_{n=0}^{\infty}\sum_{k=0}^{4} \zeta^k N_{\mathrm{D}}(k,5,5n+3) q^n = 5\phi(q)\psi(q) \tag{3.13}$$

using the fact that

$$A(q)D(q) = B(q)C(q). \tag{3.14}$$

Equation (3.13) implies that

$$N_D(0,5,5n+3) \equiv N_D(k,5,5n+3)) \pmod 5, \tag{3.15}$$

for $1 \le k \le 4$, and

$$p_{-2}(5n+3) \equiv 0 \pmod 5. \tag{3.16}$$

This last observation is due to Byungchan Kim.

4. The 5-core-birank

For ordinary partitions π the 5-*core-crank* is defined by

$$\text{5-core-crank}(\pi) = r_1 + 2r_2 - 2r_3 - r_4, \tag{4.1}$$

where r_j is the number of cells labelled j in the 5-residue diagram of π. See [15, Prop. 1, p. 7]. We define the 5-core-crank analog for bipartitions $\pi = (\pi_1, \pi_2)$ by

$$\text{5-core-birank}(\pi) = \text{5-core-crank}(\pi_1) + 2\,(\text{5-core-crank}(\pi_2)). \tag{4.2}$$

In this section we prove

Theorem 4.1. *The residue of the* 5-*core-birank* mod 5 *divides the bipartitions of n into* 5 *equal classes provided n* $\equiv$ 2, 3 *or* 4 (mod 5).

We let $N_{5C}(m, t, n)$ denote the number of bipartitions $\pi = (\pi_1, \pi_2)$ with 5-core-birank congruent to m (mod t). We illustrate Theorem 4.1 for the case $n = 3$.

Bipartitions of 3	5-core-birank (mod 5)
$(3, -)$	$3 + 0 \equiv 3$
$(2 + 1, -)$	$0 + 0 \equiv 0$
$(1 + 1 + 1, -)$	$-3 + 0 \equiv 2$
$(2, 1)$	$1 + 0 \equiv 1$
$(1 + 1, 1)$	$-1 + 0 \equiv 4$
$(1, 2)$	$0 + 2 \equiv 2$
$(1, 1 + 1)$	$0 - 2 \equiv 3$
$(-, 3)$	$0 + 6 \equiv 1$
$(-, 2 + 1)$	$0 + 0 \equiv 0$
$(-, 1 + 1 + 1)$	$0 - 6 \equiv 4$

Thus

$$N_{5C}(0, 5, 3) = N_{5C}(1, 5, 3) = N_{5C}(2, 5, 3) = N_{5C}(3, 5, 3) = N_{5C}(4, 5, 3) = 2,$$

and we see that the residue of the 5-core-birank mod 5 divides the 10 bipartitions of 3 into 5 equal classes. We note that although the Dyson-birank does not in general divide the bipartitions of $5n + 3$ into 5 equal classes the 5-core-birank does.

To prove Theorem 4.1 we need the 5-dissection of the 5-core-crank generating function when $z = \zeta$. The 5-core-crank generating function is

$$\Phi(z, q) = \sum_{\pi} z^{\text{5-core-crank}(\pi)} q^{|\pi|} = \frac{1}{E^5(q^5)} T(z, q), \tag{4.3}$$

where

$$T(z, q) := \sum_{\pi \text{ a 5-core}} z^{\text{5-core-crank}(\pi)} q^{|\pi|} = \sum_{\substack{\vec{n} \in \mathbb{Z}^5 \\ \vec{n} \cdot \vec{1} = 0}} z^{n_1 + 3n_2 + n_3} q^{\frac{5}{2}\|\vec{n}\|^2 + \vec{b} \cdot \vec{n}}, \tag{4.4}$$

where $\vec{1} = (1, 1, 1, 1, 1)$ and $\vec{b} = (0, 1, 2, 3, 4)$. Equation (4.3) can be proved combinatorially and in a straightforward manner using Bijections 1 and 2 from [15, pp. 2–3] and [15, (4.2), p. 6].

We need the 5-dissection of $T(\zeta, q)$:

$$T(\zeta, q) = W(q^5)(1 + qR(q^5) + q^2(\zeta^2 + \zeta^3)R(q^5)^2 - q^3(\zeta^2 + \zeta^3)R(q^5)^3), \tag{4.5}$$

where

$$W(q) := J_{2,5}(q)^3(J_{10,25}(q) - q(1 + \zeta^2 + \zeta^3)J_{5,25}(q)), \tag{4.6}$$

and

$$R(q) := \frac{J_{1,5}(q)}{J_{2,5}(q)}. \tag{4.7}$$

We will prove (4.5) in the next section. Theorem 4.1 follows easily from (4.5).

The 5-core-birank generating function is

$$\sum_{\pi=(\pi_1,\pi_2)} z^{\text{5-core-birank}(\pi)} q^{|\pi|} = \frac{1}{E^{10}(q^5)} T(z, q)T(z^2, q), \tag{4.8}$$

where $T(z, q)$ is the generating function for the 5-core-crank of partitions that are 5-cores given in (4.4). Thus we have

$$\sum_{n=0}^{\infty} \sum_{k=0}^{4} \zeta^k N_{5C}(k, 5, n)q^n = \frac{1}{E^{10}(q^5)} T(\zeta, q)T(\zeta^2, q). \tag{4.9}$$

From (4.5) we find that

$$T(\zeta, q)T(\zeta^2, q) = Z(q^5)(1 + 2q^5 R^5(q^5) + qR(q^5)(2 - q^5 R^5(q^5)), \tag{4.10}$$

where

$$Z(q) = J_{2,5}(q)^6(J_{10,25}(q)^2 - qJ_{10,25}(q)J_{5,25}(q) - q^2 J_{5,25}(q)^2). \tag{4.11}$$

Since the coefficient of q^n in the q-expansion of $T(\zeta, q)T(\zeta^2, q)$ is zero when $n \equiv 2$, 3 or 4 (mod 5), Theorem 4.1 then follows from (4.9).

5. A theta-function identity

In this section we will prove the following theta-function identity.

$$\begin{aligned} U(z, q) = {} & F_0(q)S_0(z, q) + F_1(q)S_1(z, q) + F_2(q)S_2(z, q) \\ & + F_3(q)S_3(z, q) + F_4(q)S_4(z, q), \end{aligned} \tag{5.1}$$

where

$$F_0(q) = W(q^{10})(1 + q^2 R(q^{10}) + (\zeta^2 + \zeta^3)q^4 R(q^{10})^2 - (\zeta^2 + \zeta^3)q^6 R(q^{10})^3), \tag{5.2}$$

$$F_1(q) = W(q^{10})(\zeta^4 + \zeta q^2 R(q^{10}) + (1+\zeta)q^4 R(q^{10})^2 - (\zeta^2+\zeta^3)q^6 R(q^{10})^3), \tag{5.3}$$

$$F_2(q) = W(q^{10})(1 + \zeta^4 q^2 R(q^{10}) + (1+\zeta)q^4 R(q^{10})^2 - (1+\zeta^4)q^6 R(q^{10})^3), \tag{5.4}$$

$$F_3(q) = W(q^{10})(1 + \zeta q^2 R(q^{10}) + (1+\zeta^4)q^4 R(q^{10})^2 - (1+\zeta)q^6 R(q^{10})^3), \tag{5.5}$$

$$F_4(q) = W(q^{10})(\zeta + \zeta^4 q^2 R(q^{10}) + (1+\zeta^4)q^4 R(q^{10})^2 - (\zeta^2+\zeta^3)q^6 R(q^{10})^3), \tag{5.6}$$

$$S_0(z,q) = \sum_{n=-\infty}^{\infty} z^{5n} q^{25n^2+20n}, \tag{5.7}$$

$$S_1(z,q) = \sum_{n=-\infty}^{\infty} z^{5n+1} q^{25n^2+30n+5}, \tag{5.8}$$

$$S_2(z,q) = \sum_{n=-\infty}^{\infty} z^{5n+2} q^{25n^2+40n+12}, \tag{5.9}$$

$$S_3(z,q) = \sum_{n=-\infty}^{\infty} z^{5n+3} q^{25n^2+50n+21}, \tag{5.10}$$

$$S_4(z,q) = \sum_{n=-\infty}^{\infty} z^{5n+4} q^{25n^2+60n+32}, \tag{5.11}$$

and

$$U(z,q) = \sum_{\vec{n}\in\mathbb{Z}^5} z^{\vec{n}\cdot\vec{1}} \zeta^{n_1+3n_2+n_3} q^{5\|\vec{n}\|^2+2\vec{b}\cdot\vec{n}}, \tag{5.12}$$

with $W(q)$ and $R(q)$ defined in (4.6) and (4.7) respectively, and the vectors $\vec{n} = (n_0, n_1, n_2, n_3, n_4)$, $\vec{1} = (1,1,1,1,1)$ and $\vec{b} = (0,1,2,3,4)$ as before. We note that (4.5) follows from (5.1) by taking the coefficient of z^0 and replacing q by $q^{1/2}$. Equation (4.5) was the crucial identity needed in the proof of Theorem 4.1.

We prove the identity (5.1) using standard techniques. We show that both sides satisfy the same functional equation and both sides agree for enough values of the parameter z. Most of these evaluations can be proved by elementary means using Jacobi's triple product. For one evaluation we will need the theory of modular functions.

We define the following Jacobi theta function

$$\Theta(z,q) = \sum_{n=-\infty}^{\infty} z^n q^{n^2}, \tag{5.13}$$

for $z \neq 0$ and $|q| < 1$. We will need Jacobi's triple product identity

$$\sum_{n=-\infty}^{\infty} z^n q^{n^2} = (-zq;q^2)_\infty(-z^{-1}q;q^2)_\infty(q^2;q^2)_\infty, \tag{5.14}$$

and the well-known functional equation

$$\Theta(zq^2,q) = z^{-1}q^{-1}\Theta(z,q), \tag{5.15}$$

for $z \neq 0$ and $0 < |q| < 1$. From the definition (5.12) we have

$$U(z,q) = \Theta(z\zeta^4,q^5)\Theta(zq^2\zeta,q^5)\Theta(zq^4,q^5)\Theta(z\zeta q^6,q^5)\,\Theta(z\zeta^4q^8,q^5). \tag{5.16}$$

From (5.15) and (5.16) we have

$$U(zq^{10},q) = z^{-5}q^{-45}U(z,q), \tag{5.17}$$

and

$$U(z,q) = 0 \quad \text{for} \quad z = -q^5\zeta, \quad -q^3\zeta^4, -q, -\zeta^4q^9, -\zeta q^7. \tag{5.18}$$

Let $V(z,q)$ denote the function of the right side of (5.1). Each $S_j(z,q)$ can be written in terms of the theta function $\Theta(z,q)$ and we find that $S_j(zq^{10},q) = z^{-5}q^{-45}S_j(z,q)$ for each j so that

$$V(zq^{10},q) = z^{-5}q^{-45}V(z,q). \tag{5.19}$$

Hence the left and right sides of (5.1) satisfy the same functional equation (i.e. (5.17), (5.19)). In view of [4, Lemma 2] or [19, Lemma 1], it suffices to show that (5.1) holds for 6 distinct values of z with $|q|^{10} < |z| \leq 1$. We claim that

$$V(z,q) = 0 \quad \text{for} \quad z = -q^5\zeta, \quad -q^3\zeta^4, -q, -\zeta^4q^9, -\zeta q^7. \tag{5.20}$$

Using (5.14) we can easily evaluate each $S_j(z,q)$ for these values of z.

$$S_0(-\zeta q^5,q) = -q^{-20}J_{2,5}(q^{10}), \tag{5.21}$$

$$S_1(-\zeta q^5,q) = \zeta q^{-20}J_{2,5}(q^{10}), \tag{5.22}$$

$$S_2(-\zeta q^5,q) = -\zeta^2q^{-18}J_{1,5}(q^{10}), \tag{5.23}$$

$$S_3(-\zeta q^5,q) = 0, \tag{5.24}$$

$$S_4(-\zeta q^5,q) = \zeta^4q^{-18}J_{1,5}(q^{10}), \tag{5.25}$$

$$S_0(-q^3\zeta^4,q) = -q^{-10}J_{1,5}(q^{10}), \tag{5.26}$$

$$S_1(-q^3\zeta^4,q) = \zeta^4q^{-12}J_{2,5}(q^{10}), \tag{5.27}$$

$$S_2(-q^3\zeta^4,q) = -\zeta^3q^{-12}J_{2,5}(q^{10}), \tag{5.28}$$

$$S_3(-q^3\zeta^4, q) = \zeta^2 q^{-10} J_{1,5}(q^{10}), \tag{5.29}$$

$$S_4(-q^3\zeta^4, q) = 0, \tag{5.30}$$

$$S_0(-q, q) = 0, \tag{5.31}$$

$$S_1(-q, q) = q^{-4} J_{1,5}(q^{10}), \tag{5.32}$$

$$S_2(-q, q) = -q^{-6} J_{2,5}(q^{10}), \tag{5.33}$$

$$S_3(-q, q) = q^{-6} J_{2,5}(q^{10}), \tag{5.34}$$

$$S_4(-q, q) = -q^{-4} J_{1,5}(q^{10}), \tag{5.35}$$

$$S_0(-\zeta^4 q^9, q) = -q^{-40} J_{1,5}(q^{10}), \tag{5.36}$$

$$S_1(-\zeta^4 q^9, q) = 0, \tag{5.37}$$

$$S_2(-\zeta^4 q^9, q) = \zeta^3 q^{-40} J_{1,5}(q^{10}), \tag{5.38}$$

$$S_3(-\zeta^4 q^9, q) = -\zeta^2 q^{-42} J_{2,5}(q^{10}), \tag{5.39}$$

$$S_4(-\zeta^4 q^9, q) = \zeta q^{-42} J_{2,5}(q^{10}), \tag{5.40}$$

$$S_0(-\zeta q^7, q) = -q^{-30} J_{2,5}(q^{10}), \tag{5.41}$$

$$S_1(-\zeta q^7, q) = \zeta q^{-28} J_{1,5}(q^{10}), \tag{5.42}$$

$$S_2(-\zeta q^7, q) = 0, \tag{5.43}$$

$$S_3(-\zeta q^7, q) = -\zeta^3 q^{-28} J_{1,5}(q^{10}), \tag{5.44}$$

$$S_4(-\zeta q^7, q) = \zeta^4 q^{-30} J_{2,5}(q^{10}). \tag{5.45}$$

The verification of (5.20) is just a routine calculation.

Thus both sides of (5.1) agree for 5 distinct values of z in the region $|q|^{10} < |z| \le 1$. We show that both sides agree for $z = -1$, and then our identity (5.1) will follow. To achieve this we use the theory of modular functions. Since this is a standard technique we just sketch some of the details.

First, we calculate the 5-dissection of each theta function on the right side of (5.16) when $z = -1$. By (5.14) we find that

$$\begin{aligned}\Theta(-\zeta^4, q^5) = {} & J_{1,2}(q^{125}) + (1 + \zeta^2 + \zeta^3) q^5 J_{3,10}(q^{25}) \\ & + (\zeta^2 + \zeta^3) q^{20} J_{1,10}(q^{25}),\end{aligned} \tag{5.46}$$

$$\begin{aligned}\Theta(-\zeta q^2, q^5) = {} & J_{23,50}(q^5) + \zeta^3 q^{16} J_{7,50}(q^5) - \zeta q^7 J_{13,50}(q^5) - \zeta^4 q^3 J_{17,50}(q^5) \\ & + q^{24} \zeta^2 J_{3,50}(q^5),\end{aligned} \tag{5.47}$$

$$\begin{aligned}\Theta(-q^4, q^5) = {} & J_{21,50}(q^5) - q\, J_{19,50}(q^5) + q^{12} J_{9,50}(q^5) + q^{28} J_{1,50}(q^5) \\ & - q^9 J_{11,50}(q^5),\end{aligned} \tag{5.48}$$

$$q\,\Theta(-\zeta q^6,q^5) = -\zeta^4 J_{21,50}(q^5) + q\,J_{19,50}(q^5) - \zeta q^{12} J_{9,50}(q^5) - \zeta^2 q^{28} J_{1,50}(q^5) + \zeta^3 q^9 J_{11,50}(q^5), \tag{5.49}$$

$$q^3\Theta(-\zeta^4 q^8,q^5) = -\zeta J_{23,50}(q^5) - \zeta^4 q^{16} J_{7,50}(q^5) + \zeta^2 q^7 J_{13,50}(q^5) + q^3 J_{17,50}(q^5) - \zeta^3 q^{24} J_{3,50}(q^5). \tag{5.50}$$

Next, we evaluate each $S_j(-1,q)$ using (5.14)

$$S_0(-1,q) = S_1(-1,q) = J_{1,10}(q^5), \tag{5.51}$$

$$S_2(-1,q) = S_4(-1,q) = -q^{-3} J_{3,10}(q^5), \tag{5.52}$$

$$S_3(-1,q) = q^{-4} J_{1,2}(q^{25}). \tag{5.53}$$

For $0 \le r \le 4$, we define the operator $\mathcal{U}_{r,5}$ by

$$\mathcal{U}_{r,5}\left(\sum_n a(n)q^n\right) = \sum_n a(5n+r)q^n. \tag{5.54}$$

To show that (5.1) holds for $z=-1$ we need to prove 5 identities

$$\mathcal{U}_{r,5}(q^4\,U(-1,q)) = \mathcal{U}_{r,5}(q^4\,V(-1,q)), \tag{5.55}$$

for $0 \le r \le 4$. It turns out that each of these identities is equivalent to a modular function identity for the group $\Gamma_1(50)$. We provide some detail for the case $r=0$. Using (5.51)–(5.53) we find that

$$\begin{aligned}\mathcal{U}_{0,5}(q^4\,V(-1,q)) &= (J_{2,5}(q^{10}) - q^2(1+\zeta^2+\zeta^3)J_{1,5}(q^{10}))\\ &\quad\times (J_{1,2}(q^5)J_{4,10}^3(q) + q(-1+\zeta^2+\zeta^3)\\ &\quad\times J_{2,10}^2(q)J_{3,10}(q)J_{4,10}(q) - 2q^2(\zeta^2+\zeta^3)J_{2,10}^3 J_{1,10}(q)).\end{aligned} \tag{5.56}$$

We can utilize (5.46)–(5.50) to write the left side of the $r=0$ case of (5.55) as a sum of 135 explicit theta products

$$\begin{aligned}\mathcal{U}_{0,5}(q^4\,U(-1,q)) &= q^{10}J_{3,50}^2(q)J_{19,50}^2(q)J_{25,50}(q) + \cdots\\ &\quad + 2q^{10}(\zeta^2+\zeta^3)J_{1,50}(q)J_{9,50}(q)J_{13,50}(q)J_{17,50}(q)J_{25,50}(q).\end{aligned} \tag{5.57}$$

We have to prove that the right side of (5.56) equals the right side of (5.57). After dividing both sides by $J_{2,5}(q^{10})J_{1,2}(q^5)J_{4,10}^3(q)$ we find that this is equivalent to showing that a certain linear combination of 140 generalized eta-quotients simplifies to the constant 1

$$\begin{aligned}&(1+\zeta^2+\zeta^3)\eta_{50,10}\eta_{50,20}^{-1} + \cdots\\ &\quad + \eta_{50,4}^{-3}\eta_{50,5}^{-2}\eta_{50,6}^{-3}\eta_{50,10}^{-4}\eta_{50,14}^{-3}\eta_{50,15}^{-2}\eta_{50,16}^{-3}\eta_{50,20}^{-5}\eta_{50,24}^{-3}\eta_{50,23}^{2}\eta_{50,21}^{2} = 1.\end{aligned} \tag{5.58}$$

Here

$$\eta_{n,m} = \eta_{n,m}(\tau) = \exp(\pi i P_2(m/n)n\tau) \prod_{k\equiv \pm m \pmod{n}} (1 - \exp(2\pi i k\tau))$$
$$= q^{nP_2(m/n)/2} J_{m,n}(q), \tag{5.59}$$

where $P_2(t) = \{t\}^2 - \{t\} + \frac{1}{6}$, and $q = \exp(2\pi i \tau)$. Using [8, Theorem 2.9, p. 7], [25, Theorem 3, p. 126] we check that each generalised eta-quotient in (5.58) is indeed a modular function on $\Gamma_1(50)$. As usual we need the valence formula

$$\sum_{z\in\mathcal{F}} \mathrm{ORD}\,(f; z, \Gamma) = 0, \tag{5.60}$$

provided f is a nontrivial modular function on Γ, and $\mathcal{F}$ is a fundamental set for Γ. Using MAGMA, the following is a complete set of inequivalent cusps for $\Gamma_1(50)$

$$\mathcal{C} = \Big\{\infty, \frac{0}{1}, \frac{1}{10}, \frac{1}{9}, \frac{2}{17}, \frac{3}{25}, \frac{1}{8}, \frac{2}{15}, \frac{3}{22}, \frac{4}{29}, \frac{5}{36}, \frac{6}{43}, \frac{7}{50}, \frac{3}{20}, \frac{2}{13}, \frac{3}{19}, \frac{4}{25},$$
$$\frac{1}{6}, \frac{6}{35}, \frac{9}{52}, \frac{13}{75}, \frac{4}{23}, \frac{7}{39}, \frac{9}{50}, \frac{7}{38}, \frac{3}{16}, \frac{5}{26}, \frac{1}{5}, \frac{27}{125}, \frac{7}{32}, \frac{11}{50}, \frac{12}{53}, \frac{17}{75},$$
$$\frac{6}{25}, \frac{1}{4}, \frac{13}{50}, \frac{53}{200}, \frac{67}{250}, \frac{27}{100}, \frac{11}{40}, \frac{18}{65}, \frac{7}{25}, \frac{3}{10}, \frac{7}{20}, \frac{9}{25}, \frac{11}{30}, \frac{19}{50}, \frac{49}{125},$$
$$\frac{2}{5}, \frac{21}{50}, \frac{11}{25}, \frac{11}{20}, \frac{26}{45}, \frac{3}{5}, \frac{59}{85}, \frac{7}{10}\Big\}, \tag{5.61}$$

with corresponding widths

$$\{1, 50, 5, 50, 50, 2, 25, 10, 25, 50, 25, 50, 1, 5, 50, 50, 2, 25, 10, 25, 2, 50, 50, 1,$$
$$25, 25, 25, 10, 2, 25, 1, 50, 2, 2, 25, 1, 1, 1, 1, 5, 10, 2, 5, 5, 2, 5, 1, 2, 10, 1,$$
$$2, 5, 10, 10, 10, 5\}. \tag{5.62}$$

Using known results for the invariant order of generalized eta-quotients at cusps [8, (2.3), p. 7], [25, pp. 127–128] we have calculated the order at each cusp of every function in (5.58). As a check we verified that the total order of each function is zero. With $\mathcal{J}$ being the set of generalized eta-quotients occurring in (5.58) we calculated

$$\sum_{c\in\mathcal{C}\setminus\{\infty\}} \min_{f\in\mathcal{J}}(\mathrm{ORD}\,(f; c; \Gamma_1(50)), 0) = -145. \tag{5.63}$$

Hence, by the valence formula (5.60) it suffices to verify (5.58) (or equivalently (5.55) with $r = 0$) up to q^{145}, since generalized eta-quotients have no poles or zeros in the upper-half plane. We have actually verified the result up to q^{200}. All calculations, except for (5.61), and (5.62), were done using MAPLE. The calculations needed to verify (5.55) for $r = 1, 2, 3, 4$ are similar and have been carried out. This completes our proof of (5.1).

6. The Andrews bicrank and extensions

For a partition π, let $\ell(\pi)$ denote the largest part of π, $\varpi(\pi)$ denote the number of ones in π, and $\mu(\pi)$ denote the number of parts of π larger than $\varpi(\pi)$. The crank of π is given by

$$\text{crank}(\pi) = \begin{cases} \ell(\pi), & \text{if } \varpi(\pi) = 0, \\ \mu(\pi) - \varpi(\pi), & \text{if } \varpi(\pi) > 0. \end{cases} \tag{6.1}$$

The crank gives a combinatorial interpretation of Ramanujan's partition congruences mod 5, 7 and 11 and solves a problem of Dyson [9], [10, p. 52]. See [2].

In [1], Andrews gave a combinatorial interpretation of the congruence

$$p_{-2}(5n+3) \equiv 0 \pmod{5}, \tag{6.2}$$

in terms of the crank. This result is a crank analog of the Dyson-birank but is more complicated since it involves positive and negative weights. This complication is there because of the nature of the generating function for the crank. Let $M(m,n)$ denote the number of partitions of n with crank m. Then

$$\sum_{n\ge 0}\sum_{m} M(m,n) z^m q^n = (1-z)q + \frac{(q;q)_\infty}{(zq;q)_\infty (z^{-1}q;q)_\infty}. \tag{6.3}$$

Define $M'(m,n)$ by

$$\begin{aligned} \sum_{n\ge 0}\sum_{m} M'(m,n) z^m q^n &= \frac{(q;q)_\infty}{(zq;q)_\infty (z^{-1}q;q)_\infty} \\ &= 1 + (z - 1 + z^{-1})q + (z^2 + z^{-2})q^2 + \cdots . \end{aligned} \tag{6.4}$$

We need to interpret $M'(m,n)$ combinatorially. To do this we need to extend the definition of partition. To the set of partitions we need to add two additional partitions of 1 which we denote by 1_a and 1_b. We call this new set $\mathcal{E}$, the set of extended partitions.

$$\mathcal{E} = \{(-), 1_a, 1_b, 1, 2, 1+1, 3, 2+1, 1+1+1, \ldots\}. \tag{6.5}$$

We have $|1_a| = |1_b| = 1$. Here as usual $(-)$ is the empty partition of 0. For these extended partitions define a weight function $w(\pi)$ defined by

$$w(\pi) = \begin{cases} -1, & \text{if } \pi = 1_b. \\ |\pi|, & \text{otherwise.} \end{cases} \tag{6.6}$$

Thus for the three extended partitions of 1 we have $w(1) = w(1_a) = 1$, and $w(1_b) = -1$, and the total weight is still $p(1) = 1$. Therefore

$$\sum_{\substack{\pi\in\mathcal{E} \\ |\pi|=n}} w(\pi) = p(n). \tag{6.7}$$

We also extend the definition of crank by $\text{crank}(1_a) = 1$, and $\text{crank}(1_b) = 0$. Recall that for ordinary partition of 1 we have $\text{crank}(1) = -1$. We now have our desired combinatorial interpretation of $M'(m,n)$.

$$F(z,q) = \sum_{\pi \in \mathcal{E}} w(\pi) z^{\text{crank}(\pi)} q^{|\pi|} = \sum_{n \geq 0} \sum_{m} M'(m,n) z^m q^n = \frac{(q;q)_\infty}{(zq;q)_\infty (z^{-1}q;q)_\infty}. \tag{6.8}$$

In other words,

$$M'(m,n) = \sum_{\substack{\pi \in \mathcal{E} \\ |\pi|=n,\ \text{crank}(\pi)=m}} w(\pi). \tag{6.9}$$

We note that the function $F(z,q)$ (at least as an infinite product) occurred in Ramanujan's Lost Notebook.

We define the set of extended bipartitions by $\mathcal{E} \times \mathcal{E}$, i.e. an extended bipartition is simply a pair of extended partitions. For an extended bipartition $\pi = (\pi_1, \pi_2)$ we define a sum of parts function and a weight function in the natural way

$$|\pi| = |\pi_1| + |\pi_2|, \quad \text{and} \quad w(\pi) = w(\pi_1) w(\pi_2). \tag{6.10}$$

We denote Andrews's bicrank function by bicrank_1. We give a variant which we call bicrank_2. For an extended bipartition $\pi = (\pi_1, \pi_2)$ we define

$$\text{bicrank}_1(\pi) = \text{crank}(\pi_1) + \text{crank}(\pi_2), \tag{6.11}$$

and

$$\text{bicrank}_2(\pi) = \text{crank}(\pi_1) + 2\,\text{crank}(\pi_2). \tag{6.12}$$

Amazingly together these two bicrank functions give a new interpretation for all three congruences in (1.1). For $j = 1, 2$ we define $M_j(m,t,n)$ by

$$M_j(m,t,n) = \sum_{\substack{\pi \in \mathcal{E} \times \mathcal{E} \\ |\pi|=n,\ \text{bicrank}_j(\pi) \equiv m \pmod{t}}} w(\pi). \tag{6.13}$$

In other words, $M_j(m,t,n)$ is the number of extended bipartitions of n with bicrank_j congruent to m mod t counted by the weight w.

In this section we prove

Theorem 6.1. *We have the following two results:*

(i) *The residue of the* $\text{bicrank}_1(\pi)$ mod 5 *divides the extended bipartitions of* n *into* 5 *classes of equal weight provided* $n \equiv 3 \pmod{5}$.
(ii) *The residue of the* $\text{bicrank}_2(\pi)$ mod 5 *divides the extended bipartitions of* n *into* 5 *classes of equal weight provided* $n \equiv 2$ *or* $4 \pmod{5}$.

We illustrate the first case of Theorem 6.1 (i). There are 18 extended bipartitions of 3 giving a total weight of $p_{-2}(3) = 10$.

Extended bipartitions of 3	$\text{bicrank}_1 \pmod 5$	weight $= w$
$(2+1, -)$	$0+0 \equiv 0$	1
$(-, 2+1)$	$0+0 \equiv 0$	1
$(2, 1)$	$2-1 \equiv 1$	1
$(1, 2)$	$-1+2 \equiv 1$	1
$(1+1+1, -)$	$-3+0 \equiv 2$	1
$(2, 1_b)$	$2+0 \equiv 2$	-1
$(1+1, 1)$	$-2-1 \equiv 2$	1
$(1_b, 2)$	$0+2 \equiv 2$	-1
$(1, 1+1)$	$-1-2 \equiv 2$	1
$(-, 1+1+1)$	$0-3 \equiv 2$	1
$(3, -)$	$3+0 \equiv 3$	1
$(2, 1_a)$	$2+1 \equiv 3$	1
$(1+1, 1_b)$	$-2+0 \equiv 3$	-1
$(1_a, 2)$	$1+2 \equiv 3$	1
$(1_b, 1+1)$	$0-2 \equiv 3$	-1
$(-, 3)$	$0+3 \equiv 3$	1
$(1+1, 1_a)$	$-2+1 \equiv 4$	1
$(1_a, 1+1)$	$1-2 \equiv 4$	1

Thus

$$M_1(0,5,3) = M_1(1,5,3) = M_1(2,5,3) = M_1(3,5,3) = M_1(4,5,3) = 2,$$

and we see that the residue of the bicrank_1 mod 5 divides the 18 bipartitions of 3 into 5 classes of equal total weight 2.

We illustrate the first case of Theorem 6.1 (ii). There are 13 extended bipartitions of 2 giving a total weight of $p_{-2}(2) = 5$.

Extended bipartitions of 2	$\text{bicrank}_2 \pmod 5$	weight $= w$
$(2, -)$	$2+0 \equiv 2$	1
$(1+1, -)$	$-2+0 \equiv 3$	1
$(1, 1)$	$-1-2 \equiv 2$	1
$(1, 1_a)$	$-1+2 \equiv 1$	1
$(1, 1_b)$	$-1+0 \equiv 4$	-1
$(1_a, 1)$	$1-2 \equiv 4$	1
$(1_a, 1_a)$	$1+2 \equiv 3$	1

$(1_a, 1_b)$	$1+0 \equiv 1$	-1
$(1_b, 1)$	$0-2 \equiv 3$	-1
$(1_b, 1_a)$	$0+2 \equiv 2$	-1
$(1_b, 1_b)$	$0+0 \equiv 0$	1
$(-, 2)$	$0+4 \equiv 4$	1
$(-, 1+1)$	$0-4 \equiv 1$	1

Thus

$$M_2(0,5,2) = M_2(1,5,2) = M_2(2,5,2) = M_2(3,5,2) = M_2(4,5,2) = 1,$$

and we see that the residue of the bicrank$_2$ mod 5 divides the 13 bipartitions of 2 into 5 classes of equal total weight 1.

Theorem 6.1 (i) is due to Andrews [1]. Theorem 6.1 (ii) is a natural extension, and its proof is analogous. For completeness we include a sketch of the proof. We define

$$M'(r,t,n) = \sum_{m \equiv r \pmod t} M'(m,n), \tag{6.14}$$

which is the number of ordinary partitions of n with crank congruent to r mod t when $n \neq 1$. When $n = 1$ it is counting extended partitions. Then

$$\begin{aligned} F(\zeta,q) = \sum_{n=0}^{\infty}\sum_{k=0}^{4} \zeta^k M'(k,5,n)q^n &= \frac{(q;q)_\infty}{(\zeta q;q)_\infty(\zeta^{-1}q;q)_\infty} \\ &= A(q^5) - q(\zeta+\zeta^4)^2 B(q^5) + q^2(\zeta^2+\zeta^3)C(q^5) - q^3(\zeta+\zeta^4)D(q^5), \end{aligned} \tag{6.15}$$

where $F(z,q)$ is given in (6.8), $A(q)$, $B(q)$, $C(q)$, and $D(q)$ are given in (3.4)–(3.7). Equation (6.15) appears in Ramanujan's Lost Notebook [24, p. 20] and is proved in [13, (1.30)].

The two bicrank generating functions are given by

$$\sum_{\pi=(\pi_1,\pi_2)} z^{\text{bicrank}_1(\pi)} q^{|\pi|} = F(z,q)^2, \tag{6.16}$$

$$\sum_{\pi=(\pi_1,\pi_2)} z^{\text{bicrank}_2(\pi)} q^{|\pi|} = F(z,q)F(z^2,q), \tag{6.17}$$

where $F(z,q)$ is the generating function for the crank of extended partitions (6.8). Thus we have

$$\sum_{n=0}^{\infty}\sum_{k=0}^{4} \zeta^k M_1(k,5,n)q^n = F(\zeta,q)^2, \tag{6.18}$$

$$\sum_{n=0}^{\infty}\sum_{k=0}^{4} \zeta^k M_1(k,5,n)q^n = F(\zeta,q)F(\zeta^2,q). \tag{6.19}$$

Using only (6.15) and equations (3.12) and (3.14) we easily find that the coefficient of q^n in the q-expansion of $F(\zeta, q)^2$ is zero if $n \equiv 3 \pmod 5$, and that the coefficient of q^n in the q-expansion of $F(\zeta, q)F(\zeta^2, q)$ is zero if $n \equiv 2$, or $4 \pmod 5$. Both parts of Theorem 6.1 then follow from equations (6.18) and (6.19).

7. A multirank analog of the Hammond-Lewis birank

Let $\mathcal{P}$ denote the set of partitions. A multipartition with r components or an r-colored partition of n is simply an r-tuple

$$\vec{\pi} = (\pi_1, \pi_2, \ldots, \pi_r) \in \mathcal{P} \times \mathcal{P} \times \cdots \times \mathcal{P} = \mathcal{P}^r, \tag{7.1}$$

where

$$\sum_{k=1}^{r} |\pi_k| = n. \tag{7.2}$$

It is clear that the number of r-colored partitions of n is $p_{-r}(n)$ where

$$\sum_{n \geq 0} p_{-r}(n) q^n = \frac{1}{E(q)^r}. \tag{7.3}$$

Congruences for $p_{-r}(n)$ were first considered by Ramanujan. In [23], Ramanthan includes a copy of letter from Ramanujan to Hardy in which Ramanujan alludes to congruences for $p_{-r}(n)$. Ramanathan [23, pp. 153–4] also mentions some of these congruences that appear in the Lost Notebook. Below in Theorem 7.1 we state two elementary and well-known congruences. See for instance [12]. We note that deeper congruences have been found by Atkin [3], Boylan [6], Forbes [11], Newman [21] and Ramanathan [22].

Theorem 7.1. *Let $t > 3$ be prime.*

(i) *If $24n + 1$ is a quadratic nonresidue* mod t, *then*

$$p_{1-t}(n) \equiv 0 \pmod t. \tag{7.4}$$

(ii) *If $8n + 1$ is not a quadratic residue* mod t, *then*

$$p_{3-t}(n) \equiv 0 \pmod t. \tag{7.5}$$

These results follow easily from identities of Euler and Jacobi. Theorem 7.1 (i) follows from

$$\sum_{n \geq 0} p_{1-t}(n) q^{24n+1} = \frac{qE(q^{24})}{E(q^{24})^t} \equiv \frac{1}{E(q^{24t})} \sum_{n=-\infty}^{\infty} (-1)^n q^{(6n+1)^2} \pmod t. \tag{7.6}$$

Here we have used Euler's Pentagonal Number Theorem [18, Thm 353]

$$E(q) = \sum_{n=-\infty}^{\infty} (-1)^n q^{n(3n+1)/2}. \tag{7.7}$$

Theorem 7.1 (ii) follows from

$$\sum_{n\geq 0} p_{3-t}(n)q^{8n+1} = \frac{qE^3(q^8)}{E(q^8)^t} \equiv \frac{1}{E(q^{8t})}\sum_{n\geq 0}(-1)^n(2n+1)q^{(2n+1)^2} \pmod{t}, \quad (7.8)$$

where we have used Jacobi's Identity [18, Thm 237]

$$E(q)^3 = \sum_{n\geq 0}(-1)^n(2n+1)q^{n(n+1)/2}. \quad (7.9)$$

Theorem 7.2 (ii) is Theorem 1 in [1].

In this section we construct analogs of the Hammond-Lewis birank to combinatorially explain the two congruences in Theorem 7.1. Andrews's bicrank [1] (see also equation (6.11)) gave a combinatorial interpretation of Theorem 7.1 (ii) for the case $t = 5$, and $n \equiv 3 \pmod 5$. The Hammond-Lewis birank gave a combinatorial interpretation of Theorem 7.1 (ii) for the case $t = 5$, and all relevant n.

For even r, we define the generalized Hammond-Lewis multirank by

$$\text{gHL-multirank}(\vec{\pi}) = \sum_{k=1}^{r/2} k(\#(\pi_k) - \#(\pi_{r+1-k})), \quad (7.10)$$

for $\vec{\pi} = (\pi_1, \pi_2, \ldots, \pi_r)$ a multipartition with r components. The $r = 2$ case corresponds to the Hammond-Lewis birank. The $r = 4$ case was considered by Hammond [16, p. 67], who proved the $t = 5$ case of Theorem 7.2 (i) below.

In this section we prove

Theorem 7.2. *Let $t > 3$ be prime.*

(i) *The residue of the generalized-Hammond-Lewis-multirank* mod t *divides the multipartitions of n with $r = t - 1$ components into t equal classes provided $24n + 1$ is a quadratic nonresidue* mod t.

(ii) *The residue of the generalized-Hammond-Lewis-multirank* mod t *divides the multipartitions of n with $r = t - 3$ components into t equal classes provided $8n + 1$ is not a quadratic residue* mod t.

We illustrate Theorem 7.2 (ii) for $t = 7$ and $n = 2$.

Multipartitions of 2 with 4 components	generalized-HL-multirank (mod 7)
$(-, -, -, 1+1)$	$-2 \equiv 5$
$(-, -, -, 2)$	$-1 \equiv 6$
$(-, -, 1, 1)$	$-3 \equiv 4$
$(-, -, 1+1, -)$	$-4 \equiv 3$
$(-, -, 2, -)$	$-2 \equiv 5$
$(-, 1, -, 1)$	$1 \equiv 1$

$(-,1,1,-)$	$0\equiv 0$
$(-,1+1,-,-)$	$4\equiv 4$
$(-,2,-,-)$	$2\equiv 2$
$(1,-,-,1)$	$0\equiv 0$
$(1,-,1,-)$	$-1\equiv 6$
$(1,1,-,-)$	$3\equiv 3$
$(1+1,-,-,-)$	$2\equiv 2$
$(2,-,-,-)$	$1\equiv 1$

We see that the residue of generalized-Hammond-Lewis-multirank mod 7 divides the 14 4-colored partitions of 2 into 7 equal classes.

Both parts of Theorem 7.2 are easy to prove. For (i), we need only Euler's pentagonal number theorem (7.7). We let ζ_t be a primitive t-th root of unity. We have

$$\begin{aligned}\sum_{\vec{\pi}\in\mathcal{P}^{t-1}} \zeta_t^{\text{gHL-multirank}(\vec{\pi})} q^{|\vec{\pi}|} &= \prod_{k=1}^{(t-1)/2} \frac{1}{(\zeta_t^k q;q)_\infty(\zeta_t^{-k}q;q)_\infty}\\ &= \frac{(q;q)_\infty}{(q^t;q^t)_\infty}.\end{aligned}\tag{7.11}$$

From (7.7) we have

$$\sum_{\vec{\pi}\in\mathcal{P}^{t-1}} \zeta_t^{\text{gHL-multirank}(\vec{\pi})} q^{24|\vec{\pi}|+1} = \frac{1}{(q^{24t};q^{24t})_\infty}\sum_{n=-\infty}^{\infty}(-1)^n q^{(6n+1)^2}.\tag{7.12}$$

We see that in the q-expansion on the right side of (7.12) the coefficient of q^n is zero when n is a quadratic nonresidue mod t. Theorem 7.2 (i) follows.

For part (ii) of Theorem 7.2 we only need Jacobi's triple product identity (5.14). We have

$$\begin{aligned}&\sum_{\vec{\pi}\in\mathcal{P}^{t-3}} \zeta_t^{\text{gHL-multirank}(\vec{\pi})} q^{|\vec{\pi}|}\\ &\quad= \prod_{k=1}^{(t-3)/2} \frac{1}{(\zeta_t^k q;q)_\infty(\zeta_t^{-k}q;q)_\infty}\\ &\quad= \frac{(\zeta_t^{(t-1)/2}q;q)_\infty(\zeta_t^{-(t-1)/2}q;q)_\infty(q;q)_\infty}{(q^t;q^t)_\infty}\\ &\quad= \frac{\sum_{m=-\infty}^{\infty}(-1)^{m+1}(\zeta_t^{(m+1)(t-1)/2}-\zeta_t^{-m(t-1)/2})q^{m(m+1)/2}}{(1-\zeta_t^{(t-1)/2})(q^t;q^t)_\infty},\end{aligned}\tag{7.13}$$

and

$$\sum_{\vec{\pi} \in \mathcal{P}^{t-3}} \zeta_t^{\text{gHL-multirank}(\vec{\pi})} q^{8|\vec{\pi}|+1}$$

$$= \frac{\sum_{m=-\infty}^{\infty} (-1)^{m+1} \zeta_t^{-m(t-1)/2} (\zeta_t^{(2m+1)(t-1)/2} - 1) q^{(2m+1)^2}}{(1 - \zeta_t^{(t-1)/2})(q^{8t}; q^{8t})_\infty}. \tag{7.14}$$

We see that in the q-expansion on the right side of (7.14) the coefficient of q^n is zero when n is not a quadratic residue mod t, i.e. when n is either a quadratic nonresidue or $n \equiv 0 \pmod{t}$. Theorem 7.2 (ii) follows.

8. Multicranks

In this section we give some extensions of the bicrank to multipartitions and provide alternative interpretations for some of the congruences given in Theorem 7.1. We define two multicranks. These multicranks are defined in terms of cartesian products of extended partitions and ordinary partitions. In Section 6, we defined the set of extended partitions $\mathcal{E}$ and its associated crank and weight function. Recall from Section 7 that $\mathcal{P}$ denotes the set of ordinary partitions, and $\mathcal{P} \subset \mathcal{E}$. Let r be a positive even integer. For an extended multipartition

$$\vec{\pi} = (\pi_1, \pi_2, \ldots, \pi_r) \in \mathcal{E} \times \cdots \times \mathcal{E} \times \mathcal{P} \times \cdots \times \mathcal{P} = \mathcal{E}^{r/2} \times \mathcal{P}^{r/2}, \tag{8.1}$$

we define multicrank-I by

$$\text{multicrank-I}(\vec{\pi}) = \sum_{k=1}^{r/2} k \cdot \text{crank}(\pi_k). \tag{8.2}$$

For an extended multipartition

$$\vec{\pi} = (\pi_1, \pi_2, \ldots, \pi_r) \in \mathcal{E} \times \mathcal{E} \times \mathcal{P} \times \cdots \times \mathcal{P} = \mathcal{E} \times \mathcal{E} \times \mathcal{P}^{r-2}, \tag{8.3}$$

we define multicrank-II by

$$\text{multicrank-II}(\vec{\pi}) = \sum_{k=1}^{2} k \cdot \text{crank}(\pi_k) + \sum_{k=3}^{r} k(\#(\pi_k) - \#(\pi_{r-k+3})). \tag{8.4}$$

We note that the bicrank$_2$ corresponds to the multicrank-II when $r = 2$.

For both types of extended multipartitions we define a sum of parts function and a weight function in the natural way

$$|\vec{\pi}| = \sum_{k=1}^{r} |\pi_k|, \quad \text{and} \quad w(\vec{\pi}) = \prod_{k=1}^{r} w(\pi_k). \tag{8.5}$$

We have

$$\sum_{\substack{\vec{\pi} \in \mathcal{E}^{r/2} \times \mathcal{P}^{r/2} \\ |\vec{\pi}|=n}} w(\vec{\pi}) = \sum_{\substack{\vec{\pi} \in \mathcal{E}^{2} \times \mathcal{P}^{r-2} \\ |\vec{\pi}|=n}} w(\vec{\pi}) = p_{-r}(n). \tag{8.6}$$

Theorem 8.1. *Let $t > 3$ be prime.*

(i) *The residue of the multicrank-I* mod t *divides the extended multipartitions of n from $\mathcal{E}^{(t-1)/2} \times \mathcal{P}^{(t-1)/2}$ into t equal classes of equal weight provided $24n + 1$ is a quadratic nonresidue* mod t.

(ii) *The residue of the multicrank-I* mod t *divides the extended multipartitions of n from $\mathcal{E}^{(t-3)/2} \times \mathcal{P}^{(t-3)/2}$ into t equal classes of equal weight provided $8n + 1$ is not a quadratic residue* mod t.

(iii) *The residue of the multicrank-II* mod t *divides the extended multipartitions of n from $\mathcal{E}^{2} \times \mathcal{P}^{t-5}$ into t equal classes of equal weight provided $8n+1$ is a quadratic nonresidue* mod t.

In view of (8.6), Theorem 8.1 (i), (ii) provides alternative combinatorial interpretations of our congruences for multipartitions given in Theorem 7.1 (i), (ii). The result in part (iii) is weaker than (ii). We include it since it is a generalization of the bicrank$_2$ result.

The proof of Theorem 8.1 is very similar to Theorem 7.2. We have

$$\begin{aligned}\sum_{\vec{\pi} \in \mathcal{E}^{(t-1)/2} \times \mathcal{P}^{(t-1)/2}} \zeta_t^{\text{multicrank-I}(\vec{\pi})} w(\vec{\pi}) q^{|\vec{\pi}|} &= \left(\prod_{k=1}^{(t-1)/2} F(\zeta_t^k, q) \right) \frac{1}{E^{(t-1)/2}(q)} \\ &= \prod_{k=1}^{(t-1)/2} \frac{1}{(\zeta_t^k q; q)_\infty (\zeta_t^{-k} q; q)_\infty} \\ &= \frac{(q;q)_\infty}{(q^t;q^t)_\infty},\end{aligned} \tag{8.7}$$

where $F(z, q)$ is the crank generating function given in (6.8). Theorem 8.1 (i) then follows from (7.12).

Similarly,

$$\begin{aligned}\sum_{\vec{\pi} \in \mathcal{E}^{(t-3)/2} \times \mathcal{P}^{(t-3)/2}} \zeta_t^{\text{multicrank-I}(\vec{\pi})} w(\vec{\pi}) q^{|\vec{\pi}|} &= \left(\prod_{k=1}^{(t-3)/2} F(\zeta_t^k, q) \right) \frac{1}{E^{(t-3)/2}(q)} \\ &= \prod_{k=1}^{(t-3)/2} \frac{1}{(\zeta_t^k q; q)_\infty (\zeta_t^{-k} q; q)_\infty}.\end{aligned} \tag{8.8}$$

Theorem 8.1 (ii) then follows from (7.13), (7.14).

We have

$$\sum_{\vec{\pi}\in\mathcal{E}^2\times\mathcal{P}^{t-5}} \zeta_t^{\text{multicrank-II}(\vec{\pi})} w(\vec{\pi}) q^{|\vec{\pi}|}$$

$$= F(\zeta_t, q)\, F(\zeta_t^2, q) \prod_{k=3}^{(t-1)/2} \frac{1}{(\zeta_t^k q; q)_\infty (\zeta_t^{-k} q; q)_\infty}.$$

$$= \frac{(q)_\infty^3}{(q^t; q^t)_\infty}. \tag{8.9}$$

Then

$$\sum_{\vec{\pi}\in\mathcal{E}^2\times\mathcal{P}^{t-5}} \zeta_t^{\text{multicrank-II}(\vec{\pi})} w(\vec{\pi}) q^{8|\vec{\pi}|+1} = \frac{1}{E(q^{8t})} \sum_{n\ge 0} (-1)^n (2n+1) q^{(2n+1)^2}, \tag{8.10}$$

and Theorem 8.1 (iii) follows.

9. Concluding remarks

The two main results of this paper are the combinatorial interpretations of the 2-colored partition congruences (1.1) in terms of the Dyson-birank and the 5-core-birank. The author has been unable to extend these two results to higher dimensional multipartitions. The extensions of the Hammond-Lewis birank and Andrews bicrank are much easier because the generating functions involved are simple infinite products.

It seems unlikely that a combinatorial proof of (4.5) is possible. This identity gives the 5-dissection of the 5-core-crank generating function when $z = \zeta_5$. The proof given in the paper relies on a heavy use of the theory of modular functions. A more elementary proof is desirable. In [15], a combinatorial proof is given that the residue of 5-core-crank mod 5 divides the 5-cores of $5n+4$ into 5 equal classes. It would interesting to see if the methods of [15] could be extended to give a combinatorial proof of Theorem 4.1, which is our result for the 5-core-birank.

It is clear that the generalized-Hammond-Lewis multiranks and our multicranks are related. For instance, from equations (7.11), (8.7), (7.13), and (8.8) we have

$$\sum_{\vec{\pi}\in\mathcal{P}^{t-1}} \zeta_t^{\text{gHL-multirank}(\vec{\pi})} q^{|\vec{\pi}|} = \sum_{\vec{\pi}\in\mathcal{E}^{(t-1)/2}\times\mathcal{P}^{(t-1)/2}} \zeta_t^{\text{multicrank-I}(\vec{\pi})} w(\vec{\pi}) q^{|\vec{\pi}|}, \tag{9.1}$$

$$\sum_{\vec{\pi}\in\mathcal{P}^{t-3}} \zeta_t^{\text{gHL-multirank}(\vec{\pi})} q^{|\vec{\pi}|} = \sum_{\vec{\pi}\in\mathcal{E}^{(t-3)/2}\times\mathcal{P}^{(t-3)/2}} \zeta_t^{\text{multicrank-I}(\vec{\pi})} w(\vec{\pi}) q^{|\vec{\pi}|}. \tag{9.2}$$

It would interesting to find a combinatorial proof of these identities. However what would be more interesting is to find bijective proofs of Theorems 6.1 and 7.2. This is a reasonable problem since the generating functions involved are simple infinite products.

Throughout this paper we have assumed t is prime. Houser [20] has considered the analogoue of Theorem 7.1 for prime powers. Although he does not give rank-type results he does find combinatorial proofs. Dennis Eichhorn has observed that equations (7.12), (7.13), and (7.14) also hold when t is composite. In this case, there will be relations between the number of multipartitions in each gHL-multirank class mod t.

Acknowledgement

I would like to thank Byungchan Kim, Paul Hammond and Dennis Eichhorn for their comments and suggestions.

References

[1] G. E. Andrews, A survey of multipartitions: congruences and identities, in Surveys in number theory, *Dev. Math.*, **17** (Springer, New York, 2008) 1–19.
[2] G. E. Andrews and F. G. Garvan, Dyson's crank of a partition, *Bull. Amer. Math. Soc. (N.S.)*, **18** (1988) 167–171.
[3] A. O. L. Atkin, Ramanujan congruences for $p_{-k}(n)$, *Canad. J. Math.*, **20** (1968) 67–78; corrigendum, ibid., **21** (1968) 256.
[4] A. O. L. Atkin and P. Swinnerton-Dyer, Some properties of partitions, *Proc. London Math. Soc.*, **4** (1954) 84–106.
[5] A. Berkovich and F. G. Garvan, The BG-rank of a partition and its applications, *Adv. in Appl. Math.*, **40** (2008) 377–400.
[6] M. Boylan, Exceptional congruences for powers of the partition function, *Acta Arith.*, **111** (2004) 187–203.
[7] W. Y. C. Chen, K. Q. Ji and H. S. Wilf, BG-ranks and 2-cores, *Electron. J. Combin.*, **13** (2006) Note 18, 5 pp. (electronic).
[8] Y. S. Choi, Tenth order mock theta functions in Ramanujan's lost notebook III, *Proc. Lond. Math. Soc. (3)*, **94** (2007) 26–52.
[9] F. J. Dyson, Some guesses in the theory of partitions, Eureka (Cambridge) **8** (1944) 10–15.
[10] F. J. Dyson, Selected papers of Freeman Dyson with commentary, *Amer. Math. Soc.*, Providence, RI (1996).
[11] A. D. Forbes, Congruence properties of functions related to the partition function, *Pacific J. Math.*, **158** (1993) 145–156.
[12] J. M. Gandhi, Congruences for $p_r(n)$ and Ramanujan's τ function, *Amer. Math. Monthly*, **70** (1963) 265–274.
[13] F. G. Garvan, New combinatorial interpretations of Ramanujan's partition congruences mod 5, 7 and 11, *Trans. Amer. Math. Soc.*, **305** (1988), 47–77.
[14] F. G. Garvan, Combinatorial interpretations of Ramanujan's partition congruences in Ramanujan Revisited: Proc. of the Centenary Conference, Univ. of Illinois at Urbana-Champaign, June 1–5, 1987, Acad. Press, San Diego (1988).
[15] F. Garvan, D. Kim and D. Stanton, Cranks and t-cores, *Invent. Math.*, **101** (1990) 1–17.
[16] P. R. Hammond, Combinatorial Aspects of the Theory of q-Series, D.Phil. thesis, University of Sussex (2006), 78 pp.
[17] P. Hammond and R. Lewis, Congruences in ordered pairs of partitions, *Int. J. Math. Math. Sci.*, (2004) nos. 45–48, 2509–2512.
[18] G. H. Hardy and E. M. Wright, An Introduction to the Theory of Numbers, Oxford Univ. Press, London (1979).

[19] M. Hirschhorn, F. Garvan and J. Borwein, Cubic analogues of the Jacobian theta function $\theta(z,q)$, *Canad. J. Math.*, **45** (1993) 673–694.

[20] G. Houser, Combinatorial Proofs of Classical Congruences for $C_h(n)$, the Number of Partitions of n in h Colors, M.S. Thesis, The University of Arizona (2002), 34 pp.

[21] M. Newman, Some theorems about $p_r(n)$, *Canad. J. Math.*, **9** (1957) 68–70.

[22] K. G. Ramanathan, Identities and congruences of the Ramanujan type, *Canadian J. Math.*, **2** (1950) 168–178.

[23] K. G. Ramanathan, Ramanujan and the congruence properties of partitions, *Proc. Indian Acad. Sci. Sect. A Math. Sci.*, **89** (1980) 133–157.

[24] S. Ramanujan, The lost notebook and other unpublished papers, Springer-Verlag, Berlin (1988).

[25] S. Robins, Generalized Dedekind η-products, in The Rademacher legacy to mathematics (eds. G. E. Andrews, D. M. Bressoud and L. A. Parson), *Contemp. Math.*, **166** (Amer. Math. Soc., Providence, RI, 1994) 119–128.

[26] N. Robbins, Some properties of two-color partitions, *Ars Combin.*, **65** (2002) 149–154.

Ramanujan Rediscovered No. 14, 2010, pp. 113–124.

The Number of Representations of a Number by Various Forms Involving Triangles, Squares, Pentagons and Octagons

Michael D. Hirschhorn

School of Mathematics and Statistics, UNSW, Sydney, Australia 2052
e-mail: m.hirschhorn@unsw.edu.au

Abstract. In a recent paper, I began with four classical results due to Jacobi, Dirichlet and Lorenz which give the number of representations of a number by various forms involving squares in terms of divisor functions. I deduced another sixteen similar representation theorems involving triangular numbers and/or squares, including celebrated results of Legendre and Ramanujan (and omitted a seventeenth, given here).

In this paper I deduce a further twenty–nine representation theorems involving triangular numbers, squares, pentagonal numbers and octagonal numbers.

Keywords. Representations, triangular numbers, squares, pentagonal numbers, octagonal numbers.

Dedicated to the memory of K. Venkatachaliengar on the occasion of the Centenary of his birth

1. Introduction

In a recent paper [1], I began with four classical results due to Jacobi, Dirichlet and Lorenz which give the number of representations of a number by various forms involving squares in terms of divisor functions. I deduced another sixteen similar representation theorems involving triangular numbers and/or squares, including celebrated results of Legendre and Ramanujan (and omitted a seventeenth, given here as (3.21)).

In this paper I deduce a further twenty–nine representation theorems involving triangular numbers, squares, pentagonal numbers (numbers of the form $(3k^2 \pm k)/2$) and octagonal numbers (numbers of the form $3k^2 \pm 2k$). These formulas all follow easily from results in [1].

Since the appearance of [1], similar results have been given by H. Y. Lam [2] and R. S. Melham [3]. For more than 300 further results (most as yet unproven) the reader is referred to Melham [4].

For integral x, let $d_{r,m}(x)$ denote the number of divisors d of x with $d \equiv r \pmod{m}$, $\sigma(x)$ the sum of the divisors of x, both 0 if x is not an integer.

In what follows, for example, $r\{\Pi + 2\Pi\}(n)$ means the number of representations of n as the sum of a pentagonal number and twice a pentagonal number. We prove the following results: (Note that Π stands for 'pentagonal number' and Ω for 'octagonal number'.) For $n \geq 0$,

$$r\{\Delta + \Pi\}(n) = d_{1,6}(6n+1) - d_{5,6}(6n+1), \tag{1.1}$$

$$\begin{aligned}
r\{\Delta+4\Pi\}(n) &= d_{1,24}(24n+7)+d_{19,24}(24n+7)-d_{5,24}(24n+7)\\
&\quad -d_{23,24}(24n+7), \qquad (1.2)\\
r\{3\Delta+\Pi\}(n) &= d_{1,12}(12n+5)-d_{11,12}(12n+5), \qquad (1.3)\\
r\{3\Delta+2\Pi\}(n) &= d_{1,8}(24n+11)-d_{7,8}(24n+11), \qquad (1.4)\\
r\{4\Delta+\Pi\}(n) &= d_{1,24}(24n+13)+d_{19,24}(24n+13)-d_{5,24}(24n+13)\\
&\quad -d_{23,24}(24n+13), \qquad (1.5)\\
r\{6\Delta+\Pi\}(n) &= d_{1,8}(24n+19)-d_{7,8}(24n+19), \qquad (1.6)\\
r\{\Delta+2\Omega\}(n) &= d_{1,24}(24n+19)+d_{13,24}(24n+19)-d_{11,24}(24n+19)\\
&\quad -d_{23,24}(24n+19), \qquad (1.7)\\
r\{2\Delta+\Omega\}(n) &= d_{1,12}(12n+7)-d_{11,12}(12n+7), \qquad (1.8)\\
r\{3\Delta+\Omega\}(n) &= d_{1,24}(24n+17)+d_{19,24}(24n+17)-d_{5,24}(24n+17)\\
&\quad -d_{23,24}(24n+17), \qquad (1.9)\\
r\{12\Delta+\Omega\}(n) &= (d_{1,8}(6n+11)+d_{3,8}(6n+11)-d_{5,8}(6n+11)\\
&\quad -d_{7,8}(6n+11))/2, \qquad (1.10)\\
r\{\Box+2\Pi\}(n) &= d_{1,3}(12n+1)-d_{2,3}(12n+1), \qquad (1.11)\\
r\{2\Box+\Pi\}(n) &= d_{1,3}(24n+1)-d_{2,3}(24n+1), \qquad (1.12)\\
r\{3\Box+\Pi\}(n) &= d_{1,8}(24n+1)+d_{3,8}(24n+1)-d_{5,8}(24n+1)\\
&\quad -d_{7,8}(24n+1), \qquad (1.13)\\
r\{3\Box+4\Pi\}(n) &= d_{1,8}(6n+1)+d_{3,8}(6n+1)-d_{5,8}(6n+1)-d_{7,8}(6n+1), \qquad (1.14)\\
r\{\Box+\Omega\}(n) &= d_{1,3}(3n+1)-d_{2,3}(3n+1)\\
&\quad +2(d_{1,3}((3n+1)/4)-d_{2,3}((3n+1)/4)), \qquad (1.15)\\
r\{3\Box+\Omega\}(n) &= d_{1,4}(3n+1)-d_{3,4}(3n+1), \qquad (1.16)\\
r\{3\Box+2\Omega\}(n) &= d_{1,8}(3n+2)+d_{3,8}(3n+2)-d_{5,8}(3n+2)-d_{7,8}(3n+2), \qquad (1.17)\\
r\{6\Box+\Omega\}(n) &= d_{1,8}(3n+1)+d_{3,8}(3n+1)-d_{5,8}(3n+1)-d_{7,8}(3n+1), \qquad (1.18)\\
r\{\Pi+\Pi\}(n) &= d_{1,4}(12n+1)-d_{3,4}(12n+1), \qquad (1.19)\\
r\{\Pi+2\Pi\}(n) &= d_{1,8}(24n+3)-d_{7,8}(24n+3)\\
&\quad -(d_{1,8}((8n+1)/3)-d_{7,8}((8n+1)/3)), \qquad (1.20)\\
r\{\Pi+4\Pi\}(n) &= d_{1,8}(24n+5)-d_{7,8}(24n+5), \qquad (1.21)
\end{aligned}$$

$$\begin{aligned} r\{\Pi+\Omega\}(n) &= (d_{1,8}(24n+9)+d_{3,8}(24n+9)-d_{5,8}(24n+9) \\ &\quad - d_{7,8}(24n+9)-(d_{1,8}((8n+3)/3)+d_{3,8}((8n+3)/3) \\ &\quad - d_{5,8}((8n+3)/3)-d_{7,8}((8n+3)/3)))/2, \end{aligned} \tag{1.22}$$

$$\begin{aligned} r\{\Pi+2\Omega\}(n) &= d_{1,24}(24n+17)+d_{13,24}(24n+17)-d_{11,24}(24n+17) \\ &\quad - d_{23,24}(24n+17), \end{aligned} \tag{1.23}$$

$$r\{2\Pi+\Omega\}(n) = d_{1,12}(12n+5)-d_{11,12}(12n+5), \tag{1.24}$$

$$\begin{aligned} r\{4\Pi+\Omega\}(n) &= (d_{1,8}(6n+3)+d_{3,8}(6n+3)-d_{5,8}(6n+3)-d_{7,8}(6n+3) \\ &\quad - (d_{1,8}((2n+1)/3)+d_{3,8}((2n+1)/3)-d_{5,8}((2n+1)/3) \\ &\quad - d_{7,8}((2n+1)/3)))/2, \end{aligned} \tag{1.25}$$

$$r\{\Omega+\Omega\}(n) = d_{1,4}(3n+2)-d_{3,4}(3n+2), \tag{1.26}$$

$$\begin{aligned} r\{\Omega+2\Omega\}(n) &= (d_{1,8}(3n+3)+d_{3,8}(3n+3)-d_{5,8}(3n+3)-d_{7,8}(3n+3) \\ &\quad - (d_{1,8}((n+1)/3)+d_{3,8}((n+1)/3)-d_{5,8}((n+1)/3) \\ &\quad - d_{7,8}((n+1)/3)))/2, \end{aligned} \tag{1.27}$$

$$r\{3\triangle+3\triangle+\Pi+\Pi\}(n) = \sigma(6n+5)/6, \tag{1.28}$$

$$r\{3\square+3\square+\Omega+\Omega\}(n) = (\sigma(3n+2)-4\sigma((3n+2)/4))/3. \tag{1.29}$$

2. Preliminary results

As usual, let

$$\phi(q)=\sum_{-\infty}^{\infty} q^{n^2}, \quad \psi(q)=\sum_{n\geq 0} q^{(n^2+n)/2}.$$

We need the easy lemmas

$$\phi(q)=\phi(q^9)+2q\Omega(q^3), \tag{2.1}$$

$$\psi(q)=\Pi(q^3)+q\psi(q^9) \tag{2.2}$$

where

$$\Omega(q)=\sum_{-\infty}^{\infty} q^{3n^2+2n}, \quad \Pi(q)=\sum_{-\infty}^{\infty} q^{(3n^2+n)/2}.$$

Note that the powers appearing in $\Pi(q)$ are the pentagonal numbers and the powers appearing in $\Omega(q)$ are the octagonal numbers.

3. Proofs of theorems

We start with the results from [1]. They all hold for $n \geq 0$, with the exception that (3.15), (3.16), (3.17) and (3.20) hold only for $n \geq 1$.

$$r\{\triangle+\triangle\}(n) = d_{1,4}(4n+1)-d_{3,4}(4n+1), \tag{3.1}$$

$$r\{\Delta+2\Delta\}(n)=d_{1,8}(8n+3)-d_{7,8}(8n+3), \tag{3.2}$$

$$r\{\Delta+3\Delta\}(n)=d_{1,3}(2n+1)-d_{2,3}(2n+1), \tag{3.3}$$

$$r\{\Delta+4\Delta\}(n)=d_{1,8}(8n+5)-d_{7,8}(8n+5), \tag{3.4}$$

$$r\{\Delta+12\Delta\}(n)=(d_{1,3}(8n+13)-d_{2,3}(8n+13))/2, \tag{3.5}$$

$$r\{3\Delta+4\Delta\}(n)=(d_{1,3}(8n+7)-d_{2,3}(8n+7))/2, \tag{3.6}$$

$$r\{\Delta+\Box\}(n)=d_{1,8}(8n+1)+d_{3,8}(8n+1)-d_{5,8}(8n+1)-d_{7,8}(8n+1), \tag{3.7}$$

$$r\{\Delta+2\Box\}(n)=d_{1,4}(8n+1)-d_{3,4}(8n+1), \tag{3.8}$$

$$r\{\Delta+6\Box\}(n)=d_{1,3}(8n+1)-d_{2,3}(8n+1), \tag{3.9}$$

$$r\{2\Delta+\Box\}(n)=d_{1,4}(4n+1)-d_{3,4}(4n+1), \tag{3.10}$$

$$r\{2\Delta+3\Box\}(n)=d_{1,3}(4n+1)-d_{2,3}(4n+1), \tag{3.11}$$

$$r\{3\Delta+2\Box\}(n)=d_{1,3}(8n+3)-d_{2,3}(8n+3), \tag{3.12}$$

$$r\{4\Delta+\Box\}(n)=d_{1,8}(2n+1)+d_{3,8}(2n+1)-d_{5,8}(2n+1)-d_{7,8}(2n+1), \tag{3.13}$$

$$r\{6\Delta+\Box\}(n)=d_{1,3}(4n+3)-d_{2,3}(4n+3), \tag{3.14}$$

$$r\{\Box+\Box\}(n)=4(d_{1,4}(n)-d_{3,4}(n)), \tag{3.15}$$

$$r\{\Box+2\Box\}(n)=2(d_{1,8}(n)+d_{3,8}(n)-d_{5,8}(n)-d_{7,8}(n)), \tag{3.16}$$

$$r\{\Box+3\Box\}(n)=2(d_{1,3}(n)-d_{2,3}(n))+4(d_{1,3}(n/4)-d_{2,3}(n/4)), \tag{3.17}$$

$$r\{\Delta+\Delta+\Delta+\Delta\}(n)=\sigma(2n+1), \tag{3.18}$$

$$r\{\Delta+\Delta+\Box+\Box\}(n)=\sigma(4n+1), \tag{3.19}$$

$$r\{\Box+\Box+\Box+\Box\}(n)=8(\sigma(n)-4\sigma(n/4)), \tag{3.20}$$

to which we should add

$$r\{\Delta+\Delta+2\Delta+2\Delta\}=\sigma(4n+3)/4. \tag{3.21}$$

(3.1) is equivalent to

$$\psi(q)^2=\sum_{n\geq 0}(d_{1,4}(4n+1)-d_{3,4}(4n+1))q^n,$$

or,

$$\Pi(q^3)^2+2q\Pi(q^3)\psi(q^9)+q^2\psi(q^9)^2=\sum_{n\geq 0}(d_{1,4}(4n+1)-d_{3,4}(4n+1))q^n.$$

If we extract those terms in which the power of q is congruent to 0 modulo 3 and replace q^3 by q we obtain

$$\Pi(q)^2=\sum_{n\geq 0}(d_{1,4}(12n+1)-d_{3,4}(12n+1))q^n$$

which yields (1.19), while if we extract those terms in which the power of q is congruent to 1 modulo 3, divide by $2q$ and replace q^3 by q, we obtain

$$\psi(q^3)\Pi(q) = \sum_{n\geq 0}(d_{1,4}(12n+5) - d_{3,4}(12n+5))/2\, q^n$$

which gives (1.3).

(3.2) is equivalent to

$$(\Pi(q^3) + q\psi(q^9))(\Pi(q^6) + q^2\psi(q^{18})) = \sum_{n\geq 0}(d_{1,8}(8n+3) - d_{7,8}(8n+3))q^n.$$

If we extract those terms in which the power of q is congruent to 0 modulo 3, and replace q^3 by q, we obtain

$$\Pi(q)\Pi(q^2) + q\psi(q^3)\psi(q^6) = \sum_{n\geq 0}(d_{1,8}(24n+3) - d_{7,8}(24n+3))q^n,$$

while

$$q\psi(q^3)\psi(q^6) = \sum_{n\geq 0}(d_{1,8}(8n+3) - d_{7,8}(8n+3))q^{3n+1}$$

$$= \sum_{n\geq 0}(d_{1,8}((8n+1)/3) - d_{7,8}((8n+1)/3))q^n.$$

By subtraction, we obtain

$$\Pi(q)\Pi(q^2) = \sum_{n\geq 0}(d_{1,8}(24n+3) - d_{7.8}(24n+3)$$

$$- (d_{1,8}((8n+1)/3) - d_{7,8}((8n+1)/3)))q^n,$$

which gives (1.20).

If from the above we extract those terms in which the power of q is congruent to 1 modulo 3, divide by q and replace q^3 by q we obtain

$$\psi(q^3)\Pi(q^2) = \sum_{n\geq 0}(d_{1,8}(24n+11) - d_{7,8}(24n+11))q^n$$

which gives (1.4), while if we extract those terms in which the power of q is 2 modulo 3, divide by q^2 and replace q^3 by q, we obtain

$$\psi(q^6)\Pi(q) = \sum_{n\geq 0}(d_{1,8}(24n+19) - d_{7,8}(24n+19))q^n$$

which gives (1.6).

(3.3) is equivalent to

$$(\Pi(q^3) + q\psi(q^9))\psi(q^3) = \sum_{n\geq 0}(d_{1,3}(2n+1) - d_{2,3}(2n+1))q^n.$$

If we extract those terms in which the power of q is congruent to 0 modulo 3 and replace q^3 by q, we obtain

$$\psi(q)\Pi(q) = \sum_{n\geq 0}(d_{1,3}(6n+1) - d_{2,3}(6n+1))q^n,$$

which gives (1.1).

(3.4) is equivalent to

$$(\Pi(q^3) + q\psi(q^9))(\Pi(q^{12}) + q^4\psi(q^{36})) = \sum_{n\geq 0}(d_{1,8}(8n+5) - d_{7,8}(8n+5))q^n.$$

If we extract those terms in which the power of q is congruent to 0 modulo 3 and replace q^3 by q we obtain

$$\Pi(q)\Pi(q^4) = \sum_{n\geq 0}(d_{1,8}(24n+5) - d_{7,8}(24n+5))q^n,$$

which gives (1.21).

(3.5) is equivalent to

$$(\Pi(q^3) + q\psi(q^9))\psi(q^{12}) = \sum_{n\geq 0}(d_{1,3}(8n+13) - d_{2,3}(8n+13))/2\, q^n.$$

If we extract those terms in which the power of q is congruent to 0 modulo 3 and replace q^3 by q, we obtain

$$\psi(q^4)\Pi(q) = \sum_{n\geq 0}(d_{1,3}(24n+13) - d_{2,3}(24n+13))/2\, q^n,$$

from which (1.5) follows.

(3.6) is equivalent to

$$\psi(q^3)(\Pi(q^{12}) + q^4\psi(q^{36})) = \sum_{n\geq 0}(d_{1,3}(8n+7) - d_{2,3}(8n+7))/2\, q^n.$$

If we extract those terms in which the power of q is 0 modulo 3 and replace q^3 by q, we obtain

$$\psi(q)\Pi(q^4) = \sum_{n\geq 0}(d_{1,3}(24n+7) - d_{2,3}(24n+7))/2\, q^n,$$

which gives (1.2).

(3.7) is equivalent to

$$\begin{aligned}&(\Pi(q^3) + q\psi(q^9))(\phi(q^9) + 2q\Omega(q^3))\\ &\quad = \sum_{n\geq 0}(d_{1,8}(8n+1) + d_{3,8}(8n+1) - d_{5,8}(8n+1) - d_{7,8}(8n+1))q^n,\end{aligned}$$

If we extract those terms in which the power of q is congruent to 0 modulo 3 and replace q^3 by q, we obtain

$$\phi(q^3)\Pi(q) = \sum_{n\geq 0}(d_{1,8}(24n+1)+d_{3,8}(24n+1)-d_{5,8}(24n+1)-d_{7,8}(24n+1))q^n,$$

which gives (1.13), while if we extract those terms in which the power of q is congruent to 2 modulo 3, divide by $2q^2$ and replace q^3 by q, we obtain

$$\psi(q^3)\Omega(q) = \sum_{n\geq 0}(d_{1,8}(24n+17)+d_{3,8}(24n+17) - d_{5,8}(24n+17)-d_{7,8}(24n+17))/2\, q^n,$$

which gives (1.9).

If we extract those terms in which the power of q is congruent to 1 modulo 3, divide by q and replace q^3 by q, we obtain

$$2\Pi(q)\Omega(q) + \psi(q^3)\phi(q^3) - \sum_{n\geq 0}(d_{1,8}(24n+9)+d_{3,8}(24n+9)-d_{5,8}(24n+9)-d_{7,8}(24n+9))q^n$$

while

$$\psi(q^3)\phi(q^3) = \sum_{n\geq 0}(d_{1,8}(8n+1)+d_{3,8}(8n+1)-d_{5,8}(8n+1)-d_{7,8}(8n+1))q^{3n}$$

$$= \sum_{n\geq 0}(d_{1,8}((8n+3)/3)+d_{3,8}((8n+3)/3)-d_{5,8}((8n+3)/3) - d_{7,8}((8n+3)/3))q^n.$$

If we subtract the second from the first and divide by 2, we obtain

$$\Pi(q)\Omega(q) = \sum_{n\geq 0}(d_{1,8}(24n+9)+d_{3,8}(24n+9)-d_{5,8}(24n+9)-d_{7,8}(24n+9) - (d_{1,8}((8n+3)/3)+d_{3,8}((8n+3)/3)-d_{5,8}((8n+3)/3) - d_{7,8}((8n+3)/3)))/2\, q^n,$$

which gives (1.22).

(3.8) is equivalent to

$$(\Pi(q^3)+q\psi(q^9))(\phi(q^{18})+2q^2\Omega(q^6)) = \sum_{n\geq 0}(d_{1,4}(8n+1)-d_{3,4}(8n+1))q^n.$$

If we extract those terms in which the power of q is congruent to 2 modulo 3, divide by $2q^2$ and replace q^3 by q, we obtain

$$\Pi(q)\Omega(q^2) = \sum_{n\geq 0}(d_{1,4}(24n+17)-d_{3,4}(24n+17))/2\, q^n,$$

which gives (1.23).

(3.9) is equivalent to

$$(\Pi(q^3)+q\psi(q^9))\phi(q^6)=\sum_{n\geq 0}(d_{1,3}(8n+1)-d_{2,3}(8n+1))q^n.$$

If we extract those terms in which the power of q is congruent to 0 modulo 3 and replace q^3 by q, we obtain

$$\phi(q^2)\Pi(q)=\sum_{n\geq 0}(d_{1,3}(24n+1)-d_{2,3}(24n+1))q^n,$$

which gives (1.12).

(3.10) is equivalent to

$$(\Pi(q^6)+q^2\psi(q^{18}))(\phi(q^9)+2q\Omega(q^3))=\sum_{n\geq 0}(d_{1,4}(4n+1)-d_{3,4}(4n+1))q^n.$$

If we extract those terms in which the power of q is congruent to 1 modulo 3, divide by $2q$ and replace q^3 by q, we obtain

$$\Pi(q^2)\Omega(q)=\sum_{n\geq 0}(d_{1,4}(12n+5)-d_{3,4}(12n+5))q^n,$$

which gives (1.24).

(3.11) is equivalent to

$$(\Pi(q^6)+q^2\psi(q^{18}))\phi(q^3)=\sum_{n\geq 0}(d_{1,3}(4n+1)-d_{2,3}(4n+1))q^n.$$

If we extract those terms in which the power of q is congruent to 0 modulo 3 and replace q^3 by q, we obtain

$$\phi(q)\Pi(q^2)=\sum_{n\geq 0}(d_{1,3}(12n+1)-d_{2,3}(12n+1))q^n,$$

which gives (1.11).

(3.12) is equivalent to

$$\psi(q^3)(\phi(q^{18})+2q^2\Omega(q^6))=\sum_{n\geq 0}(d_{1,3}(8n+3)-d_{2,3}(8n+3))q^n.$$

If we extract those terms in which the power of q is congruent to 2 modulo 3, divide by $2q^2$ and replace q^3 by q, we obtain

$$\psi(q)\Omega(q^2)=\sum_{n\geq 0}(d_{1,3}(24n+19)-d_{2,3}(24n+19))/2\;q^n,$$

which gives (1.7).

(3.13) is equivalent to

$$(\Pi(q^{12}) + q^4\psi(q^{36}))(\phi(q^9) + 2q\Omega(q^3))$$
$$= \sum_{n\geq 0}(d_{1,8}(2n+1) + d_{3,8}(2n+1) - d_{5,8}(2n+1) - d_{7,8}(2n+1))q^n.$$

If we extract those terms in which the power of q is congruent to 0 modulo 3 and replace q^3 by q, we obtain

$$\phi(q^3)\Pi(q^4) = \sum_{n\geq 0}(d_{1,8}(6n+1) + d_{3,8}(6n+1) - d_{5,8}(6n+1) - d_{7,8}(6n+1))q^n,$$

which gives (1.14), while if we extract those terms in which the power of q is congruent to 2 modulo 3, divide by $2q^5$ and replace q^3 by q, we obtain

$$\psi(q^{12})\Omega(q) = \sum_{n\geq 0}(d_{1,8}(6n+11)+d_{3,8}(6n+11)-d_{5,8}(6n+11)-d_{7,8}(6n+11))/2\, q^n$$

which gives (1.10).

If we extract those terms in which the power of q is 1 modulo 3, divide by q and replace q^3 by q, we obtain

$$2\Pi(q^4)\Omega(q) + q\psi(q^{12})\phi(q^3)$$
$$= \sum_{n\geq 0}(d_{1,8}(6n+3) + d_{3,8}(6n+3) - d_{5,8}(6n+3) - d_{7,8}(6n+3))q^n,$$

while

$$q\psi(q^{12})\phi(q^3)$$
$$= \sum_{n\geq 0}(d_{1,8}(2n+1) + d_{3,8}(2n+1) - d_{5,8}(2n+1) - d_{7,8}(2n+1))q^{3n+1}$$
$$= \sum_{n\geq 0}(d_{1,8}((2n+1)/3) + d_{3,8}((2n+1)/3) - d_{5,8}((2n+1)/3)$$
$$- d_{7,8}((2n+1)/3))q^n.$$

If we subtract the second from the first and divide by 2, we obtain

$$\Pi(q^4)\Omega(q)$$
$$= \sum_{n\geq 0}(d_{1,8}(6n+3) + d_{3,8}(6n+3) - d_{5,8}(6n+3) - d_{7,8}(6n+3)$$
$$- (d_{1,8}((2n+1)/3) + d_{3,8}((2n+1)/3) - d_{5,8}((2n+1)/3)$$
$$- d_{7,8}((2n+1)/3)))/2\, q^n,$$

which gives (1.25).

(3.14) is equivalent to

$$\psi(q^6)(\phi(q^9) + 2q\Omega(q^3)) = \sum_{n\geq 0}(d_{1,3}(4n+3) - d_{2,3}(4n+3))q^n.$$

If we extract those terms in which the power of q is congruent to 1 modulo 3, divide by $2q$ and replace q^3 by q, we obtain

$$\psi(q^2)\Omega(q) = \sum_{n\geq 0}(d_{1,3}(12n+7) - d_{2,3}(12n+7))/2\, q^n,$$

which gives (1.8).

(3.15) is (essentially) equivalent to

$$(\phi(q^9) + 2q\Omega(q^3))^2 = 1 + 4\sum_{n\geq 1}(d_{1,4}(n) - d_{3,4}(n))q^n.$$

If we extract those terms in which the power of q is congruent to 1 modulo 3, divide by $4q$ and replace q^3 by q, we obtain

$$\phi(q^3)\Omega(q) = \sum_{n\geq 0}(d_{1,4}(3n+1) - d_{3,4}(3n+1))q^n,$$

which gives (1.16), while if we extract those terms in which the power of q is congruent to 2 modulo 3, divide by $4q^2$ and replace q^3 by q, we obtain

$$\Omega(q)^2 = \sum_{n\geq 0}(d_{1,4}(3n+2) - d_{3,4}(3n+2))q^n,$$

which gives (1.26).

(3.16) is (essentially) equivalent to

$$\begin{aligned}(\phi(q^9) &+ 2q\Omega(q^3))(\phi(q^{18}) + 2q^2\Omega(q^6))\\ &= 1 + 2\sum_{n\geq 1}(d_{1,8}(n) + d_{3,8}(n) - d_{5,8}(n) - d_{7,8}(n))q^n.\end{aligned}$$

If we extract those terms in which the power of q is congruent to 1 modulo 3, divide by $2q$ and replace q^3 by q, we obtain

$$\phi(q^6)\Omega(q) = \sum_{n\geq 0}(d_{1,8}(3n+1) + d_{3,8}(3n+1) - d_{5,8}(3n+1) - d_{7,8}(3n+1))q^n,$$

which gives (1.18), while if we extract those terms in which the power of q is congruent to 2 modulo 3, divide by $2q^2$ and replace q^3 by q, we obtain

$$\phi(q^3)\Omega(q^2) = \sum_{n\geq 0}(d_{1,8}(3n+2) + d_{3,8}(3n+2) - d_{5,8}(3n+2) - d_{7,8}(3n+2))q^n,$$

which gives (1.17).

If we extract those terms in which the power of q is congruent to 0 modulo 3 and replace q^3 by q, we obtain

$$\phi(q^3)\phi(q^6)+4q\Omega(q)\Omega(q^2) = 1+2\sum_{n\geq 1}(d_{1,8}(3n)+d_{3,8}(3n)-d_{5,8}(3n)-d_{7,8}(3n))q^n,$$

while

$$\begin{aligned}\phi(q^3)\phi(q^6) &= 1 + 2\sum_{n\geq 1}(d_{1,8}(n) + d_{3,8}(n) - d_{5,8}(n) - d_{7,8}(n))q^{3n} \\ &= 1 + 2\sum_{n\geq 0}(d_{1,8}(n/3) + d_{3,8}(n/3) - d_{5,8}(n/3) - d_{7,8}(n/3))q^n.\end{aligned}$$

If we subtract the second from the first and divide by $4q$, we obtain

$$\begin{aligned}&\Omega(q)\Omega(q^2) \\ &\quad = \sum_{n\geq 0}(d_{1,8}(3n+3) + d_{3,8}(3n+3) - d_{5,8}(3n+3) - d_{7,8}(3n+3) \\ &\quad - (d_{1,8}((n+1)/3) + d_{3,8}((n+1)/3) - d_{5,8}((n+1)/3) \\ &\quad - d_{7,8}((n+1)/3)))/2\, q^n,\end{aligned}$$

which gives (1.27).

(3.17) is (essentially) equivalent to

$$\begin{aligned}&(\phi(q^9) + 2q\Omega(q^3))\phi(q^3) \\ &\quad = 1 + 2\sum_{n\geq 1}(d_{1,3}(n) - d_{2,3}(n))q^n + 4\sum_{n\geq 1}(d_{1,3}(n/4) - d_{2,3}(n/4))q^n.\end{aligned}$$

If we extract those terms in which the power of q is congruent to 1 modulo 3, divide by $2q$ and replace q^3 by q, we obtain

$$\begin{aligned}\phi(q)\Omega(q) &= \sum_{n\geq 0}(d_{1,3}(3n+1) - d_{2,3}(3n+1))q^n \\ &\quad + 2\sum_{n\geq 0}(d_{1,3}((3n+1)/4) - d_{2,3}((3n+1)/4))q^n,\end{aligned}$$

which gives (1.15).

(3.18) is equivalent to

$$(\Pi(q^3) + q\psi(q^9))^4 = \sum_{n\geq 0}\sigma(2n+1)q^n.$$

If we extract those terms in which the power of q is congruent to 2 modulo 3, divide by $6q^2$ and replace q^3 by q, we obtain

$$\psi(q^3)^2\Pi(q)^2 = \sum_{n\geq 0}\sigma(6n+5)/6\, q^n,$$

which gives (1.28).

(3.20) is (essentially) equivalent to

$$(\phi(q^9) + 2q\Omega(q^3))^4 = 1 + 8\sum_{n\geq 1}(\sigma(n) - 4\sigma(n/4))q^n.$$

If we extract those terms in which the power of q is congruent to 2 modulo 3, divide by $24q^2$ and replace q^3 by q, we obtain

$$\phi(q^3)^2\Omega(q)^2 = \sum_{n\geq 0}(\sigma(3n+2) - 4\sigma((3n+2)/4))/3\, q^n,$$

which gives (1.29).

References

[1] M. D. Hirschhorn, The number of representations of a number by various forms, *Discrete Math.*, **298** (2005) 205–211.
[2] H. Y. Lam, The number of representations by sums of squares and triangular numbers, *Integers*, **7(1)** (2007) #A28.
[3] R. S. Melham, Analogues of two classical theorems on the representations of a number, *Integers*, **8(1)** (2008) #A51.
[4] R. S. Melham, Analogues of Jacobi's two–square theorem, http://maths.science.uts.edu.au/maths/wiki/RayMelham

Ramanujan Rediscovered No. 14, 2010, pp. 125–138.

On the 2-, 3-, 4- and 6-Dissections of Ramanujan's Cubic Continued Fraction and its Reciprocal

Michael D. Hirschhorn[1] and Roselin[2]

[1]School of Mathematics and Statistics, UNSW, Sydney, Australia 2052
[2]Department of Mathematics and Statistics, Sam Higginbottom Institute of Agriculture, Technology & Sciences, (Formerly Allahabad Agricultural Institute), Allahabad 211 007, Uttar Pradesh, India
e-mail: m.hirschhorn@unsw.edu.au; roselinmaths@gmail.com

Abstract. We give 2-, 3-, 4- and 6-dissections of Ramanujan's cubic continued fraction, with each component a single product or if not a single product then a sum of two products.

Keywords. Ramanujan, cubic continued fraction, dissection.

2000 Subject Classification: 30B70

Dedicated to the memory of K. Venkatachaliengar
on the occasion of the Centenary of his birth

1. Introduction

The Rogers–Ramanujan continued fraction is

$$R(q) := \frac{1}{1} \, \frac{q}{+\,1} \, \frac{q^2}{+\,1} \, \frac{q^3}{+\,1} \, + \cdots = \frac{(q, q^4; q^5)_\infty}{(q^2, q^3; q^5)_\infty}, \quad |q| < 1.$$

Ramanujan [13, p. 50] gave the 2-dissections of this continued fraction and its reciprocal,

$$R(q) = \frac{(q^4, q^4, q^{16}, q^{16}; q^{20})_\infty}{(q^2, q^{10}, q^{10}, q^{18}; q^{20})_\infty} - q\frac{(q^4, q^6, q^{14}, q^{16}; q^{20})_\infty}{(q^8, q^{10}, q^{10}, q^{12}; q^{20})_\infty}$$

and

$$1/R(q) = \frac{(q^8, q^8, q^{12}, q^{12}; q^{20})_\infty}{(q^6, q^{10}, q^{10}, q^{14}; q^{20})_\infty} + q\frac{(q^2, q^8, q^{12}, q^{18}; q^{20})_\infty}{(q^4, q^{10}, q^{10}, q^{16}; q^{20})_\infty}$$

and these were first proved by Andrews [2].

Here we use the standard notation

$$(a; q)_\infty = \prod_{k=0}^{\infty}(1 - aq^k), \quad (a_1, \cdots, a_n; q)_\infty = \prod_{k=1}^{n}(a_k; q)_\infty.$$

Ramanujan [13, p. 50] also gave 5-dissections of $R(q)$ and its reciprocal, and these results were improved upon and proved by Hirschhorn [8]. In so doing, he was able to demonstrate the periodic behaviour of the sign of the coefficients in the series expansion of $R(q)$ and its reciprocal, first observed by Szekeres [16]. (It would be surprising if Ramanujan did not also observe this periodicity, given the tables in [13, p. 49].) In the same paper, Hirschhorn conjectured formulas for the 4-dissection of $R(q)$ and its reciprocal, and these were first proved by Lewis and Liu [12].

Gordon's continued fraction is

$$G(q) := \frac{1}{1+q} \, \frac{}{+} \, \frac{q^2}{1+q^3} \, \frac{}{+} \, \frac{q^4}{1+q^5} \, \frac{}{+} \cdots = \frac{(q,q^7;q^8)_\infty}{(q^3,q^5;q^8)_\infty}, \qquad |q|<1.$$

Hirschhorn [10] found the 8-dissection of $G(q)$ and its reciprocal, thereby demonstrating the periodicity of the sign of the coefficients in the expansion of $G(q)$ and its reciprocal, and in particular that certain coefficients are zero, a phenomenon first observed and proved by Richmond and Szekeres [14]. These themes have been generalised by Andrews and Bressoud [3], Alladi and Gordon [1] and Chan and Yesilyurt [5].

Ramanujan's cubic continued fraction is

$$RC(q) := \frac{1}{1} \, \frac{}{+} \, \frac{q+q^2}{1} \, \frac{}{+} \, \frac{q^2+q^4}{1} \, \frac{}{+} \cdots, \qquad |q|<1.$$

Ramanujan [13, p. 44] states that

$$RC(q) = \frac{(q,q^5;q^6)_\infty}{(q^3,q^3;q^6)_\infty}$$

and, indeed, this is a special case of a general result of Ramanujan [13, p. 41], proved by Hirschhorn [7], as is $R(q)$, as is also

$$\frac{1}{1} \, \frac{}{+} \, \frac{q+q^2}{1} \, \frac{}{+} \, \frac{q^4}{1} \, \frac{}{+} \, \frac{q^3+q^6}{1} \, \frac{}{+} \cdots = \frac{(q,q^7;q^8)_\infty}{(q^3,q^5;q^8)_\infty} = G(q).$$

The object of this note is to establish the following results. The first two are the 2- and 3-dissections of $RC(q)$ and its reciprocal. In these dissections each component is a single product. The third and fourth results are 4- and 6-dissections of $RC(q)$ and its reciprocal. In these dissections, not all components are single products. Those components that are not single products are shown to be expressible as sums of two products.

Note that the 2-and 4-dissections of the continued fraction $1/RC(q)$ were given by B. Srivastava [15].

Theorem 1.1. *The 2-dissections of $RC(q)$ and its reciprocal are given by*

$$RC(q) = \frac{(q^4;q^4)_\infty^2 (q^{12};q^{12})_\infty^2}{(q^6;q^6)_\infty^4} - q\frac{(q^2;q^2)_\infty^2 (q^{12};q^{12})_\infty^6}{(q^4;q^4)_\infty^2 (q^6;q^6)_\infty^6},$$

$$1/RC(q) = \frac{(q^4;q^4)_\infty^3}{(q^2;q^2)_\infty (q^6;q^6)_\infty (q^{12};q^{12})_\infty} + q\frac{(q^2;q^2)_\infty (q^{12};q^{12})_\infty^3}{(q^4;q^4)_\infty (q^6;q^6)_\infty^3}.$$

Theorem 1.2. *The 3-dissections of $RC(q)$ and its reciprocal are given by*

$$\begin{aligned} RC(q) &= \frac{(q^6;q^6)_\infty (q^{21},q^{33},q^{54};q^{54})_\infty}{(q^3;q^3)_\infty^2} \\ &\quad - q\frac{(q^6;q^6)_\infty (q^{15},q^{39},q^{54};q^{54})_\infty}{(q^3;q^3)_\infty^2} \\ &\quad - q^5\frac{(q^6;q^6)_\infty (q^{3},q^{51},q^{54};q^{54})_\infty}{(q^3;q^3)_\infty^2}, \\ 1/RC(q) &= \frac{(q^3;q^3)_\infty (-q^{12},-q^{15},q^{27};q^{27})_\infty}{(q^6;q^6)_\infty^2} \\ &\quad + q\frac{(q^3;q^3)_\infty (-q^{6},-q^{21},q^{27};q^{27})_\infty}{(q^6;q^6)_\infty^2} \\ &\quad + q^2\frac{(q^3;q^3)_\infty (-q^{3},-q^{24},q^{27};q^{27})_\infty}{(q^6;q^6)_\infty^2}. \end{aligned}$$

Theorem 1.3. *The 4-dissections of $RC(q)$ and its reciprocal are given by*

$$\begin{aligned} &RC(q) \\ &= \frac{(q^4;q^4)_\infty^2 (q^{24};q^{24})_\infty^{14}}{(q^{12};q^{12})_\infty^{12} (q^{48};q^{48})_\infty^4} \\ &\quad - q\left(\frac{(q^8;q^8)_\infty^4 (q^{24};q^{24})_\infty^4}{(q^{12};q^{12})_\infty^8} + q^4\frac{(q^4;q^4)_\infty^4 (q^{24};q^{24})_\infty^{12}}{(q^8;q^8)_\infty^4 (q^{12};q^{12})_\infty^{12}}\right) \\ &\quad + 4q^6\frac{(q^4;q^4)_\infty^2 (q^{24};q^{24})_\infty^2 (q^{48};q^{48})_\infty^4}{(q^{12},q^{12})_\infty^8} \\ &\quad + 2q^3\frac{(q^4;q^4)_\infty^2 (q^{24};q^{24})_\infty^8}{(q^{12};q^{12})_\infty^{10}}, \\ &1/RC(q) \\ &= \frac{(q^4;q^4)_\infty (q^{16};q^{16})_\infty^2 (q^{24};q^{24})_\infty^5}{(q^8;q^8)_\infty (q^{12};q^{12})_\infty^5 (q^{48};q^{48})_\infty^2} + q\frac{(q^8;q^8)_\infty^2 (q^{24};q^{24})_\infty^2}{(q^{12};q^{12})_\infty^4} \\ &\quad + q^2\frac{(q^8;q^8)_\infty^5 (q^{48};q^{48})_\infty^2}{(q^4;q^4)_\infty (q^{12};q^{12})_\infty^3 (q^{16};q^{16})_\infty^2 (q^{24};q^{24})_\infty} \\ &\quad - q^3\frac{(q^4;q^4)_\infty^2 (q^{24};q^{24})_\infty^6}{(q^8;q^8)_\infty^2 (q^{12};q^{12})_\infty^6}. \end{aligned}$$

Theorem 1.4. *The 6-dissections of $RC(q)$ and its reciprocal are given by*

$$RC(q) = \left(\frac{(q^{12};q^{12})_\infty^3 (q^{18},q^{18})_\infty^2 (q^{24},q^{30},q^{54};q^{54})_\infty}{(q^6;q^6)_\infty^5 (q^{36};q^{36})_\infty} \right.$$

$$\left. -q^6 \frac{(q^{12};q^{12})_\infty^2 (q^{36};q^{36})_\infty^2 (q^6,q^{48},q^{54};q^{54})_\infty}{(q^6;q^6)_\infty^4 (q^{18};q^{18})_\infty} \right)$$

$$-q\left(\frac{(q^{12};q^{12})_\infty^2 (q^{42},q^{66},q^{108};q^{108})_\infty^2}{(q^6;q^6)_\infty^4} \right.$$

$$\left. +2q^{12} \frac{(q^{12};q^{12})_\infty^2 (q^6,q^{30},q^{78},q^{102},q^{108},q^{108};q^{108})_\infty}{(q^6;q^6)_\infty^4} \right)$$

$$-q^2\left(\frac{(q^{12};q^{12})_\infty^3 (q^{18};q^{18})_\infty^2 (q^{12},q^{42},q^{54};q^{54})_\infty}{(q^6;q^6)_\infty^5 (q^{36};q^{36})_\infty} \right.$$

$$\left. -\frac{(q^{12};q^{12})_\infty^2 (q^{36};q^{36})_\infty^2 (q^{24},q^{30},q^{54};q^{54})_\infty}{(q^6;q^6)_\infty^4 (q^{18};q^{18})_\infty} \right)$$

$$+q^3\left(2\frac{(q^{12};q^{12})_\infty^2 (q^{30},q^{42},q^{66},q^{78},q^{108},q^{108};q^{108})_\infty}{(q^6;q^6)_\infty^4} \right.$$

$$\left. -q^{18} \frac{(q^{12};q^{12})_\infty^2 (q^6,q^{102},q^{108};q^{108})_\infty^2}{(q^6;q^6)_\infty^4} \right)$$

$$-q^4\left(\frac{(q^{12};q^{12})_\infty^3 (q^{18};q^{18})_\infty^2 (q^6,q^{48}.q^{54};q^{54})_\infty}{(q^6;q^6)_\infty^5 (q^{36};q^{36})_\infty} \right.$$

$$\left. +\frac{(q^{12};q^{12})_\infty^2 (q^{36};q^{36})_\infty^2 (q^{12},q^{42},q^{54};q^{54})_\infty}{(q^6;q^6)_\infty^4 (q^{18};q^{18})_\infty} \right)$$

$$-q^5\left(\frac{(q^{12};q^{12})_\infty^2 (q^{30},q^{78},q^{108};q^{108})_\infty^2}{(q^6;q^6)_\infty^4} \right.$$

$$\left. -2q^6 \frac{(q^{12};q^{12})_\infty^2 (q^6,q^{42},q^{66},q^{102},q^{108},q^{108};q^{108})_\infty}{(q^6;q^6)_\infty^4} \right),$$

$$1/RC(q) = \left(\frac{(q^{18};q^{18})_\infty (q^{48},q^{60},q^{108};q^{108})_\infty}{(q^6;q^6)_\infty^2 (q^{36};q^{36})_\infty} \right.$$

$$\left. -q^6 \frac{(q^{36};q^{36})_\infty^2 (q^{24},q^{84},q^{108};q^{108})_\infty}{(q^6;q^6)_\infty (q^{12};q^{12})_\infty (q^{18};q^{18})_\infty} \right)$$

$$+q\frac{(q^{12};q^{12})_\infty(q^{42},q^{66},q^{108};q^{108})_\infty}{(q^6;q^6)_\infty^2}$$

$$+q^2\left(\frac{(q^{36};q^{36})_\infty^2(q^{48},q^{60},q^{108};q^{108})_\infty}{(q^6;q^6)_\infty(q^{12};q^{12})_\infty(q^{18};q^{18})_\infty}\right.$$

$$\left.-q^6\frac{(q^{18};q^{18})_\infty(q^{12},q^{96},q^{108};q^{108})_\infty}{(q^6;q^6)_\infty^2(q^{36};q^{36})_\infty}\right)$$

$$-q^3\frac{(q^{12};q^{12})_\infty(q^{30},q^{78},q^{108};q^{108})_\infty}{(q^6;q^6)_\infty^2}$$

$$-q^4\left(\frac{(q^{18};q^{18})_\infty(q^{24},q^{84},q^{108};q^{108})_\infty}{(q^6;q^6)_\infty^2(q^{36};q^{36})_\infty}\right.$$

$$\left.+q^6\frac{(q^{36};q^{36})_\infty^2(q^{12},q^{96},q^{108};q^{108})_\infty}{(q^6;q^6)_\infty(q^{12};q^{12})_\infty(q^{18};q^{18})_\infty}\right)$$

$$-q^{11}\frac{(q^{12};q^{12})_\infty(q^{6},q^{102},q^{108};q^{108})_\infty}{(q^6;q^6)_\infty^2}.$$

Theorem 1.5. *If we write*

$$RC(q) =: \sum_{n\geq 0} a_n q^n, \quad 1/RC(q) =: \sum_{n\geq 0} b_n q^n,$$

then the sign of the a_n is periodic with period 3, and the sign of the b_n is periodic with period 6. Indeed,

$$b_{6n} > 0, \;\; a_{3n} = b_{6n+1} > 0, \;\; b_{6n+2} > 0, \;\; a_{3n+1} = b_{6n+3} < 0, \;\; b_{6n+4} < 0,$$

$$a_{3n+2} = b_{6n+5} < 0$$

except $a_2 = b_5 = b_8 = 0$.

2. Preliminaries

Lemma 2.1. *The following 2-dissections hold:*

$$\frac{(q^3;q^3)_\infty^3}{(q;q)_\infty} = \frac{(q^4;q^4)_\infty^3(q^6;q^6)_\infty^2}{(q^2;q^2)_\infty^2(q^{12};q^{12})_\infty} + q\frac{(q^{12};q^{12})_\infty^3}{(q^4;q^4)_\infty},$$

$$\frac{(q;q)_\infty}{(q^3;q^3)_\infty^3} = \frac{(q^2;q^2)_\infty(q^4;q^4)_\infty^2(q^{12};q^{12})_\infty^2}{(q^6;q^6)_\infty^7} - q\frac{(q^2;q^2)_\infty^3(q^{12};q^{12})_\infty^6}{(q^4;q^4)_\infty^2(q^6;q^6)_\infty^9}.$$

Proof. Let

$$c^*(q) = \frac{1}{3} \sum_{m,n=-\infty}^{\infty} q^{m^2+mn+n^2+m+n}.$$

It is shown in Hirschhorn, Garvan and Borwein [11], using nothing deeper than Jacobi's triple product identity, that [11, (1.7), (1.36)],

$$c^*(q) = \frac{(q^3;q^3)_\infty^3}{(q;q)_\infty} = \frac{(q^4;q^4)_\infty^3 (q^6;q^6)_\infty^2}{(q^2;q^2)_\infty^2 (q^{12};q^{12})_\infty} + q\, c^*(q^4).$$

This is the first identity. Also,

$$\begin{aligned}
\frac{(q;q)_\infty}{(q^3;q^3)_\infty^3} &= \frac{1}{c^*(q)} = \frac{c^*(-q)}{c^*(q)c^*(-q)} \\
&= \frac{(q^2;q^2)_\infty^3 (q^{12};q^{12})_\infty^3}{(q^4;q^4)_\infty (q^6;q^6)_\infty^9} \left\{ \frac{(q^4;q^4)_\infty^3 (q^6;q^6)_\infty^2}{(q^2;q^2)_\infty^2 (q^{12};q^{12})_\infty} - q\frac{(q^{12};q^{12})_\infty^3}{(q^4;q^4)_\infty} \right\} \\
&= \frac{(q^2;q^2)_\infty (q^4;q^4)_\infty^2 (q^{12};q^{12})_\infty^2}{(q^6;q^6)_\infty^7} - q\frac{(q^2;q^2)_\infty^3 (q^{12};q^{12})_\infty^6}{(q^4;q^4)_\infty^2 (q^6;q^6)_\infty^9},
\end{aligned}$$

which is the second identity. □

Lemma 2.2. *The following 3-dissections hold:*

$$\begin{aligned}
\frac{(q^2;q^2)_\infty}{(q;q)_\infty} &= \frac{(q^6;q^6)_\infty}{(q^3;q^3)_\infty^2}\{(-q^{12},-q^{15},q^{27};q^{27})_\infty + q(-q^6,-q^{21},q^{27};q^{27})_\infty \\
&\quad + q^2(-q^3,-q^{24},q^{27};q^{27})_\infty\}, \\
\frac{(q;q)_\infty}{(q^2;q^2)_\infty} &= \frac{(q^3;q^3)_\infty}{(q^6;q^6)_\infty^2}\{(q^{21},q^{33},q^{54};q^{54})_\infty - q(q^{15},q^{39},q^{54};q^{54})_\infty \\
&\quad - q^5(q^3,q^{51},q^{54};q^{54})_\infty\}.
\end{aligned}$$

Proof. The first identity follows from Entry 31 [4, p. 48] by setting $a = q$, $b = q^2$ and $n = 3$, the second by setting $a = -q$, $b = -q^5$ and $n = 3$. □

Lemma 2.3. *The following 2-dissection holds:*

$$\frac{1}{(q;q)_\infty^4} = \frac{(q^4;q^4)_\infty^{14}}{(q^2;q^2)_\infty^{14}(q^8;q^8)_\infty^4} + 4q\frac{(q^4;q^4)_\infty^2 (q^8;q^8)_\infty^4}{(q^2;q^2)_\infty^{10}}.$$

Proof. With

$$\phi(q) = \sum_{-\infty}^{\infty} q^{n^2} = \frac{(q^2;q^2)_\infty^5}{(q;q)_\infty^2 (q^4;q^4)_\infty^2} \quad \text{and} \quad \psi(q) = \sum_{n\ge 0} q^{(n^2+n)/2} = \frac{(q^2;q^2)_\infty^2}{(q;q)_\infty}$$

it is easy to show that (Berndt [4, p. 40], Hirschhorn [9])

$$\phi(q)^2 = \phi(q^2)^2 + 4q\psi(q^4)^2.$$

That is,

$$\frac{(q^2;q^2)_\infty^{10}}{(q;q)_\infty^4(q^4;q^4)_\infty^4} = \frac{(q^4;q^4)_\infty^{10}}{(q^2;q^2)_\infty^4(q^8;q^8)_\infty^4} + 4q\frac{(q^8;q^8)_\infty^4}{(q^4;q^4)_\infty^2}.$$

The result follows. □

Lemma 2.4. *The following 2-dissection holds:*

$$\frac{1}{(q;q)_\infty(q^3;q^3)_\infty} = \frac{(q^8;q^8)_\infty^2(q^{12};q^{12})_\infty^5}{(q^2;q^2)_\infty^2(q^4;q^4)_\infty(q^6;q^6)_\infty^4(q^{24};q^{24})_\infty^2} + q\frac{(q^4;q^4)_\infty^5(q^{24};q^{24})_\infty^2}{(q^2;q^2)_\infty^4(q^6;q^6)_\infty^2(q^8;q^8)_\infty^2(q^{12};q^{12})_\infty}.$$

Proof. It can be shown that (Berndt [4, (36.8)], Cooper and Hirschhorn [6, lemma (xxxiii)])

$$\psi(q)\psi(q^3) = \phi(q^6)\psi(q^4) + q\phi(q^2)\psi(q^{12}).$$

That is,

$$\frac{(q^2;q^2)_\infty^2}{(q;q)_\infty}\frac{(q^6;q^6)_\infty^2}{(q^3;q^3)_\infty} = \frac{(q^{12};q^{12})_\infty^5}{(q^6;q^6)_\infty^2(q^{24};q^{24})_\infty^2}\frac{(q^8;q^8)_\infty^2}{(q^4;q^4)_\infty} + q\frac{(q^4;q^4)_\infty^5}{(q^2;q^2)_\infty^2(q^8;q^8)_\infty^2}\frac{(q^{24};q^{24})_\infty^2}{(q^{12};q^{12})_\infty}.$$

The result follows. □

Lemma 2.5. *The following 3-dissections hold:*

$$\frac{(q^2;q^2)_\infty^2}{(q;q)_\infty} = \frac{(q^6;q^6)_\infty(q^9;q^9)_\infty^2}{(q^3;q^3)_\infty(q^{18};q^{18})_\infty} + q\frac{(q^{18};q^{18})_\infty^2}{(q^9;q^9)_\infty},$$

$$(q,q)_\infty = (q^{12},q^{15},q^{27};q^{27})_\infty - q(q^6,q^{21},q^{27};q^{27})_\infty - q^2(q^3,q^{24},q^{27};q^{27})_\infty.$$

Proof. The first identity follows from the first equality in Corollary (ii) [4, p. 49], the second from Entry 31 [4, p. 48] by setting $a=-q, b=-q^2$ and $n=3$. □

3. Proofs of theorems

Proof (Theorem (1.1)). We have, by Lemma (2.1),

$$
\begin{aligned}
RC(q) &= \frac{(q;q^2)_\infty}{(q^3;q^6)^3_\infty} = \frac{(q;q)_\infty(q^6;q^6)^3_\infty}{(q^2;q^2)_\infty(q^3;q^3)^3_\infty}\\
&= \frac{(q^6;q^6)^3_\infty}{(q^2;q^2)_\infty}\left\{\frac{(q^2;q^2)_\infty(q^4;q^4)^2_\infty(q^{12};q^{12})^2_\infty}{(q^6;q^6)^7_\infty} - q\frac{(q^2;q^2)^3_\infty(q^{12};q^{12})^6_\infty}{(q^4;q^4)^2_\infty(q^6;q^6)^9_\infty}\right\}\\
&= \frac{(q^4;q^4)^2_\infty(q^{12};q^{12})^2_\infty}{(q^6;q^6)^4_\infty} - q\frac{(q^2;q^2)^2_\infty(q^{12};q^{12})^6_\infty}{(q^4;q^4)^2_\infty(q^6;q^6)^6_\infty}
\end{aligned}
$$

and

$$
\begin{aligned}
1/RC(q) &= \frac{(q^2;q^2)_\infty(q^3;q^3)^3_\infty}{(q^6;q^6)^3_\infty(q;q)_\infty}\\
&= \frac{(q^2;q^2)_\infty}{(q^6;q^6)^3_\infty}\left\{\frac{(q^4;q^4)^3_\infty(q^6;q^6)^2_\infty}{(q^2;q^2)^2_\infty(q^{12};q^{12})_\infty} + q\frac{(q^{12};q^{12})^3_\infty}{(q^4;q^4)_\infty}\right\}\\
&= \frac{(q^4;q^4)^3_\infty}{(q^2;q^2)_\infty(q^6;q^6)_\infty(q^{12};q^{12})_\infty} + q\frac{(q^2;q^2)_\infty(q^{12};q^{12})^3_\infty}{(q^4;q^4)_\infty(q^6;q^6)^3_\infty}. \qquad \square
\end{aligned}
$$

Proof (Theorem (1.2)). We have, by Lemma (2.2),

$$
\begin{aligned}
RC(q) &= \frac{(q^6;q^6)^3_\infty(q;q)_\infty}{(q^3;q^3)^3_\infty(q^2;q^2)_\infty}\\
&= \frac{(q^6;q^6)_\infty}{(q^3;q^3)^2_\infty}\{(q^{21},q^{33},q^{54};q^{54})_\infty - q(q^{15},q^{39},q^{54};q^{54})_\infty\\
&\quad - q^5(q^3,q^{51},q^{54};q^{54})_\infty\}\\
&= \frac{(q^6;q^6)_\infty(q^{21},q^{33},q^{54};q^{54})_\infty}{(q^3;q^3)^2_\infty} - q\frac{(q^6;q^6)_\infty(q^{15},q^{39},q^{54};q^{54})_\infty}{(q^3;q^3)^2_\infty}\\
&\quad - q^5\frac{(q^6;q^6)_\infty(q^3,q^{51},q^{54};q^{54})_\infty}{(q^3;q^3)^2_\infty}
\end{aligned}
$$

and

$$
1/RC(q) = \frac{(q^3;q^3)^3_\infty(q^2;q^2)_\infty}{(q^6;q^6)^3_\infty(q;q)_\infty}
$$

$$= \frac{(q^3;q^3)_\infty}{(q^6;q^6)_\infty^2}\{(-q^{12},-q^{15},q^{27};q^{27})_\infty + q(-q^6,-q^{21},q^{27};q^{27})_\infty$$
$$+ q^2(-q^3,-q^{24},q^{27};q^{27})_\infty\}$$
$$= \frac{(q^3;q^3)_\infty(-q^{12},-q^{15},q^{27};q^{27})_\infty}{(q^6;q^6)_\infty^2} + q\frac{(q^3;q^3)_\infty(-q^6,-q^{21},q^{27};q^{27})_\infty}{(q^6;q^6)_\infty^2}$$
$$+ q^2\frac{(q^3;q^3)_\infty(-q^3,-q^{24},q^{27};q^{27})_\infty}{(q^6;q^6)_\infty^2}. \qquad \square$$

Proof (Theorem (1.3)). We have, by Theorem (1.1) and Lemma (2.3) with q^6 for q and Lemma (2.1) with q^2 for q,

$$RC(q)$$
$$= \frac{(q^4;q^4)_\infty^2(q^{12};q^{12})_\infty^2}{(q^6;q^6)_\infty^4} - q\frac{(q^2;q^2)_\infty^2(q^{12};q^{12})_\infty^6}{(q^4;q^4)_\infty^2(q^6;q^6)_\infty^6}$$
$$= (q^4;q^4)_\infty^2(q^{12};q^{12})_\infty^2$$
$$\times\left(\frac{(q^{24};q^{24})_\infty^{14}}{(q^{12};q^{12})_\infty^{14}(q^{48};q^{48})_\infty^4} + 4q^6\frac{(q^{24};q^{24})_\infty^2(q^{48};q^{48})_\infty^4}{(q^{12};q^{12})_\infty^{10}}\right) - q\frac{(q^{12};q^{12})_\infty^6}{(q^4;q^4)_\infty^2}$$
$$\times\left(\frac{(q^4;q^4)_\infty(q^8;q^8)_\infty^2(q^{24};q^{24})_\infty^2}{(q^{12};q^{12})_\infty^7} - q^2\frac{(q^4;q^4)_\infty^3(q^{24};q^{24})_\infty^6}{(q^8;q^8)_\infty^2(q^{12};q^{12})_\infty^9}\right)^2$$
$$= \frac{(q^4;q^4)_\infty^2(q^{24};q^{24})_\infty^{14}}{(q^{12};q^{12})_\infty^{12}(q^{48};q^{48})_\infty^4}$$
$$- q\left(\frac{(q^8;q^8)_\infty^4(q^{24};q^{24})_\infty^4}{(q^{12};q^{12})_\infty^8} + q^4\frac{(q^4;q^4)_\infty^4(q^{24};q^{24})_\infty^{12}}{(q^8;q^8)_\infty^4(q^{12};q^{12})_\infty^{12}}\right)$$
$$+ 4q^6\frac{(q^4;q^4)_\infty^2(q^{24};q^{24})_\infty^2(q^{48};q^{48})_\infty^4}{(q^{12},q^{12})_\infty^8} + 2q^3\frac{(q^4;q^4)_\infty^2(q^{24};q^{24})_\infty^8}{(q^{12};q^{12})_\infty^{10}}$$

and, by Theorem (1.1) and Lemma (2.4) with q^2 for q and Lemma (2.1) with q^2 for q,

$$1/RC(q) = \frac{(q^4;q^4)_\infty^3}{(q^2;q^2)_\infty(q^6;q^6)_\infty(q^{12};q^{12})_\infty} + q\frac{(q^2;q^2)_\infty(q^{12};q^{12})_\infty^3}{(q^4;q^4)_\infty(q^6;q^6)_\infty^3}$$
$$= \frac{(q^4;q^4)_\infty^3}{(q^{12};q^{12})_\infty}\left\{\frac{(q^{16};q^{16})_\infty^2(q^{24};q^{24})_\infty^5}{(q^4;q^4)_\infty^2(q^8;q^8)_\infty(q^{12};q^{12})_\infty^4(q^{48};q^{48})_\infty^2}\right.$$
$$\left.+q^2\frac{(q^8;q^8)_\infty^5(q^{48};q^{48})_\infty^2}{(q^4;q^4)_\infty^4(q^{12};q^{12})_\infty^2(q^{16};q^{16})_\infty^2(q^{24};q^{24})_\infty}\right\}$$

$$
\begin{aligned}
&+q\frac{(q^{12};q^{12})_\infty^3}{(q^4;q^4)_\infty}\left\{\frac{(q^4;q^4)_\infty(q^8;q^8)_\infty^2(q^{24};q^{24})_\infty^2}{(q^{12};q^{12})_\infty^7}\right.\\
&\left.-q^2\frac{(q^4;q^4)_\infty^3(q^{24};q^{24})_\infty^6}{(q^8;q^8)_\infty^2(q^{12};q^{12})_\infty^9}\right\}\\
&=\frac{(q^4;q^4)_\infty(q^{16};q^{16})_\infty^2(q^{24};q^{24})_\infty^5}{(q^8;q^8)_\infty(q^{12};q^{12})_\infty^5(q^{48};q^{48})_\infty^2}+q\frac{(q^8;q^8)_\infty^2(q^{24};q^{24})_\infty^2}{(q^{12};q^{12})_\infty^4}\\
&+q^2\frac{(q^8;q^8)_\infty^5(q^{48};q^{48})_\infty^2}{(q^4;q^4)_\infty(q^{12};q^{12})_\infty^3(q^{16};q^{16})_\infty^2(q^{24};q^{24})_\infty}\\
&-q^3\frac{(q^4;q^4)_\infty^2(q^{24};q^{24})_\infty^6}{(q^8;q^8)_\infty^2(q^{12};q^{12})_\infty^6}.
\end{aligned}
$$

□

Proof (Theorem (1.4)). We have, by Theorem (1.1) and Lemmas (2.2) and (2.5),

$$
\begin{aligned}
RC(q)&=\frac{(q^{12};q^{12})_\infty^2}{(q^6;q^6)_\infty^4}\left(\frac{(q^{12};q^{12})_\infty(q^{18};q^{18})_\infty^2}{(q^6;q^6)_\infty(q^{36};q^{36})_\infty}+q^2\frac{(q^{36};q^{36})_\infty^2}{(q^{18};q^{18})_\infty}\right)\\
&\times((q^{24},q^{30},q^{54};q^{54})_\infty-q^2(q^{12},q^{42},q^{54};q^{54})_\infty\\
&-q^4(q^6,q^{48},q^{54};q^{54})_\infty)-q\frac{(q^{12};q^{12})_\infty^6}{(q^6;q^6)_\infty^6}\\
&\times\left(\frac{(q^6;q^6)_\infty}{(q^{12};q^{12})_\infty^2}\{(q^{42},q^{66},q^{108};q^{108})_\infty\right.\\
&\left.-q^2(q^{30},q^{78},q^{108};q^{108})_\infty-q^{10}(q^6,q^{102},q^{108};q^{108})_\infty\}\right)^2\\
&=\left(\frac{(q^{12};q^{12})_\infty^3(q^{18},q^{18})_\infty^2(q^{24},q^{30},q^{54};q^{54})_\infty}{(q^6;q^6)_\infty^5(q^{36};q^{36})_\infty}\right.\\
&\left.-q^6\frac{(q^{12};q^{12})_\infty^2(q^{36};q^{36})_\infty^2(q^6,q^{48},q^{54};q^{54})_\infty}{(q^6;q^6)_\infty^4(q^{18};q^{18})_\infty}\right)\\
&-q\left(\frac{(q^{12};q^{12})_\infty^2(q^{42},q^{66},q^{108};q^{108})_\infty^2}{(q^6;q^6)_\infty^4}\right.\\
&\left.+2q^{12}\frac{(q^{12};q^{12})_\infty^2(q^6,q^{30},q^{78},q^{102},q^{108},q^{108};q^{108})_\infty}{(q^6;q^6)_\infty^4}\right)
\end{aligned}
$$

$$- q^2\left(\frac{(q^{12};q^{12})^3_\infty(q^{18};q^{18})^2_\infty(q^{12},q^{42},q^{54};q^{54})_\infty}{(q^6;q^6)^5_\infty(q^{36};q^{36})_\infty}\right.$$

$$\left.-\frac{(q^{12};q^{12})^2_\infty(q^{36};q^{36})^2_\infty(q^{24},q^{30},q^{54};q^{54})_\infty}{(q^6;q^6)^4_\infty(q^{18};q^{18})_\infty}\right)$$

$$+q^3\left(2\frac{(q^{12};q^{12})^2_\infty(q^{30},q^{42},q^{66},q^{78},q^{108},q^{108};q^{108})_\infty}{(q^6;q^6)^4_\infty}\right.$$

$$\left.-q^{18}\frac{(q^{12};q^{12})^2_\infty(q^6,q^{102},q^{108};q^{108})^2_\infty}{(q^6;q^6)^4_\infty}\right)$$

$$- q^4\left(\frac{(q^{12};q^{12})^3_\infty(q^{18};q^{18})^2_\infty(q^6,q^{48}.q^{54};q^{54})_\infty}{(q^6;q^6)^5_\infty(q^{36};q^{36})_\infty}\right.$$

$$\left.+\frac{(q^{12};q^{12})^2_\infty(q^{36};q^{36})^2_\infty(q^{12},q^{42},q^{54};q^{54})_\infty}{(q^6;q^6)^4_\infty(q^{18};q^{18})_\infty}\right)$$

$$- q^5\left(\frac{(q^{12};q^{12})^2_\infty(q^{30},q^{78},q^{108};q^{108})^2_\infty}{(q^6;q^6)^4_\infty}\right.$$

$$\left.-2q^6\frac{(q^{12};q^{12})^2_\infty(q^6,q^{42},q^{66},q^{102},q^{108},q^{108};q^{108})_\infty}{(q^6;q^6)^4_\infty}\right),$$

and

$$1/RC(q)=\frac{1}{(q^6;q^6)_\infty(q^{12};q^{12})_\infty}\left(\frac{(q^{12};q^{12})_\infty(q^{18};q^{18})^2_\infty}{(q^6;q^6)_\infty(q^{36};q^{36})_\infty}+q^2\frac{(q^{36};q^{36})^2_\infty}{(q^{18};q^{18})_\infty}\right)$$

$$\times((q^{48},q^{60},q^{108};q^{108})_\infty-q^4(q^{24},q^{84},q^{108};q^{108})_\infty$$

$$-q^8(q^{12},q^{96},q^{108};q^{108})_\infty)+q\frac{(q^{12};q^{12})^3_\infty}{(q^6;q^6)^3_\infty}$$

$$\times\left(\frac{(q^6;q^6)_\infty}{(q^{12};q^{12})^2_\infty}\{(q^{42},q^{66},q^{108};q^{108})_\infty-q^2(q^{30},q^{78},q^{108};q^{108})_\infty\right.$$

$$\left.-q^{10}(q^6,q^{102},q^{108};q^{108})_\infty\}\right)$$

$$= \left(\frac{(q^{18};q^{18})_\infty^2 (q^{48},q^{60},q^{108};q^{108})_\infty}{(q^6;q^6)_\infty^2 (q^{36};q^{36})_\infty} \right.$$

$$\left. -q^6 \frac{(q^{36};q^{36})_\infty^2 (q^{24},q^{84},q^{108};q^{108})_\infty}{(q^6;q^6)_\infty (q^{12};q^{12})_\infty (q^{18};q^{18})_\infty} \right)$$

$$+ q \frac{(q^{12};q^{12})_\infty (q^{42},q^{66},q^{108};q^{108})_\infty}{(q^6;q^6)_\infty^2}$$

$$+ q^2 \left(\frac{(q^{36};q^{36})_\infty^2 (q^{48},q^{60},q^{108};q^{108})_\infty}{(q^6;q^6)_\infty (q^{12};q^{12})_\infty (q^{18};q^{18})_\infty} \right.$$

$$\left. -q^6 \frac{(q^{18};q^{18})_\infty^2 (q^{12},q^{96},q^{108};q^{108})_\infty}{(q^6;q^6)_\infty^2 (q^{36};q^{36})_\infty} \right)$$

$$- q^3 \frac{(q^{12};q^{12})_\infty (q^{30},q^{78},q^{108};q^{108})_\infty}{(q^6;q^6)_\infty^2}$$

$$- q^4 \left(\frac{(q^{18};q^{18})_\infty^2 (q^{24},q^{84},q^{108};q^{108})_\infty}{(q^6;q^6)_\infty^2 (q^{36};q^{36})_\infty} \right.$$

$$\left. +q^6 \frac{(q^{36};q^{36})_\infty^2 (q^{12},q^{96},q^{108};q^{108})_\infty}{(q^6;q^6)_\infty (q^{12};q^{12})_\infty (q^{18};q^{18})_\infty} \right)$$

$$- q^{11} \frac{(q^{12};q^{12})_\infty (q^{6},q^{102},q^{108};q^{108})_\infty}{(q^6;q^6)_\infty^2}. \qquad \square$$

Proof (Theorem (1.5)). We have, from Theorems (1.2) and (1.4),

$$\sum_{n\ge 0} a_{3n} q^n = \sum_{n\ge 0} b_{6n+1} q^n$$

$$= 1/(q,q,q^2,q^3,q^3,q^4,q^5,q^5,q^6,q^7,q^8,q^9,q^9,q^{10},$$

$$q^{11},q^{12},q^{13},q^{13},q^{14},q^{15},q^{15},q^{16},q^{17},q^{17};q^{18})_\infty,$$

$$\sum_{n\ge 0} a_{3n+1} q^n = \sum_{n\ge 0} b_{6n+3} q^n$$

$$= -1/(q,q,q^2,q^3,q^3,q^4,q^5,q^6,q^7,q^7,q^8,q^9,q^9,q^{10},$$

$$q^{11},q^{11},q^{12},q^{13},q^{14},q^{15},q^{15},q^{16},q^{17},q^{17};q^{18})_\infty,$$

$$\sum_{n\geq 0} a_{3n+2}q^n = \sum_{n\geq 0} b_{6n+5}q^n$$
$$= -q/(q,q^2,q^3,q^3,q^4,q^5,q^5,q^6,q^7,q^7,q^8,q^9,q^9,q^{10},$$
$$q^{11},q^{11},q^{12},q^{13},q^{13},q^{14},q^{15},q^{15},q^{16},q^{17};q^{18})_\infty.$$

Also, from Theorem (1.2),

$$\sum_{n\geq 0} b_{3n}q^n = \frac{(-q^4,-q^5,q^9;q^9)_\infty}{\psi(q)} = \frac{(-q^4,-q^5,q^9;q^9)_\infty}{(-q,-q^2,q^4;q^4)_\infty}$$
$$= \frac{(-q^4,-q^{14},q^{18};q^{18})_\infty}{(q^4;q^4)_\infty}\frac{(-q^5,q^9,-q^{13};q^{18})_\infty}{(-q;q^2)_\infty},$$

so

$$\sum_{n\geq 0}(-1)^n b_{3n}q^n = \frac{(-q^4,-q^{14},q^{18};q^{18})_\infty}{(q^4;q^4)_\infty}\frac{(q^5,-q^9,q^{13};q^{18})_\infty}{(q;q^2)_\infty}$$
$$= \frac{(-q^9;q^{18})_\infty \sum_{-\infty}^{\infty} q^{9n^2-5n}}{(q^4;q^4)_\infty(q,q^3,q^7,q^9,q^{11},q^{15},q^{17};q^{18})_\infty}$$
$$= f(q^4,q^{14})f(q^9,q^{27})/(q,q^3,q^4,q^7,q^8,q^9,q^{11},q^{12},$$
$$q^{15},q^{16},q^{17},q^{19},q^{20},q^{21},q^{24},q^{25},q^{27},q^{28},q^{29},$$
$$q^{32},q^{33},q^{35},q^{36},q^{36};q^{36})_\infty,$$

and similarly,

$$\sum_{n\geq 0}(-1)^n b_{3n+1}q^n$$
$$= f(q^2,q^{16})f(q^9,q^{27})/(q,q^3,q^4,q^5,q^8,q^9,q^{12},q^{13},q^{15},q^{16},q^{17},$$
$$q^{19},q^{20},q^{21},q^{23},q^{24},q^{27},q^{28},q^{31},q^{32},q^{33},q^{35},q^{36},q^{36};q^{36})_\infty,$$

$$\sum_{n\geq 0}(-1)^n b_{3n+2}q^n$$
$$= f(q^8,q^{10})f(q^9,q^{27})/(q,q^3,q^4,q^5,q^7,q^8,q^9,q^{11},q^{12},q^{13},q^{15},q^{16},$$
$$q^{20},q^{21},q^{23},q^{24},q^{25},q^{27},q^{28},q^{29},q^{31},q^{32},q^{33},q^{36},q^{36};q^{36})_\infty.$$

The results follow. □

Acknowledgement

The latter part of the proof of Theorem (1.5) is essentially the proof of Chan and Yesilyurt [5, §4].

References

[1] K. Alladi and B. Gordon, Vanishing coefficients in the expansion of products of Rogers–Ramanujan type, The Rademacher Legacy to Mathematics, *Contemporary Mathematics*, **166** (1994) 129–139.

[2] G. E. Andrews, Ramanujan's "lost" notebook III: The Rogers–Ramanujan continued fraction, *Advances in Mathematics*, **41** (1981) 186–208.

[3] G. E. Andrews and D. Bressoud, Vanishing coefficients in infinite product expansions, *Journal of the Australian Mathematical Society Series A*, **27** (1979) 199–202.

[4] B. C. Berndt, Ramanujan's Notebooks, Part III, Springer–Verlag, New York (1990).

[5] S. H. Chan and H. Yesilyurt, The periodicity of the signs of the coefficients of certain infinite products, *Pacific Journal of Mathematics*, **225** (2006) 13–32.

[6] S. Cooper and M. D. Hirschhorn, Results of Hurwitz type for three squares, *Discrete Mathematics*, **274** (2004) 9–24.

[7] M. D. Hirschhorn, A continued fraction of Ramanujan, *Journal of the Australian Mathematical Society Series A*, **29** (1980) 80–86.

[8] M. D. Hirschhorn, On the expansion of Ramanujan's continued fraction, *The Ramanujan Journal*, **2** (1998) 521–527.

[9] M. D. Hirschhorn, Arithmetic consequences of Jacobi's two–square theorem, *The Ramanujan Journal*, **4** (2000) 51–57.

[10] M. D. Hirschhorn, On the expansion of a continued fraction of Gordon, *The Ramanujan Journal*, **5** (2001) 369–375.

[11] M. D. Hirschhorn, F. Garvan and J. Borwein, Cubic analogues of the Jacobian theta function $\theta(z, q)$, *Canadian Journal of Mathematics*, **45** (1993) 673–694.

[12] R. P. Lewis and Z.-G. Liu, A conjecture of Hirschhorn on the 4-dissection of Ramanujan's continued fraction, *The Ramanujan Journal*, **4** (2000) 347–352.

[13] S. Ramanujan, The Lost Notebook and Other Unpublished Papers, Narosa, Delhi (1988).

[14] B. Richmond and G. Szekeres, The Taylor coefficients of certain infinite products, *Acta Sci. Math. (Szeged)*, **40** (1978) 347–369.

[15] B. Srivastava, On 2-dissection and 4-dissection of Ramanujan's cubic continued fraction and identities, *Tamsui Oxford Journal of Mathematical Sciences*, **23(3)** (2007) 305–315.

[16] G. Szekeres, Private communication (1969).

Ramanujan Rediscovered No. 14, 2010, pp. 139–156.

Coupled Systems of Differential Equations for Modular forms of Level *n*

Tim Huber

Department of Mathematics, University of Texas-Pan American, Edinburg, TX, USA
e-mail: hubertj@utpa.edu

Abstract. In this paper, we derive a general analogue of Ramanujan's differential equations for the classical Eisenstein series. Several such analogues were studied by V. Ramamani in the 1970s and have been more recently investigated by R. Maier in [32]. We extend the work of Ramamani and Maier by proving that an infinite class of modular forms on subgroups of the full modular group satisfy a coupled system of nonlinear differential equations analogous to the differential equations satisfied by the Eisenstein series on $SL(2, \mathbb{Z})$. Some consequences of these differential equations are given, motivated by results that hold for the classical Eisenstein series.

Keywords. Eisenstein series, differential equations, elliptic modular forms, cubic theta functions.

2000 Subject Classification: Primary: 30C15; Secondary: 11F11

In memory of Prof. K. Venkatachaliengar on the centenary of his birth

1. Introduction

Define, for $|q| < 1$,

$$P(q) := 1 - 24\sum_{n=1}^{\infty}\frac{nq^n}{1-q^n}, \quad Q(q) := 1 + 240\sum_{n=1}^{\infty}\frac{n^3q^n}{1-q^n},$$

$$R(q) := 1 - 504\sum_{n=1}^{\infty}\frac{n^5q^n}{1-q^n}.$$

In his famous paper [38], Ramanujan employed the trigonometric series identities

$$\left(\frac{1}{8}\cot^2\theta + \frac{1}{12} + \sum_{n=1}^{\infty}\frac{nq^n}{1-q^n}(1-\cos(2n\theta))\right)^2$$
$$= \left(\frac{1}{8}\cot^2\theta + \frac{1}{12}\right)^2 + \frac{1}{12}\sum_{n=1}^{\infty}\frac{n^3q^n}{1-q^n}(5+\cos(2n\theta)) \qquad (1.1)$$

and

$$\left(\frac{1}{4}\cot\theta+\sum_{n=1}^{\infty}\frac{q^n\sin(2n\theta)}{1-q^n}\right)^2$$
$$=\left(\frac{1}{4}\cot\theta\right)^2+\sum_{n=1}^{\infty}\frac{q^n\cos(2n\theta)}{(1-q^n)^2}+\frac{1}{2}\sum_{n=1}^{\infty}\frac{nq^n}{1-q^n}(1-\cos(2n\theta)) \tag{1.2}$$

to give a beautiful proof that $P(q)$, $Q(q)$, and $R(q)$ satisfy the differential equations

$$q\frac{dP}{dq}=\frac{P^2-Q}{12},\quad q\frac{dQ}{dq}=\frac{PQ-R}{3},\quad q\frac{dR}{dq}=\frac{PR-Q^2}{2}. \tag{1.3}$$

The three series $P(q)$, $Q(q)$, and $R(q)$ and their differential equations (1.3) are of fundamental importance in many areas of number theory and Ramanujan's own work. We cite just a small number of books and papers where the differential equations (1.3) arise: [5, Chapters 11–16], [6, Chapter 15], [7, pp. 484–486], [8, Chapters 4–6], [9–13,15–17,29,30,36,38]. Furthermore, Q and R are the building blocks in the theory of modular forms; e.g., see R. A. Rankin's text [40] for a development of the theory of Eisenstein series from the viewpoint of modular forms.

Given the extensive applications of the differential equations (1.3), a natural question to ask is whether or not similar differential equations hold for Eisenstein series on subgroups of the modular group and if these differential equations can be applied to derive classical results such as those that follow by employing (1.3). The question was answered in the affirmative by V. Ramamani, a student of Venkatacheliangar. In her Ph.D. dissertation [34,35], Ramamani considered

$$e(q):=1+24\sum_{n=1}^{\infty}\frac{nq^n}{1+q^n},\quad \mathcal{P}(q):=1+8\sum_{n=1}^{\infty}\frac{(-1)^{n-1}nq^n}{1-q^n}, \tag{1.4}$$

$$\mathcal{Q}(q):=1-16\sum_{n=1}^{\infty}\frac{(-1)^{n-1}n^3q^n}{1-q^n}. \tag{1.5}$$

The Lambert series $e(q)$ and $\mathcal{Q}(q)$ are normalized modular forms on $\Gamma_0(2)$ of weight 2 and 4, respectively [10, Lemma 3.3], [19, Theorem 1.1], where

$$\Gamma_0(2):=\left\{\begin{pmatrix}a & b\\ c & d\end{pmatrix}\ \middle|\ a,b,c,d\in\mathbb{Z};\ ad-bc=1; c\text{ even}\right\}. \tag{1.6}$$

The series $\mathcal{P}(q)$ is called a quasimodular form [25]. Ramamani showed that these series satisfy differential equations similar to (1.3):

$$q\frac{de}{dq}=\frac{e\mathcal{P}-\mathcal{Q}}{2},\quad q\frac{d\mathcal{P}}{dq}=\frac{\mathcal{P}^2-\mathcal{Q}}{4},\quad q\frac{d\mathcal{Q}}{dq}=\mathcal{P}\mathcal{Q}-e\mathcal{Q}. \tag{1.7}$$

Ramamani found an additional differential equation for another set of Lambert series. In his Ph.D. thesis [28], the current author found a further coupled system of differential equations and showed that, under renormalization, the three classes of series studied by Ramamani and himself satisfy the same set of coupled differential equations. Moreover, the corresponding series turn out to be modular forms on subgroups conjugate to $\Gamma_0(2)$. The theorem appears in the present paper as Theorem 5.1.

More recently, R. Maier [31] was compelled by the work of Ramamani to seek out a corresponding coupled system of nonlinear differential equations for modular forms on $\Gamma_0(3)$ and $\Gamma_0(4)$. Maier derived two resulting systems of differential equations via the theory of modular forms. In the paper at hand, we desire to derive Maier's differential equations using trigonometric series identities in the spirit of Ramanujan's proof for (1.3). This strategy leads to an infinite class of differential equations encompassing Maier's differential equations.

Identity (1.1) appears in the classical literature on elliptic functions and is equivalent to a well-known differential equation satisfied by the Weierstrass-$\wp$ function. Identity (1.2) relates the square of the Weierstrass zeta function with the Weierstrass-$\wp$ function and first appeared in Ramanujan's paper [38]. In this paper, we apply (1.1) and (1.2) and additional identities from the theory of elliptic functions to derive a class of differential equations for modular forms corresponding to subgroups of the modular group. Section 2 is dedicated to proving the differential equations, presented below in Theorem 1.1. Section 3 demonstrates how Maier's differential equations for modular forms on $\Gamma_0(3)$ can be derived from Theorem 1.1. We also list a number of applications of these differential equations to prove results for the cubic theta functions analogous to classical results for Eisenstein series. In Section 4, we list applications of the differential equations on subgroups of higher level. We conclude the paper with Section 5, where we list applications of Ramamani's differential equations for modular forms on $\Gamma_0(2)$.

Throughout the sequel, we will employ the Jacobi theta function

$$\theta_1(z|q) = -iq^{1/8} \sum_{n=-\infty}^{\infty} (-1)^n q^{n(n+1)/2} \sin(2n+1)z, \qquad q = e^{2\pi i\tau}.$$

This function and its allied theta functions [42, p. 489] constitute the building blocks of elliptic functions. The quasiperiodic nature of the theta functions plays an important role in our current study. In particular, to determine the periods of elliptic functions constructed later in the paper, we will apply the easily verified transformation formulas

$$\theta_1(z+\pi|q) = -\theta_1(z|q), \quad \theta_1(z+\pi\tau|q) = -q^{-1/2}e^{-2iz}\theta_1(z|q). \tag{1.8}$$

If we denote differentiation with respect to z by $'$ and logarithmically differentiate the Jacobi triple product representation [42, p. 469] for $\theta_1(z|q)$

$$\theta_1(z|q) = 2q^{1/8}\sin z(q;q)_\infty(qe^{2iz};q)_\infty(qe^{-2iz};q)_\infty,$$

we derive

$$\frac{\theta_1'}{\theta_1}(z|q) = \cot z + 4\sum_{n=1}^{\infty} \frac{q^n}{1-q^n}\sin 2nz, \tag{1.9}$$

$$\frac{\theta_1''}{\theta_1}(z|q) = -1 + 16\sum_{n=1}^{\infty} \frac{q^n}{(1-q^n)^2} \cos 2nz + 8\sum_{n=1}^{\infty} \frac{nq^n}{1-q^n}. \tag{1.10}$$

Ramanujan's identity may then be written in compact form as

$$(\theta_1'/\theta_1)^2 = \theta_1''/\theta_1 - (\theta_1'/\theta_1)'. \tag{1.11}$$

The preceding identities motivate the main result of this paper.

Theorem 1.1. *Let*

$$e_\alpha(q) = 1 + \frac{4}{\cot(\pi\alpha)} \sum_{n=1}^{\infty} \frac{\sin(2n\pi\alpha)q^n}{1-q^n},$$

$$P_\alpha(q) = 1 - \frac{8}{\csc^2(\pi\alpha)} \sum_{n=1}^{\infty} \frac{\cos(2n\pi\alpha)nq^n}{1-q^n},$$

$$Q_\alpha(q) = 1 - \frac{8}{\cot(\pi\alpha)\csc^2(\pi\alpha)} \sum_{n=1}^{\infty} \frac{\sin(2n\pi\alpha)n^2q^n}{1-q^n}.$$

Then

$$q\frac{d}{dq}e_\alpha = \frac{\csc^2(\pi\alpha)}{4}(e_\alpha P_\alpha - Q_\alpha), \tag{1.12}$$

$$q\frac{d}{dq}P_\alpha = \frac{\csc^2(\pi\alpha)}{4}P_\alpha^2 - \frac{1}{2}\cot^2(\pi\alpha)e_\alpha Q_\alpha + \frac{1}{2}\cot(\pi\alpha)\cot(2\pi\alpha)e_{1-2\alpha}Q_\alpha, \tag{1.13}$$

$$\begin{aligned} q\frac{d}{dq}Q_\alpha = {} & \frac{1}{4}Q_\alpha P_\alpha \csc^2(\pi\alpha) + \frac{1}{2}P_{1-2\alpha}Q_\alpha\csc^2(2\pi\alpha) - \frac{1}{2}e_{1-2\alpha}^2 Q_\alpha \cot^2(2\pi\alpha) \\ & + \frac{3}{2}e_\alpha e_{1-2\alpha}Q_\alpha\cot(\pi\alpha)\cot(2\pi\alpha) - e_\alpha^2 Q_\alpha\cot^2(\pi\alpha). \end{aligned} \tag{1.14}$$

Note that

$$e_{1-2\alpha} = e_{2\alpha}, \quad Q_{1-2\alpha} = Q_{2\alpha}, \quad \text{and} \quad P_{1-2\alpha} = P_{2\alpha}. \tag{1.15}$$

Since

$$\frac{1}{\phi(m)} \sum_{\chi \,(\mathrm{mod}\, m)} \bar{\chi}(a)\chi(n) = \begin{cases} 1, & n \equiv a \pmod{m}, \\ 0, & \text{otherwise}, \end{cases}$$

where χ ranges over all characters modulo m, we see that, in each case, the N-periodic functions

$$\chi_{e,N}(n) := \frac{\sin(2n\pi/N)}{\cot(\pi/N)}, \quad \chi_{P,N}(n) := \frac{\cos(2n\pi/N)}{\csc^2(\pi/N)},$$

$$\chi_{Q,N}(n) := \frac{\sin(2n\pi/N)}{\cot(\pi/N)\csc^2(\pi/N)},$$

may be represented by an appropriate linear combination of Dirichlet characters. For each integer $N \geq 3$, the series $e_{1/N}$, $P_{1/N}$ and $Q_{1/N}$ are modular forms on $\Gamma_0(N)$ with multiplier system defined by a linear combination of Dirichlet characters.

2. Proof of Theorem 1.1

To prove Theorem 1.1, we derive an additional identity analogous to (1.2) and (1.1). The basis for our proof is the following well-known fact from the theory of elliptic functions:

Theorem 2.1. *The sum of the residues of an elliptic function on a period-parallelogram is zero.*

Lemma 2.2.

$$\left(\cot(\pi(1-2\alpha))e_{1-2\alpha} + \cot(\theta+\alpha\pi) - \cot(\theta-\alpha\pi)\right.$$

$$\left. +8\sum_{n=1}^{\infty} \frac{\sin(2n\pi\alpha)q^n}{1-q^n}\cos(2n\theta)\right)^2$$

$$= -P + \csc^2((1-2\alpha)\pi)P_{1-2\alpha} + \csc^2(\theta+\pi\alpha)$$

$$+ \csc^2(\theta-\pi\alpha) - 16\sum_{n=1}^{\infty} \frac{\cos(2n\pi\alpha)nq^n}{1-q^n}\cos(2n\theta).$$

Proof. Define

$$f(z,x,\alpha) = \frac{\theta_1(z+x+\pi\alpha|q)\theta_1(z-x+\pi\alpha|q)\theta_1(z+\pi(1-2\alpha)|q)}{\theta_1^3(z|q)}.$$

By (1.8), f is an elliptic function in the parameter z with periods π and $\pi\tau$. The only singularity of f in a period parallelogram is a pole of order 3 at $z=0$. Setting $F(z) = z^3 f(z)$ in Theorem 2.1 results in

$$0 = \operatorname{Res}(f;0) = \lim_{z\to 0}\frac{d^2}{dz^2}F(z) = \lim_{z\to 0} F(z)\left[\left(\frac{F'(z)}{F(z)}\right)^2 + \left(\frac{F'(z)}{F(z)}\right)'\right].$$

Since $\lim_{z\to 0} z^3 f(z) \neq 0$, we obtain

$$\left(\frac{F'(0)}{F(0)}\right)^2 + \left(\frac{F'(0)}{F(0)}\right)' = 0.$$

Noting that

$$\frac{F'(z)}{F(z)} = \frac{3}{z} - 3\cot z - 12\sum_{n=1}^{\infty}\frac{q^n}{1-q^n}\sin(2nz)$$
$$+ \cot(z+(1-2\alpha)\pi) + 4\sum_{n=1}^{\infty}\frac{q^n}{1-q^n}\sin(2n(z+(1-2\alpha)\pi)$$
$$+ \cot(z+x+\pi\alpha) + 4\sum_{n=1}^{\infty}\frac{q^n}{1-q^n}\sin(2n(z+x+\pi\alpha)$$
$$+ \cot(z-x+\pi\alpha) + 4\sum_{n=1}^{\infty}\frac{q^n}{1-q^n}\sin(2n(z-x+\pi\alpha),$$

we obtain the result in Lemma 2.2 by setting z equal to 0. □

We next equate coefficients of θ^0 and θ^2 on both sides of Lemma 2.2 to obtain the following corollary:

Corollary 2.3.

$$(\cot(\pi(1-2\alpha))e_{1-2\alpha} + 2\cot(\pi\alpha)e_\alpha)^2$$
$$= \csc^2((1-2\alpha)\pi)P_{1-2\alpha} + 2\csc^2(\pi\alpha)P_\alpha - P, \tag{2.1}$$
$$2(\cot(\pi(1-2\alpha))e_{1-2\alpha} + 2\cot(\pi\alpha)e_\alpha)\cot(\pi\alpha)\csc^2(\pi\alpha)Q_\alpha$$
$$= (2+\cos(2\pi\alpha))\csc^4(\pi\alpha) + 16\sum_{n=1}^{\infty}\frac{\cos(2n\pi\alpha)n^3q^n}{1-q^n}. \tag{2.2}$$

Proof of Theorem 1.1. Replace θ by $\theta+\pi\alpha$ in (1.2) and equate coefficients on both sides of the resulting identity to obtain

$$2\left(\frac{\cot\pi\alpha}{4} + \sum_{n=1}^{\infty}\frac{q^n\sin(2\pi\alpha n)}{1-q^n}\right)\left(-\frac{\csc^2\pi\alpha}{4} + 2\sum_{n=1}^{\infty}\frac{\cos(2\pi\alpha n)nq^n}{1-q^n}\right)$$
$$= -\frac{1}{8}\cot(\pi\alpha)\csc^2(\pi\alpha) + \sum_{n=1}^{\infty}\frac{\sin(2n\pi\alpha)n^2q^n}{1-q^n} - 2\sum_{n=1}^{\infty}\frac{\sin(2\pi\alpha n)nq^n}{(1-q^n)^2}. \tag{2.3}$$

With the definitions of e_α, P_α, and Q_α from Theorem 1.1, (2.3) can easily be seen to be equivalent to (1.12). To derive the differential equation for P_α we again replace θ by $\theta+\pi\alpha$ in (1.2) and equate coefficients of θ^2 in the resulting identity to obtain

$$\frac{1}{8}\cot^2(\pi\alpha)\csc^2(\pi\alpha)e_\alpha Q_\alpha + \frac{1}{16}\csc^4(\pi\alpha)P_\alpha^2$$

$$= \frac{1}{16}(2+\cos(2\pi\alpha))\csc^4(\pi\alpha) + \sum_{n=1}^{\infty}\frac{\cos(2n\pi\alpha)n^3q^n}{1-q^n} + \frac{\csc^2(\pi\alpha)}{4}q\frac{d}{dq}P_\alpha$$

$$= \frac{1}{8}(\cot(\pi(1-2\alpha))e_{1-2\alpha} + 2\cot(\pi\alpha)e_\alpha)\cot(\pi\alpha)\csc^2(\pi\alpha)Q_\alpha$$

$$+ \frac{\csc^2(\pi\alpha)}{4}q\frac{d}{dq}P_\alpha, \tag{2.4}$$

where the last equality follows by (2.2). We deduce (1.13) by equating the extreme left and right sides of (2.4) and solving for $q\frac{d}{dq}P_\alpha$. To derive the differential equation for Q_α we proceed along similar lines. Replace θ by $\theta + \pi\alpha$ in (1.2) and equate coefficients of θ^3 in the resulting identity to derive

$$\frac{1}{2}\cot(\pi\alpha)e_\alpha\left(-\frac{1}{12}(2+\cos(2\pi\alpha))\csc^4(\pi\alpha) - \frac{4}{3}\sum_{n=1}^{\infty}\frac{\cos(2n\pi\alpha)n^3q^n}{1-q^n}\right)$$

$$-\frac{1}{8}\csc^4(\pi\alpha)\cot(\pi\alpha)P_\alpha Q_\alpha$$

$$= -\frac{1}{48}(11\cos(\pi\alpha)+\cos(3\pi\alpha))\csc^5(\pi\alpha)$$

$$-\frac{2}{3}\sum_{n=1}^{\infty}\frac{\sin(2n\pi\alpha)n^4q^n}{1-q^n} - \frac{\cot(\pi\alpha)\csc^2(\pi\alpha)}{6}q\frac{d}{dq}Q_\alpha. \tag{2.5}$$

We next replace θ by $\theta + \pi\alpha$ in (1.1) and equate coefficients of θ in the resulting identity to obtain

$$2\left(-\frac{P}{24}+\frac{1}{8}\csc^2(\pi\alpha)P_\alpha\right)\left(-\frac{1}{4}\cot(\pi\alpha)\csc^2(\pi\alpha) + 2\sum_{n=1}^{\infty}\frac{\sin(2n\pi\alpha)n^2q^n}{1-q^n}\right)$$

$$= -\frac{1}{96}(5+\cos(2\pi\alpha))\cot(\pi\alpha)\csc^4(\pi\alpha) - \frac{1}{6}\sum_{n=1}^{\infty}\frac{\sin(2n\pi\alpha)n^4q^n}{1-q^n}$$

$$= -\frac{1}{192}(11\cos(\pi\alpha)+\cos(3\pi\alpha))\csc^5(\pi\alpha) - \frac{1}{6}\sum_{n=1}^{\infty}\frac{\sin(2n\pi\alpha)n^4q^n}{1-q^n}. \tag{2.6}$$

Through the use of (2.6) and (2.2), identity (2.5) can be written

$$-\frac{1}{12}(\cot(\pi(1-2\alpha))e_{1-2\alpha} + 2\cot(\pi\alpha)e_\alpha)\cot^2(\pi\alpha)\csc^2(\pi\alpha)e_\alpha Q_\alpha$$

$$-\frac{1}{8}\cot(\pi\alpha)\csc^4(\pi\alpha)P_\alpha Q_\alpha$$

$$= -2\left(-\frac{P}{24} + \frac{1}{8}\csc^2(\pi\alpha)P_\alpha\right)\cot(\pi\alpha)\csc^2(\pi\alpha)Q_\alpha$$

$$-\frac{\cot(\pi\alpha)\csc^2(\pi\alpha)}{6}q\frac{d}{dq}Q_\alpha. \qquad (2.7)$$

We complete the proof of (1.14) by solving for P in (2.1), inserting the result in (2.7), and solving for $q\frac{d}{dq}Q_\alpha$. □

3. Cubic Eisenstein series

Let $\chi : \{0, 1, \ldots, N-1\} \to \mathbb{Z}$, and denote $\sigma_k(n|\chi) = \sum_{d|n} \chi(d \mod N)d^k$. Let $\chi_3(n)$ denote the Jacobi symbol modulo -3. Define the cubic Eisenstein series as follows:

$$e_{1/3}(q) = 1 + 6\sum_{n=1}^{\infty}\sigma_0(n|\chi_3)q^n, \quad P_{1/3}(q) = 1 + 3\sum_{n=1}^{\infty}\sigma_1(n; -2, 1, 1)q^n,$$

$$Q_{1/3}(q) = 1 - 9\sum_{n=1}^{\infty}\sigma_2(n|\chi_3)q^n.$$

The series $e_{1/3}(q)$ and $Q_{1/3}(q)$ are null values of cubic theta functions, commonly referred to as $a(q)$ and $b^3(q)$ [18]. The Eisenstein series $P_{1/3}(q)$ may be obtained by logarithmically differentiating the cube of the Borwein's cubic theta function $c(q)$. The series $e_{1/3}(q)$ and $Q_{1/3}(q)$ are, respectively, weight 1 and 3 modular forms on $\Gamma_0(3)$ with multiplier system χ_{-3}. For more information on these functions from the perspective of modular forms, see [31].

The three cubic Eisenstein series satisfy a coupled system of nonlinear first order differential equations in q analogous to those satisfied by the classical Eisenstein series. Maier's paper [32] includes differential equations that are equivalent to those in the following theorem. The relations are an immediate consequence of Theorem (1.1) in the case $\alpha = \frac{1}{3}$.

Corollary 3.1. *For* $|q| < 1$,

$$q\frac{d}{dq}e_{1/3} = \frac{e_{1/3}P_{1/3} - Q_{1/3}}{3}, \quad q\frac{d}{dq}P_{1/3} = \frac{P_{1/3}^2 - e_{1/3}Q_{1/3}}{3},$$

$$q\frac{d}{dq}Q_{1/3} = P_{1/3}Q_{1/3} - e_{1/3}^2Q_{1/3}.$$

In the next corollary, we observe that the Borwein's cubic theta function identity $a^3(q) = b^3(q) + c^3(q)$ is a direct consequence of Corollary 3.1.

Corollary 3.2.

$$e_{1/3}^3 - Q_{1/3} = 27q\frac{(q^3; q^3)_\infty^9}{(q; q)_\infty^3}. \qquad (3.1)$$

Proof. The result follows by logarithmically differentiating each side of (3.1) and applying Corollary 3.1 as well as the identity

$$\frac{(q^3;q^3)^3_\infty}{(q;q)_\infty} = \frac{(q^3;q^3)^2_\infty}{(q;q^3)_\infty (q^2;q^3)_\infty}. \tag{3.2}$$

□

In 1956, Rankin [39] presented a general method for deriving identities satisfied by derivatives of modular forms. Let $'$ denote the operator $q\frac{d}{dq}$. For example, with $\Delta = \frac{Q^3-R^2}{1728}$, Rankin derived

$$4QQ'' - 5(Q')^2 = 960\Delta, \quad 6RR'' - 7(R')^2 = -3024Q\Delta.$$

In the following corollary, we see that similar relations exist for the derivatives of Eisenstein series of Level 3. Each may be proven using the differential equations from Corollary 3.1.

Corollary 3.3.

$$\begin{aligned} e''_{1/3} &- \frac{2}{3}P_{1/3}e'_{1/3} + \frac{2}{9}Q'_{1/3} \\ &= Q''_{1/3} + e^2_{1/3}Q'_{1/3} - 4Q_{1/3}P'_{1/3} + 5e_{1/3}Q_{1/3}e'_{1/3} \\ &= Q_{1/3}Q''_{1/3} - (Q'_{1/3})^2 - Q^2_{1/3}P'_{1/3} + 2e_{1/3}Q^2_{1/3}e'_{1/3} = 0, \end{aligned}$$

and

$$e_{1/3}e''_{1/3} - 2(e'_{1/3})^2 = 6qQ_{1/3}\frac{(q^3;q^3)^9_\infty}{(q;q)^3_\infty},$$

$$Q_{1/3}Q''_{1/3} - \frac{4}{3}(Q'_{1/3})^2 = -9qe_{1/3}Q^2_{1/3}\frac{(q^3;q^3)^9_\infty}{(q;q)^3_\infty},$$

$$P''_{1/3} - \frac{2}{3}P_{1/3}P'_{1/3} + \frac{4}{3}Q_{1/3}e'_{1/3} = 9qQ_{1/3}\frac{(q^3;q^3)^9_\infty}{(q;q)^3_\infty}.$$

We may also imitate a number of identities in Ramanujan's Lost Notebook using cubic Eisenstein series instead of those on the full modular group. The following integral equation is similar to an identity from Ramanujan's lost notebook [13, Thm. 3.1, pp. 82–83], [33,37]:

Corollary 3.4. *For* $0 < q < 1$,

$$\int_0^q \sqrt{e_{1/3}(t)Q_{1/3}(t)}\frac{dt}{t} = \log\left(\frac{e^{3/2}_{1/3}(q) - Q^{1/2}_{1/3}(q)}{e^{3/2}_{1/3}(q) + Q^{1/2}_{1/3}(q)}\right).$$

An analogue of the Rankin-Cohen bracket [22] may be defined using Corollary 3.1. A proof of the following theorem follows directly from the argument given in [43].

Theorem 3.5. *Let*

$$D_{1/3} = \frac{q_{1/3}P_{1/3} - Q_{1/3}}{3}\frac{\partial}{\partial e_{1/3}} + \frac{P_{1/3}^2 - e_{1/3}Q_{1/3}}{3}\frac{\partial}{\partial P_{1/3}}$$
$$+ (P_{1/3}Q_{1/3} - e_{1/3}^2 Q_{1/3})\frac{\partial}{\partial Q_{1/3}},$$

and define

$$[f,g]_{D_{1/3},n} = \sum_{r+s=n} (-1)^r \binom{n+k-1}{s}\binom{n+\ell-1}{r} D_{1/3}^r f D_{1/3}^s g.$$

Then the subalgebra generated by $e_{1/3}$ *and* $Q_{1/3}$ *is closed under the bracket operator* $[\,]_n := [\,]_{D_{1/3},n}$.

The following examples of Theorem 3.5 may be derived by employing Corollary 3.1. Denote $\Delta_{1/3} = e_{1/3}^3 - Q_{1/3}$. Then

$$[e_{1/3}, Q_{1/3}]_1 = -Q_{1/3}\Delta_{1/3}, \quad [e_{1/3}, \Delta_{1/3}]_1 = -Q_{1/3}\Delta_{1/3},$$

$$[Q_{1/3}, \Delta_{1/3}]_1 = 3e_{1/3}^2 Q_{1/3}\Delta_{1/3}, \quad [e_{1/3}, e_{1/3}]_2 = \frac{4}{9}Q_{1/3}\Delta_{1/3},$$

$$[e_{1/3}, Q_{1/3}]_2 = e_{1/3}^2 Q_{1/3}\Delta_{1/3}, \quad [Q_{1/3}Q_{1/3}]_2 = -4e_{1/3}Q_{1/3}^2\Delta_{1/3},$$

and

$$[\Delta_{1/3}, \Delta_{1/3}]_2 = -4e_{1/3}Q_{1/3}\Delta_{1/3}^2.$$

4. Eisenstein series on subgroups of higher level

Setting $\alpha = \frac{1}{4}$ in Theorem 1.1, we obtain

$$q\frac{d}{dq}e_{1/4} = \frac{e_{1/4}P_{1/4} - Q_{1/4}}{2}, \quad q\frac{d}{dq}P_{1/4} = \frac{P_{1/4}^2 - e_{1/4}Q_{1/4}}{2}, \tag{4.1}$$

$$q\frac{d}{dq}Q_{1/4} = Q_{1/4}\frac{P_{1/2} + P_{1/4} - 2e_{1/4}^2}{2}. \tag{4.2}$$

The differential equations (4.1)–(4.2) are expressed in equivalent form in Maier's paper [31] as a coupled system for Eisenstein series $\Gamma_0(4)$ in the parameters $P_{1/4}$, $e_{1/4}^2$ and $2P_{1/4} - P_{1/2}$.

For $\alpha = \frac{1}{6}$, Theorem 1.1 becomes

$$q\frac{d}{dq}e_{1/6} = e_{1/6}P_{1/6} - Q_{1/6}, \quad q\frac{d}{dq}P_{1/4} = \frac{2P_{1/6}^2 - 3Q_{1/6}e_{1/6} + Q_{1/6}e_{2/3}}{2}, \tag{4.3}$$

$$q\frac{d}{dq}Q_{1/6} = Q_{1/6}\frac{4P_{2/3} + 6P_{1/6} - 18e_{1/6}^2 + 9e_{1/6}e_{2/3} - e_{2/3}^2}{6}. \tag{4.4}$$

One consequence of (4.3) and (4.4) is the following corollary:

Corollary 4.1.

$$e_{1/6}e''_{1/6} - (e'_{1/6})^2 - Q_{1/6}e'_{1/6} - e^2_{1/6}P'_{1/6} + e_{1/6}Q'_{1/6} = 0.$$

By applying Theorem 1.1 for $\alpha = 1/5$ we obtain a corresponding system of differential equations that can be used to deduce Corollary 4.2:

Corollary 4.2. *Let* $\alpha = \frac{5+\sqrt{5}}{10}$. *Then*

$$e_{1/5}e''_{1/5} - (e'_{1/5})^2 - \alpha Q_{1/5}e'_{1/5} - \alpha e^2_{1/5}P'_{1/5} + \alpha e_{1/5}Q'_{1/5} = 0.$$

In view of the similarity of the two preceding corollaries, a natural question to ask is whether or not the parameters e_α, P_α, and Q_α satisfy the same differential equations for each α. Although the derivation is tedious, we may apply Theorem 1.1 to derive the following corollary:

Corollary 4.3. *Let* $\beta = \frac{\csc^2(\alpha\pi)}{4}$. *Then*

$$e_\alpha e''_\alpha - (e'_\alpha)^2 - \beta Q_\alpha e'_\alpha - \beta e^2_\alpha P'_\alpha + \beta e_\alpha Q'_\alpha = 0.$$

It may be interesting to determine further results like Corollary 4.3. In particular, Rankin's method [39] may be implemented to generate differential equations satisfied by e_α, P_α, and Q_α for values of α such that the series involved are modular forms. However, the differential equation in Corollary 4.3 is not a relation between modular forms for arbitrary values of α. Thus, to obtain similar results with the same generality, one may have to look beyond the theory of modular forms.

5. Differential equations for Eisenstein series on conjugate subgroups of level 2

This section concerns a number of consequences of the following theorem:

Theorem 5.1. *For* $j \in \{1, 2, 3\}$,

$$q\frac{d\mathcal{P}_j}{dq} = \frac{\mathcal{P}_j^2 - \mathcal{Q}_j}{4}, \quad q\frac{d\mathcal{Q}_j}{dq} = \mathcal{P}_j\mathcal{Q}_j - 6e_j\mathcal{Q}_j, \quad q\frac{de_j}{dq} = \frac{\mathcal{P}_je_j}{2} - \frac{\mathcal{Q}_j}{12},$$

where

$$e_1(q) = \frac{1}{6} + 4\sum_{n=1}^{\infty}\frac{nq^n}{1+q^n}, \quad -12e_j(q)$$

$$= 1 + 24(-1)^j \sum_{\substack{n\geq 1 \\ n \text{ odd}}}^{\infty} \frac{nq^{n/2}}{1+(-1)^jq^{n/2}}, \quad j = 2, 3,$$

$$\mathcal{P}_1(q) = 1 + 8\sum_{n=1}^{\infty} \frac{(-1)^{n-1} n q^n}{1-q^n}, \quad \mathcal{P}_j(q) = -8\sum_{n=1}^{\infty} \frac{(-1)^{jn} n q^{n/2}}{1-q^n}, \quad j = 2, 3,$$

$$\mathcal{Q}_1(q) = 1 - 16\sum_{n=1}^{\infty} \frac{(-1)^{n-1} n^3 q^n}{1-q^n}, \quad \mathcal{Q}_j(q) = 16\sum_{n=1}^{\infty} \frac{(-1)^{jn} n^3 q^{n/2}}{1-q^n}, \quad j = 2, 3.$$

A proof of Theorem 5.1 for the cases $j = 1, 2$ was first given by Ramamani [34,35]. H. Hahn includes a proof for the case $j = 1$ [25]. A proof of the theorem for the case $j = 3$ is given in [28]. In this section, we give some applications of Theorem 5.1. Motivated by Corollary 3.4 and the corresponding results from Ramanujan's Lost Notebook [13, Thm. 3.1, pp. 82–83], [33,37], we derive the following corollary:

Corollary 5.2. *Let* $0 < q < 1$. *For* $j = 1, 2, 3$,

$$\log\left(\frac{6e_j(q) - \mathcal{Q}_j^{1/2}(q)}{6e_j(q) + \mathcal{Q}_j^{1/2}(q)}\right) = \int_0^q \sqrt{\mathcal{Q}_j(t)}\frac{dt}{t}.$$

On pages 188 and 369 of the Lost Notebook [37, pp. 188, 369], [17,14], [41, pp. 31–32], Ramanujan derives expressions for the series

$$T_{2k}(q) := 1 + \sum_{n=1}^{\infty}(-1)^n\{(6n-1)^{2k}q^{n(3n-1)/2} + (6n+1)^{2k}q^{n(3n+1)/2}\} \quad \text{and}$$

$$F_{2k}(q) := \sum_{n=0}^{\infty}(-1)^n(2n+1)^{2k+1}q^{n(n+1)/2} \tag{5.1}$$

in terms of the Eisenstein series $P(q)$, $Q(q)$ and $R(q)$. These representations have inspired a number of analogous relations [25, pp. 32–37], [20]. We derive a result of a similar kind in the following corollary:

Corollary 5.3. *Define*

$$\varphi(q) = \sum_{n=-\infty}^{\infty} q^{n^2}, \tag{5.2}$$

$$\mathcal{U}_{2k}(q) = 2^{2k+1}\sum_{n=1}^{\infty}(-1)^n n^{2k} q^{n^2/2}, \quad \mathcal{V}_2(q) = 2^{2k+1}\sum_{n=1}^{\infty} n^{2k} q^{n^2/2}.$$

Then, for $k \geq 1$, *and* $\alpha_{\ell,m,n} \in \mathbb{Q}$,

$$\frac{\mathcal{U}_{2k}(q)}{\varphi(-q^{1/2})} = \sum_{2\ell+2m+4n=2k} \alpha_{\ell,m,n}\mathcal{P}_2^{\ell} e_2^m \mathcal{Q}_2^n, \tag{5.3}$$

$$\frac{\mathcal{V}_{2k}(q)}{\varphi(q^{1/2})} = \sum_{2\ell+2m+4n=2k} \alpha_{\ell,m,n}\mathcal{P}_3^{\ell} e_3^m \mathcal{Q}_3^n. \tag{5.4}$$

Proof. We provide a proof only for (5.3). The proof for (5.4) is similar.

First note that

$$8q\frac{d\varphi(-q^{1/2})}{dq}=8\sum_{n=1}^{\infty}(-1)^n n^2 q^{n^2/2}=\mathcal{U}_2(q). \tag{5.5}$$

Next note that, by (5.20),

$$8\frac{q\frac{d\varphi(-q^{1/2})}{dq}}{\varphi(-q^{1/2})}=\mathcal{P}_2. \tag{5.6}$$

In general, we have

$$\mathcal{U}_{2k}=\frac{8^k}{2^{k-1}}\sum_{n=1}^{\infty}(-1)^n n^{2k}q^{n^2/2}=8^k\left(q\frac{d}{dq}\right)^k\varphi(-q^{1/2})=8q\frac{d\mathcal{U}_{2k-2}}{dq}. \tag{5.7}$$

If Corollary 5.3 holds for all positive integers strictly less than k, then

$$\mathcal{U}_{2k-2}=\varphi(-q^{1/2})f(\mathcal{P}_2,\mathcal{Q}_2,e_2), \tag{5.8}$$

for some polynomial f in $\mathcal{P}_2$, $\mathcal{Q}_2$ and

$$\begin{aligned}\mathcal{U}_{2k}&=8q\frac{d}{dq}\{\varphi(-q^{1/2})f(\mathcal{P}_2,\mathcal{Q}_2)\}\\&=8\varphi(-q^{1/2})q\frac{d}{dq}\{f(\mathcal{P}_2,\mathcal{Q}_2,e_2)\}+f(\mathcal{P}_2,\mathcal{Q}_2,e_2)\varphi(-q^{1/2})\cdot 8\frac{q\frac{d}{dq}\varphi(-q^{1/2})}{\varphi(-q^{1/2})}\\&=\varphi(-q^{1/2})\left\{8q\frac{d}{dq}\{f(\mathcal{P}_2,\mathcal{Q}_2,e_2)\}+\mathcal{P}_2f(\mathcal{P}_2,\mathcal{Q}_2,e_2)\right\}.\end{aligned} \tag{5.9}$$

Applying Theorem 5.1 to the last line of (5.9), we find that $\mathcal{U}_{2k}$ is a polynomial in $\mathcal{P}_2$, $\mathcal{Q}_2$ and e_2 with rational coefficients. Therefore, (5.3) holds by induction on k. □

The first three examples of Corollary 5.3 are as follows:

$$\frac{\mathcal{U}_2(q)}{\varphi(-q^{1/2})}=\mathcal{P}_2, \tag{5.10}$$

$$\frac{\mathcal{U}_4(q)}{\varphi(-q^{1/2})}=3\mathcal{P}_2^2-2\mathcal{Q}_2, \tag{5.11}$$

$$\frac{\mathcal{U}_6(q)}{\varphi(-q^{1/2})}=13\mathcal{P}_2^3-28\mathcal{P}_2\mathcal{Q}_2+96e_2\mathcal{Q}_2. \tag{5.12}$$

Another consequence of Theorem 5.1 is the following set of identities. These relations are analogous to the identity [40, pp. 196–197], [8, p. 92]

$$Q^3-R^2=1728q(q;q)_\infty^{24}, \tag{5.13}$$

which can be derived by employing (1.3) and logarithmic differentiation.

Theorem 5.4. *Define*

$$\psi(q) := \sum_{n=0}^{\infty} q^{n(n+1)/2} \tag{5.14}$$

and let $\mathcal{P}_j$, $\mathcal{Q}_j$, e_j, $j = 1,2,3$, be defined as in Theorem 5.1. Then

$$36e_1^2 - \mathcal{Q}_1 = 64q\psi^8(q), \tag{5.15}$$

$$36e_2^2 - \mathcal{Q}_2 = \varphi^8(-q^{1/2})/4, \tag{5.16}$$

$$36e_3^2 - \mathcal{Q}_3 = \frac{\psi^{16}(q^{1/2})}{4\psi^8(q)} = \varphi^8(q^{1/2})/4. \tag{5.17}$$

Proof. Identity (5.15) is equivalent to an identity in Hahn's thesis [25, p. 20, Eq. (2.3.6)]. To prove (5.16) and (5.17), apply Theorem 5.1 to derive

$$q\frac{d}{dq}\log(36e_j^2 - \mathcal{Q}_j) = \mathcal{P}_j(q), \quad j = 1,2,3. \tag{5.18}$$

The Jacobi triple product representation for $\varphi(q)$ gives

$$\varphi^8(-q^{1/2}) = \frac{(q^{1/2};q^{1/2})_\infty^8}{(-q^{1/2};q^{1/2})_\infty^8} = \frac{(q^{1/2};q^{1/2})_\infty^{16}}{(q;q)_\infty^8}. \tag{5.19}$$

Therefore

$$\begin{aligned} q\frac{d}{dq}\log\varphi^8(-q^{1/2}) &= 8\sum_{n=1}^{\infty}\frac{nq^n}{1-q^n} - 16\sum_{n=1}^{\infty}\frac{(n/2)q^{n/2}}{1-q^{n/2}} \\ &= 8\sum_{n=1}^{\infty}\frac{nq^n}{1-q^n} - 8\sum_{n=1}^{\infty}\frac{nq^{n/2}(1+q^{n/2})}{1-q^n} = \mathcal{P}_2(q). \end{aligned} \tag{5.20}$$

Similarly, use the fact that

$$\frac{\psi^{16}(q^{1/2})}{\psi^8(q)} = \frac{(q;q)_\infty^{16}(q;q^2)_\infty^8}{(q^{1/2};q)_\infty^{16}(q^2;q^2)_\infty^8} \tag{5.21}$$

to obtain,

$$\begin{aligned} q\frac{d}{dq}\log[\psi^{16}(q^{1/2})/\psi^8(q)] &= q\frac{d}{dq}\log\frac{(q;q^2)_\infty^8}{(q^{1/2};q)_\infty^{16}} + q\frac{d}{dq}\log\frac{(q;q)_\infty^{16}}{(q^2;q^2)_\infty^8} \\ &= 8\sum_{n=1}^{\infty}\frac{(2n-1)q^{(2n-1)/2}}{1-q^{2n-1}} - 16\sum_{n=1}^{\infty}\frac{nq^n}{1-q^{2n}} = \mathcal{P}_3(q). \end{aligned} \tag{5.22}$$

The last equality of (5.22) follows by summing $\mathcal{P}_3(q)$ over odd and even indices, respectively. Set $j = 2$ in (5.18) and equate the left sides of (5.18) and (5.20). Integrate and exponentiate the result, and note that the left side of (5.16) reduces to $1/4$ as q tends to zero. This proves (5.16). The proof of (5.17) is accomplished similarly by applying (5.18) and (5.22) for $j = 3$. □

By replacing q by q^2 in (5.10) and equating coefficients of q, we obtain the following corollary:

Corollary 5.5. *Define* $\tilde{\sigma}(n) = n \sum_{\substack{d|n \\ d \text{ odd}}} \frac{1}{d}$. *Then*

$$\sum_{\substack{j+k^2=n \\ j\geq 1,\ k\in\mathbb{Z}}} (-1)^{k+1}\tilde{\sigma}(j) = \begin{cases} (-1)^n n, & \text{if } n \text{ is a square}, \\ 0, & \text{otherwise}. \end{cases} \tag{5.23}$$

Proof. First note that

$$\mathcal{P}_2 = -8\sum_{k=1}^{\infty} \frac{kq^k}{1-q^{2k}} = -8\sum_{n=0}^{\infty} \tilde{\sigma}(n)q^n. \tag{5.24}$$

After replacing q by q^2 in (5.10) and noting that $\tilde{\sigma}(0) = 0$, we obtain

$$\sum_{n=1}^{\infty}(-1)^n n^2 q^{n^2} = \left(\sum_{n=-\infty}^{\infty} (-1)^{n+1} q^{n^2}\right) \cdot \left(\sum_{n=0}^{\infty} \tilde{\sigma}(n) q^n\right). \tag{5.25}$$

Equating coefficients of q^n on both sides of (5.25), we obtain a proof of (5.23). □

We conclude this section with an application of Theorem 5.1 to a differential equation from the theory of modular forms. The classical Darboux-Halphen system

$$\frac{du_1}{dx} + \frac{du_2}{dx} = u_1u_2, \qquad \frac{du_2}{dx} + \frac{du_3}{dx} = u_2u_2, \qquad \frac{du_3}{dx} + \frac{du_1}{dx} = u_3u_1. \tag{5.26}$$

arose in connection with Darboux's analysis of triply orthogonal surfaces [23]. A solution to the Darboux-Halphen system was found by G. Halphen [26] in 1881 using null values of the theta functions. The Darboux-Halphen system is equivalent to the Chazy equation [21]

$$y''' = 2yy'' - 3(y')^2, \tag{5.27}$$

where $y := u_1+u_2+u_3$. The Chazy equation initially arose in the analysis of Painlevé type equations. Equation (5.27) is of particular importance in the study of integrable systems [1–4] and arises in a number of applications [24,27]. In her Ph. D. thesis, Hahn shows that if $y(\tau) = \pi i \mathcal{P}_1(q)$, $q = e^{2\pi i \tau}$, then y satisfies the nonlinear third-order differential equation

$$y^2\{y''' - 2yy'' + 3(y')^2\} = 2(y')^3 - 2(y'')^2 + 2y'y''. \tag{5.28}$$

We can apply Theorem 5.1 to obtain two more solutions to (5.28).

Theorem 5.6. *For* $q = e^{2\pi i \tau}$, *the functions*

$$y(\tau) = \pi i \mathcal{P}_j(q), \qquad j = 1, 2, 3 \tag{5.29}$$

each satisfy (5.28).

For a proof of Theorem 5.6, we refer the reader to Hahn's proof [25, pp. 30–32] that $\pi i \mathcal{P}_1$ satisfies (5.28). The only tools used in her proof are the differential equations satisfied by $\mathcal{P} := \mathcal{P}_1$, $\mathcal{Q} := \mathcal{Q}_1$ and $e := 6e_1$. We note that an identical argument will also show that $\pi i \mathcal{P}_j$ satisfies (5.28) for each $j = 1, 2, 3$.

References

[1] M. J. Ablowitz, S. Chakravarty and R. Halburd, The generalized Chazy equation and Schwarzian triangle functions, *Asian J. Math.*, **2(4)** (1998) 619–624. Mikio Sato: A great Japanese mathematician of the twentieth century.

[2] M. J. Ablowitz, S. Chakravarty and R. G. Halburd, The generalized Chazy equation from the self-duality equations, *Stud. Appl. Math.*, **103(1)** (1999) 75–88.

[3] M. J. Ablowitz, S. Chakravarty and R. G. Halburd, Integrable systems and reductions of the self-dual Yang-Mills equations, *J. Math. Phys.*, **44(8)** (2003) 3147–3173. Integrability, topological solitons and beyond.

[4] M. J. Ablowitz, S. Chakravarty and L. A. Takhtaian, Integrable systems, self-dual Yang-Mills equations and connections with modular forms. In Nonlinear problems in engineering and science—numerical and analytical approach (Beijing, 1991), pages 1–15. Science Press, Beijing (1992).

[5] G. E. Andrews and B. C. Berndt, Ramanujan's lost notebook, Part II, Springer, New York (2009).

[6] B. C. Berndt, Ramanujan's notebooks, Part III, Springer-Verlag, New York (1991).

[7] B. C. Berndt, Ramanujan's notebooks, Part V, Springer-Verlag, New York (1998).

[8] B. C. Berndt, Number theory in the spirit of Ramanujan, volume 34 of Student Mathematical Library, American Mathematical Society, Providence, RI (2006).

[9] B. C. Berndt, Ramanujan's congruences for the partition function, modulo 5, 7, and 11, *Internat. J. Number Thy.*, **3** (2007) 349–354.

[10] B. C. Berndt, P. R. Bialek and A. J. Yee, Formulas of Ramanujan for the power series coefficients of certain quotients of Eisenstein series, *Int. Math. Res. Not.*, **(21)** (2002) 1077–1109.

[11] B. C. Berndt and H. H. Chan, Eisenstein series and approximations to π, *Illinois J. Math.*, **45** (2001) 75–90.

[12] B. C. Berndt, H. H. Chan and S. H. Son, Eisenstein series in Ramanujan's lost notebook, *Ramanujan J.*, **4** (2000) 81–114.

[13] B. C. Berndt, H. H. Chan and S. S. Huang, Incomplete elliptic integrals in Ramanujan's lost notebook. In q-series from a contemporary perspective (South Hadley, MA, 1998), volume 254 of *Contemp. Math.*, pages 79–126. *Amer. Math. Soc.*, Providence, RI (2000).

[14] B. C. Berndt, S. H. Chan, Z.-G. Liu and H. Yesilyurt, A new identity for $(q; q)_\infty^{10}$ with an application to Ramanujan's partition congruence modulo 11, *Q. J. Math.*, **55(1)** (2004) 13–30.

[15] B. C. Berndt and A. J. Yee, Congruences for the coefficients of Eisenstein series, *Acta. Arith.*, **104** (2002) 297–308.

[16] B. C. Berndt and A. J. Yee, Ramanujan's contributions to Eisenstein series, especially in his lost notebook. In Number Theoretic Methods – Future Trends, pages 31–53. Kluwer, Dordnecht (2002).

[17] B. C. Berndt and A. J. Yee, A page on Eisenstein series in Ramanujan's Lost Notebook, *Glasg. Math. J.*, **45(1)** (2003) 123–129.

[18] J. M. Borwein and P. B. Borwein, A cubic counterpart of Jacobi's identity and the AGM, *Trans. Amer. Math. Soc.*, **323(2)** (1991) 691–701.

[19] M. Boylan, Swinnerton-Dyer type congruences for certain Eisenstein series. In q-series with applications to combinatorics, number theory, and physics (Urbana, IL, 2000), volume 291 of *Contemp. Math.*, pages 93–108. *Amer. Math. Soc.*, Providence, RI (2001).

[20] H. H. Chan, S. Cooper and P. C. Toh, Ramanujan's Eisenstein series and powers of Dedekind's eta-function, *Proc. London Math. Soc.*, **75(1)** (2007) 225–242.

[21] J. Chazy, Sur les équations différentielles du troisième ordre et d'ordre supérieur dont l'intégrale générale a ses points critiques fixes, *Acta Math.*, **34(1)** (1911) 317–385.

[22] H. Cohen, Sums involving the values at negative integers of L functions of quadratic characters. In Séminaire de Théorie des Nombres, 1974–1975 (Univ. Bordeaux I, Talence), Exp. No. 3, page 21 pp. Lab. Théorie des Nombres. Centre Nat. Recherche Sci., Talence (1975).

[23] G. Darboux, Mémoire sur la théorie des coordonnées curvilignes, et des systèmes orthogonaux, *Ann. Sci. École Norm. Sup. (2)*, **7** (1878) 101–150.

[24] G. W. Gibbons and C. N. Pope, The positive action conjecture and asymptotically Euclidean metrics in quantum gravity, *Comm. Math. Phys.*, **66(3)** (1979) 267–290.

[25] H. Hahn, Eisenstein series, analogues of the Rogers-Ramanujan functions, and partition identities, Ph.D. Thesis, University of Illinois at Urbana-Champaign (2004).

[26] G. Halphen, Sur un système d'équations différentielles, *C. R. Acad. Sci. Paris*, **92** (1881) 1101–1103.

[27] N. J. Hitchin, Monopoles, minimal surfaces and algebraic curves, volume 105 of Séminaire de Mathématiques Supérieures [Seminar on Higher Mathematics]. Presses de l'Université de Montréal, Montreal, QC (1987).

[28] T. Huber, Zeros of generalized Rogers-Ramanujan series and topics from Ramanujan's theory of elliptic functions, Ph.D. Thesis, University of Illinois at Urbana-Champaign (2007).

[29] T. Huber, Hadamard products for generalized Rogers-Ramanujan series, *J. Approx. Th.*, **151(2)** (2008) 126–154.

[30] T. Huber, Basic representations for Eisenstein series from their differential equations, *J. Math. Anal. Appl.*, **350(1)** (2009) 135–146.

[31] R. S. Maier, Algebraic hypergeometric transformations of modular origin, *Trans. Amer. Math. Soc.*, **359(8)** (2007) 3859–3885 (electronic).

[32] Robert S. Maier, Nonlinear differential equations satisfied by certain classical modular forms, http://www.citebase.org/abstract?id=oai:arXiv.org:0807.1081 (2008).

[33] S. Raghavan and S. S. Rangachari, On Ramanujan's elliptic integrals and modular identities. In Number theory and related topics (Bombay, 1988), volume 12 of *Tata Inst. Fund. Res. Stud. Math.*, pages 119–149. *Tata Inst. Fund. Res.*, Bombay (1989).

[34] V. Ramamani, Some identities conjectured by Srinivasa Ramanujan in his lithographed notes connected with partition theory and elliptic modular functions – their proofs – interconnections with various other topics in the theory of numbers and some generalizations thereon. Ph.D. Thesis, University of Mysore (1970).

[35] V. Ramamani, On Some Algebraic Identities Connected with Ramanujan's Work. In Ramanujan International Symposium on Analysis (Pune, 1987), pages 277–291. Macmillan of India, New Delhi (1989).

[36] S. Ramanujan, Congruence properties of partitions, *Math Z.*, **9** (1921) 147–153.

[37] S. Ramanujan, The lost notebook and other unpublished papers, Springer-Verlag, Berlin (1988). With an introduction by George E. Andrews.

[38] S. Ramanujan, On certain arithmetical functions [*Trans. Cambridge Philos. Soc.*, **22** (1916) no. 9, 159–184]. In Collected papers of Srinivasa Ramanujan, pages 136–162. AMS Chelsea Publ., Providence, RI (2000).

[39] R. A. Rankin, The construction of automorphic forms from the derivatives of a given form, *J. Indian Math. Soc. (N.S.)*, **20** (1956) 103–116.

[40] R. A. Rankin, Modular forms and functions, Cambridge University Press, Cambridge (1977).

[41] K. Venkatachaliengar, Development of elliptic functions according to Ramanujan, volume 2 of Technical Report. Madurai Kamaraj University Department of Mathematics, Madurai (1988). With a prologue by T. Soundararajan.

[42] E. T. Whittaker and G. N. Watson, A course of modern analysis. An introduction to the general theory of infinite processes and of analytic functions: with an account of the principal transcendental functions. Fourth edition. Reprinted. Cambridge University Press, New York (1962).

[43] D. Zagier, Modular forms and differential operators, *Proc. Indian Acad. Sci. Math. Sci.*, **104(1)** (1994) 57–75. K. G. Ramanathan memorial issue.

Ramanujan Rediscovered No. 14, 2010, pp. 157–163.

The Overcubic Partition Function mod 3

Byungchan Kim

School of Liberal Arts, Seoul National University of Science and Technology,
172 Gongreung 2 dong, Nowongu, Seoul,139-743, Korea
e-mail: bkim4@seoultech.ac.kr

Abstract. Let $\bar{a}(n)$ be the number of overcubic partitions of n. We show that $\bar{a}(3n+2) \equiv 0 \pmod 3$ and give a crank for this Ramanujan-type congruence.

Keywords. Integer partition, overcubic partition and partition congruences.

In memory of Prof. Venkatachaliengar on the centenary of his birth

1. Introduction and the statement of results

In a series of papers ([3–5]), H.-C. Chan has studied congruence properties of a certain kind of partition $a(n)$, which arises from Ramanujan's cubic continued fraction:

$$\nu(q) := \frac{q^{1/3}}{1} \, \frac{}{+} \, \frac{q+q^2}{1} \, \frac{}{+} \, \frac{q^2+q^4}{1} \, \frac{}{+} \cdots, \qquad |q| < 1.$$

This partition function $a(n)$ is defined by

$$\sum_{n=0}^{\infty} a(n)q^n = \frac{1}{(q;q)_\infty (q^2;q^2)_\infty}.$$

Here and in the sequel, we use the following standard q-series notations:

$$(a;q)_0 := 1,$$
$$(a;q)_n := (1-a)(1-aq)\cdots(1-aq^{n-1}), \qquad n \geq 1,$$

and

$$(a;q)_\infty := \lim_{n\to\infty} (a;q)_n, \qquad |q| < 1.$$

We can interpret $a(n)$ as the number of 2-color partitions of n with colors r and b subject to the restriction that the color b appears only in even parts. For example, there are 3 such partitions of 2:

$$2_r, \; 2_b, \; 1_r + 1_r.$$

Since $a(n)$ is closely related with Ramanujan's cubic continued fraction, we will say that $a(n)$ is the number of cubic partitions of n.

An overpartition of n is a non-increasing sequence of natural numbers for which the sum is n and in which the first occurrence of a part may be overlined. Let $\bar{p}(n)$ denote the number of overpartitions of an integer n. For convenience, define $\bar{p}(0) = 1$. For example, $\bar{p}(3) = 8$ because there are 8 possible overpartitions of 3:

$$3,\ \bar{3},\ 2+1,\ \bar{2}+1,\ 2+\bar{1},\ \bar{2}+\bar{1},\ 1+1+1,\ \bar{1}+1+1.$$

Since the overlined parts form a partition into distinct parts and the non-overlined parts form an ordinary partition, we have the generating function for overpartitions,

$$\bar{P}(q) = \sum_{n=0}^{\infty} \bar{p}(n)q^n = \frac{(-q;q)_\infty}{(q;q)_\infty}.$$

Overpartitions have been extensively studied by a number of mathematicians since they play an important role in interpreting or proving q-series identities combinatorially. For more information and references in this direction, see [6,7] and [20]. Recently, many arithmetic properties of the overpartition function have been investigated. For example, consult [9,12,15] and [19].

In this note, we study the overpartition analog of the cubic partition, namely the overcubic partition. An overcubic partition of n is a 2-color partition of n with colors r and b subject to the restriction that the color b appears only in even parts, and we may overline the first occurrence of a part. For example, there are 6 such partitions of 2:

$$2_r,\ \overline{2_r},\ 2_b,\ \overline{2_b},\ 1_r+1_r,\ \overline{1_r}+1_r.$$

Define $\bar{a}(n)$ as the number of overcubic partitions of n. Then, we can easily see that

$$\sum_{n=0}^{\infty} \bar{a}(n)q^n = \frac{(-q;q)_\infty(-q^2;q^2)_\infty}{(q;q)_\infty(q^2;q^2)_\infty}.$$

We will prove that this partition function satisfies the following Ramanujan type congruence.

Theorem 1. *For all non-negative integers n,*

$$\bar{a}(3n+2) \equiv 0 \pmod{3}. \tag{1}$$

This congruence can be readily derived from the following theorem:

Theorem 2. *For all non-negative integers n,*

$$\sum_{n=0}^{\infty} \bar{a}(3n+2)q^n = 6\frac{(q^3;q^3)_\infty^6(q^4;q^4)_\infty^3}{(q;q)_\infty^8(q^2;q^2)_\infty^3}.$$

Remark. This is an analog of "Ramanujan's most beautiful identity" (in the words of G. H. Hardy [18, p. xxxv]), namely,

$$\sum_{n=0}^{\infty} p(5n+4)q^n = 5\frac{(q^5;q^5)_\infty^5}{(q;q)_\infty^6},$$

where $p(n)$ is the number of ordinary partitions of n.

Once we have the congruence (1), it is natural to seek a combinatorial explanation. To explain Ramanujan congruences for the ordinary partition function, G. E. Andrews and F. G. Garvan [1] introduced the crank of a partition. For a given partition λ, the crank $c(\lambda)$ of a partition is defined as

$$c(\lambda) := \begin{cases} \ell(\lambda), & \text{if } r = 0, \\ \omega(\lambda) - r, & \text{if } r \geq 1, \end{cases}$$

where r is the number of appearances of 1's in λ, $\omega(\lambda)$ is the number of parts in λ that are strictly larger than r and $\ell(\lambda)$ is the largest part in λ. Here, we give a Andrews-Garvan type crank for (1). Let us define the crank for an overcubic partition of n as the number of even parts with color r minus the number of even parts with color b. Let $M(m, n)$ be the number of overcubic partitions of n with crank m. Then, from the fact that $\frac{1}{(zq;q)_\infty} = \sum_{m=0}^{\infty} p(m, n) z^m q^n$, where $p(m, n)$ denotes the number of partitions of n with the number of parts equals m, we can easily derive that

$$\sum_{m=-\infty}^{\infty} \sum_{n=0}^{\infty} M(m, n) z^m q^n = \frac{(-q; q)_\infty (-q^2; q^2)_\infty}{(q; q^2)_\infty (zq^2; q^2)_\infty (z^{-1}q^2; q^2)_\infty}.$$

Let $M(m, N, n)$ be the number of overcubic partitions of n with crank $\equiv m \pmod{N}$. Then, we can prove the following theorem:

Theorem 3. *For all nonnegative integers n,*

$$M(0, 3, 3n + 2) \equiv M(1, 3, 3n + 2) \equiv M(2, 3, 3n + 2) \pmod{3}.$$

Since $M(0, 3, 3n + 2) + M(1, 3, 3n + 2) + M(2, 3, 3n + 2) = \bar{a}(3n + 2)$, this immediately proves Theorem 1.

2. Preliminary results

In this section, we summarize some basic facts about q-series identities and modular functions. We define Ramanujan's general theta function $f(a, b)$ as

$$f(a, b) := \sum_{n=-\infty}^{\infty} a^{n(n+1)/2} b^{n(n-1)/2}, \quad |ab| < 1.$$

Then, Jacobi's triple product identity asserts that

$$f(a, b) = (-a; ab)_\infty (-b; ab)_\infty (ab; ab)_\infty. \tag{2}$$

We will also need Euler's identity

$$(-q; q)_\infty = \frac{1}{(q; q^2)_\infty}. \tag{3}$$

Remark. Euler's identity has the following famous partition-theoretic interpretation: The number of partitions of n into distinct parts equals the number of partitions of n into odd parts.

For the proof of the above two identities and more background on Ramanujan's theta function, the author strongly recommends the book of B. Berndt [2].

Now we give some basic properties of modular functions. For more details on this subject, consult [17]. Define $\Gamma = SL_2(\mathbb{Z})$, $\Gamma_0(N) := \left\{\left(\begin{smallmatrix} a & b \\ c & d \end{smallmatrix}\right) \in \Gamma : c \equiv 0 \pmod{N}\right\}$. For a meromorphic function f on the complex upper half plane $\mathcal{H}$, define the Slash operator by

$$f|\begin{pmatrix} a & b \\ c & d \end{pmatrix} := f\left(\frac{az+b}{cz+d}\right).$$

We say a meromorphic function f is a modular function on $\Gamma_0(N)$ if f is invariant under the Slash operator. Let $\mathcal{M}_0(\Gamma_0(N))$ denote the vector space of modular functions on $\Gamma_0(N)$. For a prime p, we define the U_p-operator as follows. If $f(q)$ has a Fourier expansion $f(q) = \sum a(n)q^n$, then

$$U_p f(z) := \sum a(pn)q^n.$$

It is well known that $U_p f(z) \in \mathcal{M}_0(\Gamma_0(Np))$ provided $f(z) \in \mathcal{M}_0(\Gamma_0(Np^2))$. Let $\eta(z) = q^{1/24}(q;q)_\infty$ be Dedekind Eta function, where $q = \exp(2\pi i z)$ and z is in the upper half complex plane $\mathcal{H}$. For a fixed N and integers r_i's, a function of the form

$$f(z) := \prod_{\substack{n|N \\ n>0}} \eta(nz)^{r_n}. \tag{4}$$

is called an η-quotient. The following theorem of Newman [16] gives a complete criterion for when an η-quotient becomes a modular function.

Theorem 4. *The η-quotient* (4) *is in $\mathcal{M}_0(\Gamma_0(N))$ if and only if*

(1) $\sum_{n|N} r_n = 0$,
(2) $\sum_{n|N} nr_n \equiv 0 \pmod{24}$,
(3) $\sum_{n|N} \frac{N}{n} r_n \equiv 0 \pmod{24}$,
(4) $\prod_{n|N} n^{r_n} \in \mathbb{Q}^2$.

We call the orbits of $\mathbb{Q} \cup \{i\infty\} \pmod{\Gamma_0(N)}$ the cusps of $\Gamma_0(N)$. It is well known that the cusps of $\Gamma_0(N)$ consist of numbers c/d, where $d|N$, $0 \le c < d$ and $(c,d) = 1$. At each cusp c/d, $f \in \mathcal{M}_0(\Gamma_0(N))$ has a Fourier expansion of the form $f(z) = \sum_{n \ge n_0} a(n)q^{n/N}$ with $a(n_0) \neq 0$. We call n_0 the order of f at the cusp c/d and denote it $ord_{c/d} f$. The following theorem of Ligozat [14] gives the order of the η-quotient f at the cusps c/d of $\Gamma_0(N)$ provided $f \in \mathcal{M}_0(\Gamma_0(N))$.

Theorem 5. *If the η-quotient $f \in \mathcal{M}_0(\Gamma_0(N))$, then its order at the cusp c/d of $\Gamma_0(N)$ is*

$$\frac{1}{24}\sum_{n|N} \frac{N(d,n)^2 r_n}{(d,N/d)dn}.$$

Recall that if $p|N$ and $f \in \mathcal{M}_0(\Gamma_0(pN))$, then $U_p f \in \mathcal{M}_0(\Gamma_0(N))$. The following theorem gives bounds on the order of $U_p f$ at cusps of $\Gamma_0(N)$ in terms of the order of f at cusps of $\Gamma_0(pN)$.

Theorem 6 (Theorem 4 of [10]). *Let p be a prime and $\pi(n)$ be the highest power of p dividing n. Suppose that $f \in \mathcal{M}_0(\Gamma_0(pN))$, where $p|N$ and $\alpha = c/d$ is a cusp of $\Gamma_0(N)$. Then,*

$$\operatorname{ord}_\alpha U_p f \geq \begin{cases} \frac{1}{p}\operatorname{ord}_{\alpha/p} f, & \text{if } \pi(d) \geq \frac{1}{2}\pi(N), \\ \operatorname{ord}_{\alpha/p} f, & \text{if } 0 < \pi(d) < \frac{1}{2}\pi(N), \\ \min_{0 \leq \beta \leq p-1} \operatorname{ord}_{(\alpha+\beta)/p} f, & \text{if } \pi(d) = 0. \end{cases}$$

3. Proofs of theorems

Since we can prove Theorem 1 by applying a very similar argument that we will use in the proof of Theorem 3, we omit the proof of Theorem 1.

For the proof of Theorem 2, we will follow the argument given in [10].

Proof of Theorem 2. Let us define $F(z)$ as

$$F(z) := \frac{\eta(4z)\eta^{15}(6z)}{\eta^2(z)\eta(2z)\eta^{10}(3z)\eta^3(12z)}.$$

By Theorem 4, we have $F(z) \in \mathcal{M}_0(\Gamma_0(36))$. Note that $U_3 f(z) \in \mathcal{M}_0(\Gamma_0(12))$. Let us define $G(z)$ as

$$G(z) := \frac{\eta^{12}(2z)\eta^6(3z)}{\eta^{18}(z)}.$$

Then, by Theorem 4, we have $G(z) \in \mathcal{M}_0(\Gamma_0(12))$. From the order at each cusp by employing Theorems 5 and 6, we see that $\frac{U_3 F}{G}$ is a holomorphic modular function, namely, a constant. From this, we can easily deduce that $U_3 F(z) = 6G(z)$. Recall that $U_p f(pz)g(z) = f(z)U_p g(z)$. Thus, we have

$$U_3\left(\sum_{n=0}^{\infty} \bar{a}(n)q^{n+1}\right)\frac{(q^2;q^2)_\infty^{15}}{(q;q)_\infty^{10}(q^4;q^4)_\infty^3} = 6q\frac{(q^2;q^2)_\infty^{12}(q^3;q^3)_\infty^6}{(q;q)_\infty^{18}},$$

or

$$\sum_{n=0}^{\infty} \bar{a}(3n+2)q^n = 6\frac{(q^3;q^3)_\infty^6(q^4;q^4)_\infty^3}{(q;q)_\infty^8(q^2;q^2)_\infty^3},$$

as desired. □

Proof of Theorem 3. Let ξ be a primitive third root of unity. Then, it is not hard to see that

$$F(\xi, q) := \sum_{n=0}^{\infty}\sum_{k=0}^{2} M(k, 3, n)\xi^k q^n$$

$$= \sum_{m=-\infty}^{\infty} \sum_{n=0}^{\infty} M(m,n)\xi^m q^n$$

$$= \frac{(-q;q)_\infty(-q^2;q^2)_\infty}{(q;q^2)_\infty(\xi q^2;q^2)_\infty(\xi^{-1}q^2;q^2)_\infty}.$$

Then, we can deduce that

$$\begin{aligned} F(\xi,q) &= \frac{(q^4;q^4)_\infty}{(q;q^2)_\infty^2(q^6;q^6)_\infty} \\ &= \frac{(q;q^2)_\infty(q^4;q^4)_\infty}{(q;q^2)_\infty^3(q^6;q^6)_\infty} \\ &\equiv \frac{\sum_{-\infty}^{\infty}(-1)^n q^{2n^2-n}}{(q^3;q^6)_\infty(q^6;q^6)_\infty} \pmod 3, \end{aligned}$$

where we have used Euler identity (3) for the first identity and Jacobi's triple product identity (2) for the last congruence. Since $2n^2 - n \equiv 0$ or $1 \pmod 3$, we arrive at

$$\sum_{k=0}^{2} M(k,3,3n+2)\xi^k \equiv 0 \pmod 3.$$

Since $1 + \xi + \xi^2$ is a minimal polynomial in $\mathbb{Z}[\xi]$, we must have

$$M(0,3,3n+2) \equiv M(1,3,3n+2) \equiv M(2,3,3n+2) \pmod 3.$$

This completes the proof. □

4. Concluding remarks

Though the work of S. Treneer [19] implies that there are infinitely many congruences for $\bar{a}(n)$ modulo N, if N is coprime to 6, it appears that $\bar{a}(3n+2) \equiv 0 \pmod 3$ is the only simple congruence. Recent work of M. Dewar [8] gives a criterion for the non-existence of simple congruence for a large class of generating functions. Secondly, we can easily see that $\bar{a}(n)$ is always even, except $\bar{a}(0) = 1$. In [15], K. Mahlburg conjectured that for all positive integers k, $\bar{p}(n) \equiv 0 \pmod{2^k}$ for almost all integers n. This conjecture was proven for $k = 6$ by Mahlburg [15] and for $k = 7$ by the author [11]. From numerical data, it appears that for all positive integers k, $\bar{a}(n) \equiv 0 \pmod{2^k}$ for almost all integers n. Unlike $a(3n+2) \equiv \bar{a}(3n+2) \equiv 0 \pmod 3$, it is not true, in general, that $\bar{p}(5n+4) \equiv 0 \pmod 5$ though $p(5n+4) \equiv 0 \pmod 5$. It would be interesting to investigate which partition functions and their overpartition analogs satisfy the same Ramanujan-type congruences.

Acknowledgments

The author would like to thank Bruce Berndt for his encouragement and help. The author also thanks to the referee for his/her careful reading and suggestions.

References

[1] G. E. Andrews and F. G. Garvan, Dyson's crank of a partition, *Bull. Am. Math. Soc.*, **18** (1988) 167–171.

[2] B. C. Berndt, Number theory in the spirit of Ramanujan, American Mathematical Society, Providence, RI (2006).

[3] H.-C. Chan, Ramanujan's cubic continued fraction and an analog of his "most beautiful identity", *Int. J. Number Thy*, **6** (2010) 673–680.

[4] H.-C. Chan, Ramanujan's cubic continued fraction and Ramanujan type congruences for a certain partition function, *Int. J. Number Thy*, **6** (2010) 819–834.

[5] H.-C. Chan, Distribution of a certain partition function modulo powers of primes, *Acta Math. Sin.*, to appear.

[6] S. Corteel and J. Lovejoy, Overpartitions, *Trans. Amer. Math. Soc.*, **356** (2004) 1623–1635.

[7] S. Corteel, J. Lovejoy and A. J. Yee, Overpartitions and generating functions for generalized Frobenius partitions, Mathematics and Computer Science III: Algorithms, Trees, Combinatorics, and Probabilities (2004) 15–24.

[8] M. Dewar, Non-existence of Ramanujan congruences in modular forms of level four, preprint.

[9] M. D. Hirschhorn and J. Sellers, An infinite family of overpartition congruences modulo 12, *Integers*, **5** (2005) A20, 4 pp.

[10] B. Gordon and K. Hughes, Ramanujan congruences for $q(n)$, Analytic number theory (Philadelphia, Pa., 1980), 333–359, Lecture Notes in Math., 899, Springer, Berlin-New York (1981).

[11] B. Kim, The overpartition function modulo 128, *Integers*, **8** (2008) A38, 8 pp.

[12] B. Kim, A short note on the overpartition function, *Discrete Math.*, **309** (2009) no. 8, 2528–2532.

[13] B. Kim, A Crank analog on a certain kind of partition function arising from the cubic continued fraction, preprint.

[14] G. Ligozat, Courbes modulaires de genre 1, *Bull. Soc. Math.*, France [Memoire 43] (1972) 1–80.

[15] K. Mahlburg, The overpartition function modulo small powers of 2, *Discrete Math.*, **286** (2004) 263–267.

[16] M. Newman, Construction and application of a class of modular functions II, *Proc. London Math. Soc. (3)*, **9** (1959) 373–387.

[17] K. Ono, Web of Modularity: arithmetic of the coefficients of modular forms and q-series, CBMS Regional Conference Series in Mathematics, 102. Published for the Conference Board of the Mathematical Sciences, Washington, DC, by the American Mathematical Society, Providence, RI (2004).

[18] S. Ramanujan, Collected Papers, Cambridge University Press, Cambridge, 1927; reprinted by Chelsea, New York, 1962; reprinted by the American Mathematical Society, Providence, RI (2000).

[19] S. Treneer, Congruences for the coefficients of weakly holomorphic modular forms, *Proc. London Math. Soc. (3)*, **93** (2006) 304–324.

[20] A. J. Yee, Combinatorial proofs of Ramanujan's ${}_1\psi_1$ summation and the q-Gauss summation, *J. Combin. Theory Ser. A*, **105** (2004) 63–77.

Ramanujan Rediscovered No. 14, 2010, pp. 165–183.

A Theta Function Identity and Applications

Zhi-Guo Liu

Department of Mathematics, East China Normal University, 500 Dongchuan Road, Shanghai 200241, P.R. China
e-mail: zgliu@math.ecnu.edu.cn; liuzg@hotmail.com

Abstract. We prove an identity relating a general entire function and theta functions using the complex variable theory of elliptic functions, which has many applications to the theory of elliptic theta functions.

Keywords. Elliptic function, theta function, entire function, Eisenstein series, quintuple product identity.

2000 Subject Classification: 30D15, 33E05, 11F11, 11F20, 11F27

1. Introduction

Throughout this paper, we set $q = \exp(2\pi i \tau)$, where $\mathrm{Im}\tau > 0$.

Definition 1.1. *The Jacobi theta functions $\theta_1, \theta_2, \theta_3$, and θ_4 are defined as*

$$\theta_1(z|\tau) = -i \sum_{n=-\infty}^{\infty} (-1)^n q^{(2n+1)^2/8} e^{(2n+1)iz}, \quad \theta_3(z|\tau) = \sum_{n=-\infty}^{\infty} q^{n^2/2} e^{2niz},$$

$$\theta_2(z|\tau) = \sum_{n=-\infty}^{\infty} q^{(2n+1)^2/8} e^{(2n+1)iz}, \quad \theta_4(z|\tau) = \sum_{n=-\infty}^{\infty} (-1)^n q^{n^2/2} e^{2niz}.$$

These series converge for all complex values of z whenever τ has positive imaginary part (equivalently, when $|q| < 1$). Furthermore, these series converge absolutely and uniformly on compact subsets and so are entire functions of z.

We use the prime to denote partial differential operator with respect to the first variable of theta functions. For brevity, we use $\vartheta_1'(\tau)$, $\vartheta_2(\tau)$, $\vartheta_3(\tau)$, and $\vartheta_4(\tau)$ to denote $\theta_1'(0|\tau), \theta_2(0|\tau), \theta_3(0|\tau)$, and $\theta_4(0|\tau)$ respectively.

We prove the following theorem with the complex theory of elliptic functions which is the main result of this paper:

Theorem 1.1. *Suppose that $f(z)$ is an entire function of z and satisfies the functional equations*

$$f(z) = f(z+\pi) = q^3 e^{12iz} f(z+\pi\tau). \tag{1.1a}$$

This work was supported by the National Science Foundation of China and the Shanghai Natural Science Foundation, grant number 10ZR1409100.

Then we have the identity

$$\frac{2(f(z)+f(-z))}{\theta_1^2(2z|\tau)} = \frac{f(0)}{\theta_1^2(z|\tau)} + \frac{f\left(\frac{\pi}{2}\right)}{\theta_2^2(z|\tau)} - \frac{q^{\frac{3}{4}} f\left(\frac{\pi+\pi\tau}{2}\right)}{\theta_3^2(z|\tau)} - \frac{q^{\frac{3}{4}} f\left(\frac{\pi\tau}{2}\right)}{\theta_4^2(z|\tau)}. \tag{1.1b}$$

Theorem 1.1 contains the following two important corollaries as well as many other results as special cases:

Corollary 1.1. *Let* $g(z)$ *be an entire function of* z *which satisfies the functional equations* $g(z) = -g(z+\pi) = -q^{5/2}e^{10iz}g(z+\pi\tau)$. *Then we have*

$$\frac{g'(0)}{\vartheta_1'(\tau)} = \frac{g\left(\frac{\pi}{2}\right)}{\vartheta_2(\tau)} - \frac{q^{5/8} g\left(\frac{\pi+\pi\tau}{2}\right)}{\vartheta_3(\tau)} - \frac{iq^{5/8} g\left(\frac{\pi\tau}{2}\right)}{\vartheta_4(\tau)}. \tag{1.2}$$

Corollary 1.2. *Suppose* $f(z)$ *is an entire function with* $f(0) \neq 0$, *which satisfies the functional equations in* $(1.1a)$. *Then we have*

$$2\left(1 - 24\sum_{n=1}^{\infty} \frac{nq^n}{1-q^n}\right) + (\log f)'(0)^2 + (\log f)''(0)$$
$$= \frac{2\vartheta_1'(\tau)^2}{f(0)}\left(\frac{f\left(\frac{\pi}{2}\right)}{\vartheta_2^2(\tau)} - \frac{q^{\frac{3}{4}} f\left(\frac{\pi+\pi\tau}{2}\right)}{\vartheta_3^2(\tau)} - \frac{q^{\frac{3}{4}} f\left(\frac{\pi\tau}{2}\right)}{\vartheta_4^2(\tau)}\right), \tag{1.3}$$

where $(\log)'(f)(z)$ *denotes the logarithmic derivative of* f *with respect to* z.

The remainder of this paper is organized as follows. In Section 2, we prove Theorem 1.1, Corollaries 1.1, and 1.2. In Section 3, we use Theorem 1.1 and Corollaries 1.1 to derive some well-known theta function identities as well as new identities. In Section 4, we provide a theta function representation of $\wp$-function, which was used to derive some Eisenstein series identities. In Section 6, we claim a general identity which includes Theorem 1.1 as special case.

2. Elliptic functions and the proof of theorem 1.1

We first introduce some basic properties of theta functions. We observe that the four Jacobi theta functions are mutually related; and starting from one of them we may obtain the other three by simple calculation. For example,

$$\theta_1(z+\pi/2|\tau) = \theta_2(z|\tau),$$
$$\theta_1(z+(\pi\tau)/2|\tau) = iq^{-1/8}e^{-iz}\theta_4(z|\tau),$$
$$\theta_1(z+(\pi+\pi\tau)/2|\tau) = q^{-1/8}e^{-iz}\theta_3(z|\tau). \tag{2.1}$$

It is easily seen that in the fundamental periodic parallelogram the zeros of $\theta_1, \theta_2, \theta_3$, and θ_4 are respectively $0, \pi/2, (\pi+\pi\tau)/2$, and $(\pi\tau)/2$.

The infinite product representations of theta functions are given by the following proposition (see for example, [20, p. 469]):

Proposition 2.1.

$$\theta_1(z|\tau) = 2q^{1/8}(\sin z)\prod_{n=1}^{\infty}(1-q^n)(1-q^n e^{2iz})(1-q^n e^{-2iz}),$$

$$\theta_2(z|\tau) = 2q^{1/8}(\cos z)\prod_{n=1}^{\infty}(1-q^n)(1+q^n e^{2iz})(1+q^n e^{-2iz}),$$

$$\theta_3(z|\tau) = \prod_{n=1}^{\infty}(1-q^n)(1+q^{n-1/2} e^{2iz})(1+q^{n-1/2} e^{-2iz}),$$

$$\theta_4(z|\tau) = \prod_{n=1}^{\infty}(1-q^n)(1-q^{n-1/2} e^{2iz})(1-q^{n-1/2} e^{-2iz}).$$

If we divide both sides of the first equation in Proposition 2.1 with z and then let $z \to 0$, we get the following proposition:

Proposition 2.2.

$$\vartheta_1'(\tau) = 2q^{1/8}\prod_{n=1}^{\infty}(1-q^n)^3 = 2\eta^3(\tau),$$

where $\eta(\tau)$ is the Dedekind eta function.

Now we begin to prove Theorem 1.1 using the theory of elliptic functions.

Proof. Suppose f is the given function satisfying the functional equation (1.1a). We temporarily assume that $0 < z < \pi$ and consider the function

$$F(u) = \frac{f(u)}{\theta_1(2u|\tau)\theta_1(u-z|\tau)\theta_1(u+z|\tau)}.$$

From the definition of θ_1, it is easy to find the functional equations

$$\theta_1(z|\tau) = -\theta_1(z+\pi|\tau) = -q^{1/2}e^{2iz}\theta_1(z+\pi\tau|\tau). \tag{2.2}$$

Using the above equations and (1.1a), we can verify that $F(u+\pi) = F(u+\pi\tau) = F(u)$. Hence $F(u)$ is an elliptic function with periods π and $\pi\tau$. It is obvious that with the (quasi) periods π and $\pi\tau$, 0 is the only zero of $\theta_1(u|\tau)$ in the period parallelogram. With this fact and using (2.2), we easily find that the zeros of $\theta_1(2u|\tau)$ in the period parallelogram are $0, \pi/2, (\pi+\pi\tau)/2, (\pi\tau)/2$. It follows that F has only poles at $u_1 := z, u_2 := -z, u_3 := 0, u_4 := \pi/2, u_5 := (\pi+\pi\tau)/2, u_6 := (\pi\tau)/2$ and all these are simple poles. It is well-known that the sum of all the residues of an elliptic function vanishes in the period parallelogram (see for example, [2, p. 6] and [9, p. 22]). Thus if we denote the residue of F at γ by $\mathrm{res}(F;\gamma)$, then we have

$$\mathrm{res}(F;u_1) + \mathrm{res}(F;u_2) + \cdots + \mathrm{res}(F;u_6) = 0. \tag{2.3}$$

Next we begin to compute the residues of F. We have the following elementary calculation:

$$\begin{aligned}\mathrm{res}(F; u_1) = \mathrm{res}(F; z) &= \lim_{u\to z}(u-z)F(u)\\ &= \lim_{u\to z}\frac{f(u)}{\theta_1(2u|\tau)\theta_1(u+z|\tau)} \times \lim_{u\to z}\frac{(u-z)}{\theta_1(u-z|\tau)}\\ &= \frac{f(z)}{\vartheta_1'(\tau)\theta_1^2(2z|\tau)}.\end{aligned}$$

In the above equation we replace z by $-z$ to deduce that

$$\mathrm{res}(F; u_2) = \frac{f(-z)}{\vartheta_1'(\tau)\theta_1^2(2z|\tau)}.$$

With the aid of (2.1) and some simple calculation, we easily find that

$$\mathrm{res}(F; u_3) = -\frac{f(0)}{2\vartheta_1'(\tau)\theta_1^2(z|\tau)},$$

$$\mathrm{res}(F; u_4) = -\frac{f\left(\frac{\pi}{2}\right)}{2\vartheta_1'(\tau)\theta_2^2(z|\tau)},$$

$$\mathrm{res}(F; u_5) = \frac{q^{\frac{3}{4}} f\left(\frac{\pi+\pi\tau}{2}\right)}{2\vartheta_1'(\tau)\theta_3^2(z|\tau)},$$

$$\mathrm{res}(F; u_6) = \frac{q^{\frac{3}{4}} f\left(\frac{\pi\tau}{2}\right)}{2\vartheta_1'(\tau)\theta_4^2(z|\tau)}.$$

Substituting the above six equations into (2.3), and simplifying, we obtain (1.1b) for $0 < z < \pi$. By analytic continuation we know (1.1b) holds for all u, and so this completes the proof of Theorem 1.1. □

Next we prove Corollary 1.1 using Theorem 1.1.

Proof. Taking $f(z) = \theta_1(z|\tau)g(z)$ in Theorem 1.1 and then letting $z \to 0$, we complete the proof of Corollary 1.1. □

Lastly, we use Theorem 1.1 to prove Corollary 1.2.

Proof. If we use $\phi(z)$ to denote the function on the left hand side of (1.1b), then it is obvious that $\phi(z)$ is even and has a pole at $z = 0$ with order 2. Thus we can assume that

$$\phi(z) = \frac{A}{z^2} + B + O(z^2). \tag{2.4}$$

Now we begin to determine constants A and B. If we multiply through the above equation with z^2, then we find that

$$A = \lim_{z\to 0} z^2\phi(z) = \frac{f(0)}{\vartheta_1'(\tau)^2} \quad \text{and} \quad B = \frac{1}{2}(z^2\phi(z))''\Big|_{z=0}. \tag{2.5}$$

Using the method of logarithmic differentiation and by direct computation, we deduce that

$$(z^2\phi(z))'' = z^2\phi(z)\left(\left(\frac{2}{z} + (\log\phi)'(z)\right)^2 + \left(\frac{2}{z} + (\log\phi)'(z)\right)'\right). \tag{2.6}$$

Substituting this equation into the right hand side of the second equation in (2.5), we conclude that

$$B = \frac{1}{2}z^2\phi(z)\left(\left(\frac{2}{z} + (\log\phi)'(z)\right)^2 + \left(\frac{2}{z} + (\log\phi)'(z)\right)'\right)\Bigg|_{z=0}. \tag{2.7}$$

Now we need to compute the logarithmic derivative of $\phi(z)$. From the definition of $\phi(z)$, we have

$$\frac{2}{z} + (\log\phi)'(z) = \frac{2}{z} + \frac{f'(z) - f'(-z)}{f(z) + f(-z)} - 4(\log\theta_1)'(2z|\tau). \tag{2.8}$$

Using the trigonometric series expansion for the logarithmic derivative of θ_1 (see for example, [20, p. 489]), we can readily find that

$$(\log\theta_1)'(z|\tau) = \frac{1}{z} - \frac{1}{3}\left(1 - 24\sum_{n=1}^{\infty}\frac{nq^n}{1-q^n}\right)z + O(z^2). \tag{2.9}$$

Substituting the above equation into (2.8), we are led to the formula

$$\frac{2}{z} + (\log\phi)'(z) = \frac{f'(z) - f'(-z)}{f(z) + f(-z)} + \frac{8}{3}\left(1 - 24\sum_{n=1}^{\infty}\frac{nq^n}{1-q^n}\right)z + O(z^2),$$

which immediately gives that

$$\left(\frac{2}{z} + (\log\phi)'(z)\right)\Bigg|_{z=0} = 0,$$

$$\left(\frac{2}{z} + (\log\phi)'(z)\right)'\Bigg|_{z=0} = \frac{8}{3}\left(1 - 24\sum_{n=1}^{\infty}\frac{nq^n}{1-q^n}\right) + (\log f)'(0)^2 + (\log f)''(0).$$

We now substitute the above two equations into (2.7) and conclude that

$$B = \frac{f(0)}{\vartheta_1'(\tau)^2}\left(\frac{4}{3}\left(1 - 24\sum_{n=1}^{\infty}\frac{nq^n}{1-q^n}\right) + \frac{(\log f)'(0)^2}{2} + \frac{(\log f)''(0)}{2}\right). \tag{2.10}$$

We next study the behavior of the first term on the right hand side of (1.1b). Proceeding through the same steps as above, we may readily find that

$$\frac{f(0)}{\theta_1^2(z|\tau)} = \frac{A}{z^2} + \frac{f(0)}{3\vartheta_1'(\tau)^2}\left(1 - 24\sum_{n=1}^{\infty}\frac{nq^n}{1-q^n}\right) + O(z^2). \tag{2.11}$$

Substituting this equation and (2.4) into (1.1b), and then multiplying both sides by $2\vartheta_1'(\tau)^2/f(0)$, we find that

$$2\left(1-24\sum_{n=1}^{\infty}\frac{nq^n}{1-q^n}\right)+(\log f)'(0)^2+(\log f)''(0)+O(z^2)$$

$$=\frac{2\vartheta_1'(\tau)^2}{f(0)}\left(\frac{f\left(\frac{\pi}{2}\right)}{\theta_2^2(z|\tau)}-\frac{q^{\frac{3}{4}}f\left(\frac{\pi+\pi\tau}{2}\right)}{\theta_3^2(z|\tau)}-\frac{q^{\frac{3}{4}}f\left(\frac{\pi\tau}{2}\right)}{\theta_4^2(z|\tau)}\right).$$

Putting $z=0$ in the above equation, we arrive at (1.3). Thus we complete the proof of Corollary 1.2. □

3. Some theta functions identities

Proposition 3.1. *Suppose $h(z)$ is an entire function of z and satisfies the functional equations $h(z)=h(z+\pi)=q^2e^{8iz}h(z+\pi\tau)$. Then we have* [17]

$$h(0)+q^{1/2}h\left(\frac{\pi+\pi\tau}{2}\right)=h\left(\frac{\pi}{2}\right)+q^{1/2}h\left(\frac{\pi\tau}{2}\right). \tag{3.1}$$

Proof. Suppose $h(z)$ is the given function in Proposition 3.1. Then we can choose $g(z)=\theta_1(z|\tau)h(z)$ in Corollary 1.1 to obtain Proposition 3.1. □

Example 3.1. When $g(z)=\theta_1^4(z|\tau)$, (3.1) immediately gives the Jacobi quartic identity

$$\vartheta_3^4(\tau)=\vartheta_2^4(\tau)+\vartheta_4^4(\tau).$$

Example 3.2. If we take $h(z)=\theta_1(3z|3\tau)\theta_1(z|\tau)$ in (3.1), then we get the Legendre identity [5, p. 230, Entry 5(ii)], [15]

$$\vartheta_2(\tau)\vartheta_2(3\tau)+\vartheta_4(\tau)\vartheta_4(3\tau)=\vartheta_3(\tau)\vartheta_3(3\tau).$$

Example 3.3. In (3.1), letting $h(z)=\theta_1(z|\tau)\theta_1\left(z+\frac{\pi}{7}|\tau\right)\theta_1\left(z+\frac{2\pi}{7}|\tau\right)\theta_1\left(z-\frac{3\pi}{7}|\tau\right)$ and performing a little reduction, we can find the Guetslaff identity [13] (see also [5, p. 314, Entry 19 (i)])

$$\sqrt{\vartheta_3(\tau)\vartheta_3(7\tau)}=\sqrt{\vartheta_2(\tau)\vartheta_2(7\tau)}+\sqrt{\vartheta_4(\tau)\vartheta_4(7\tau)}.$$

Example 3.4 (Quintuple product identity).

$$\sum_{n=-\infty}^{\infty}(-1)^n q^{n(3n+1)/2}\cos(2n+1)z=\prod_{n=1}^{\infty}(1-q^n)\frac{\theta_1(2z|\tau)}{\theta_1(z|\tau)}.$$

This identity has a very rich history and there are many different proofs of this identity in the literature. One may consult [5, p. 83] and the survey paper [12] for various proofs of this identity.

Proof. In Theorem 1.1, we are allowed to take the function $f(z)$ as follows:

$$e^{2iz}\theta_1(3z+\pi\tau|3\tau)\theta_1(2z|\tau)\theta_1^{-1}(z|\tau).$$

It is well-known that $\pi/2$, $(\pi\tau)/2$, and $(\pi+\pi\tau)/2$ are the zeros of $\theta_1(2z|\tau)$, and thus we have

$$f\left(\frac{\pi}{2}\right)=f\left(\frac{\pi\tau}{2}\right)=f\left(\frac{\pi+\pi\tau}{2}\right)=0.$$

With these facts, it is easily seen that (1.1b) can be simplified to the identity

$$e^{2iz}\theta_1(3z+\pi\tau|3\tau)+e^{-2iz}\theta_1(-3z+\pi\tau|3\tau)=\theta_1(\pi\tau|3\tau)\frac{\theta_1(2z|\tau)}{\theta_1(z|\tau)}.$$

Appealing to the infinite product expansion for θ_1, we easily find that

$$\theta_1(\pi\tau|3\tau)=iq^{-1/8}\prod_{n=1}^{\infty}(1-q^n).$$

Combining the above two equations, we conclude that

$$e^{2iz}\theta_1(3z+\pi\tau|3\tau)+e^{-2iz}\theta_1(-3z+\pi\tau|3\tau)=iq^{-1/8}\prod_{n=1}^{\infty}(1-q^n)\frac{\theta_1(2z|\tau)}{\theta_1(z|\tau)}.$$

Substituting the series expansion of θ_1 into the left hand side of the above equation and simplifying, we arrive at the quintuple product identity. □

Example 3.5.

$$\frac{4\theta_1^3(z+y|\tau)\theta_1^3(z-y|\tau)}{\theta_1^2(2z|\tau)}$$
$$=-\frac{\theta_1^6(y|\tau)}{\theta_1^2(z|\tau)}+\frac{\theta_2^6(y|\tau)}{\theta_2^2(z|\tau)}-\frac{\theta_3^6(y|\tau)}{\theta_3^2(z|\tau)}+\frac{\theta_4^6(y|\tau)}{\theta_4^2(z|\tau)}.$$

Proof. With the help of (2.2), it is easy to verify that $\theta_1^3(z+y|\tau)\theta_1^3(z-y|\tau)$ satisfies (1.1a). Thus we may choose $f(z)=\theta_1^3(z+y|\tau)\theta_1^3(z-y|\tau)$ in Theorem 1.1. After some simplification, identity (1.1b) becomes Example 3.5. □

Theorem 3.1. *Suppose $u_1+u_2+u_3+u_4+u_5=0$. Then we have the identity*

$$\sum_{r=1}^{5}(\log\theta_1)'(u_r|\tau)=\frac{\vartheta_1'(\tau)}{\vartheta_2(\tau)}\prod_{r=1}^{5}\frac{\theta_2(u_r|\tau)}{\theta_1(u_r|\tau)}-\frac{\vartheta_1'(\tau)}{\vartheta_3(\tau)}\prod_{r=1}^{5}\frac{\theta_3(u_r|\tau)}{\theta_1(u_r|\tau)}$$
$$+\frac{\vartheta_1'(\tau)}{\vartheta_4(\tau)}\prod_{r=1}^{5}\frac{\theta_4(u_r|\tau)}{\theta_1(u_r|\tau)}. \tag{3.2}$$

Proof. If $u_1+u_2+u_3+u_4+u_5=0$, then in Corollary 1.1 we can choose $f(z)$ as

$$f(z)=\theta_1(z+u_1|\tau)\theta_1(z+u_2|\tau)\theta_1(z+u_3|\tau)\theta_1(z+u_4|\tau)\theta_1(z+u_5|\tau).$$

With the help of (2.1) and some simple computation, we readily find that

$$f'(0)=\prod_{r=1}^{5}\theta_1(u_r|\tau)\left(\sum_{r=1}^{5}\log\theta_1)'(u_r|\tau)\right),\quad f\left(\frac{\pi}{2}\right)=\prod_{r=1}^{5}\theta_2(u_r|\tau),$$

$$f\left(\frac{\pi+\pi\tau}{2}\right)=q^{-5/8}\prod_{r=1}^{5}\theta_3(u_r|\tau),\quad f\left(\frac{\pi\tau}{2}\right)=i^5q^{-5/8}\prod_{r=1}^{5}\theta_4(u_r|\tau).$$

Substituting the above into (1.2), we immediately get (3.2) to complete the proof of the theorem. □

Corollary 3.1. *If $u_1+u_2+u_3=0$, then we have the identity*

$$-\sum_{r=1}^{3}\tan u_r+4\sum_{n=1}^{\infty}(-1)^n\frac{q^n}{1-q^n}\sum_{r=1}^{3}\sin 2nu_r=\frac{\vartheta_1'(\tau)\theta_1(u_1|\tau)\theta_1(u_2|\tau)\theta_1(u_3|\tau)}{\vartheta_2(\tau)\theta_2(u_1|\tau)\theta_2(u_2|\tau)\theta_2(u_3|\tau)}. \tag{3.3}$$

This identity can be found in [16], and when $u_1=\frac{\pi}{7}$, $u_2=\frac{2\pi}{7}$, and $u_3=-\frac{3\pi}{7}$, it gives the Ramanujan identity [5, p. 304]

$$1+2\sum_{n=1}^{\infty}\left(\frac{n}{7}\right)\frac{(-q)^n}{1-q^n}=\prod_{n=1}^{\infty}\left(\frac{1-q^n}{1+q^n}\right)\left(\frac{1-q^{7n}}{1+q^{7n}}\right). \tag{3.4}$$

Proof. Replacing (u_1,u_2,u_3,u_4,u_5) with $\left(u_1+\frac{\pi}{2},u_2-\frac{\pi}{2},u_3-\frac{\pi}{2},\frac{\pi+\pi\tau}{2},-\frac{\pi\tau}{2}\right)$ in Theorem 3.1, appealing to (2.1), we find that (3.2) becomes

$$\sum_{r=1}^{3}(\log\theta_2)'(u_r|\tau)=\frac{\vartheta_1'(\tau)\theta_1(u_1|\tau)\theta_1(u_2|\tau)\theta_1(u_3|\tau)}{\vartheta_2(\tau)\theta_2(u_1|\tau)\theta_2(u_2|\tau)\theta_2(u_3|\tau)}. \tag{3.5}$$

Substituting the trigonometric series expansion of the logarithmic derivative of θ_2,

$$(\log\theta_2)'(z|\tau)=-\tan z+4\sum_{n=1}^{\infty}\frac{(-q)^n}{1-q^n}\sin 2nz \tag{3.6}$$

into (3.5), we arrive at (3.3). We complete the proof of Corollary 3.1. □

Going through the same steps as in proving Corollary 3.1, replacing (u_1,u_2,u_3,u_4,u_5) with $\left(u_1+\frac{\pi\tau}{2},u_2-\frac{\pi\tau}{2},u_3-\frac{\pi\tau}{2},\frac{\pi+\pi\tau}{2},-\frac{\pi}{2}\right)$ in Theorem 3.1, we can derive the following Corollary [20, p. 490]:

Corollary 3.2. *If $u_1+u_2+u_3=0$, then we have the formula*

$$4\sum_{n=1}^{\infty}\frac{q^{n/2}}{1-q^n}\sum_{r=1}^{3}\sin 2nu_r=\frac{\vartheta_1'(\tau)\theta_1(u_1|\tau)\theta_1(u_2|\tau)\theta_1(u_3|\tau)}{\vartheta_4(\tau)\theta_4(u_1|\tau)\theta_4(u_2|\tau)\theta_4(u_3|\tau)}. \tag{3.7}$$

Now we make use of Theorem 3.1 to prove the following curious theta function identity involving the Dirichlet character modulo 11, which has appeared at the end of [16] without proof.

Proposition 3.2. *Let* $\chi_{11}(n)$ *be the Dirichlet character modulo* 11. *Then we have*

$$1+2\sum_{n=1}^{\infty}\chi_{11}(n)\frac{q^n}{1-q^n}$$
$$=-\frac{1}{11}\sqrt{\frac{\vartheta_1'(\tau)^3}{\vartheta_1'(11\tau)}}\left(\sqrt{\frac{\vartheta_2(11\tau)}{\vartheta_2^3(\tau)}}+\sqrt{\frac{\vartheta_3(11\tau)}{\vartheta_3^3(\tau)}}-\sqrt{\frac{\vartheta_4(11\tau)}{\vartheta_4^3(\tau)}}\right). \tag{3.8}$$

Proof. It is known that the trigonometric series expansion of the logarithmic derivative of θ_1 is

$$(\log\theta_1)'(z|\tau)=\cot z+4\sum_{n=1}^{\infty}\frac{q^n}{1-q^n}\sin 2nz. \tag{3.9}$$

In Theorem 3.1, we let $(u_1,u_2,u_3,u_4,u_5)=\left(\frac{\pi}{11},-\frac{2\pi}{11},\frac{3\pi}{11},\frac{4\pi}{11},-\frac{6\pi}{11}\right)$. Then with the help of the trigonometric series expansion of the logarithmic derivative of θ_1, we find that

$$C+4\sum_{n=1}^{\infty}A(n)\frac{q^n}{1-q^n}$$
$$=-\frac{\vartheta_1'(\tau)}{\vartheta_2(\tau)}\prod_{r=1}^{5}\frac{\theta_2\left(\frac{r\pi}{7}|\tau\right)}{\theta_1\left(\frac{r\pi}{7}|\tau\right)}-\frac{\vartheta_1'(\tau)}{\vartheta_3(\tau)}\prod_{r=1}^{5}\frac{\theta_3\left(\frac{r\pi}{7}|\tau\right)}{\theta_1\left(\frac{r\pi}{7}|\tau\right)}+\frac{\vartheta_1'(\tau)}{\vartheta_4(\tau)}\prod_{r=1}^{5}\frac{\theta_4\left(\frac{r\pi}{7}|\tau\right)}{\theta_1\left(\frac{r\pi}{7}|\tau\right)}, \tag{3.10}$$

where

$$C=\cot\frac{\pi}{11}-\cot\frac{2\pi}{11}+\cot\frac{3\pi}{11}+\cot\frac{4\pi}{11}-\cot\frac{6\pi}{11},$$
$$A(n)=\sin\frac{2n\pi}{11}-\sin\frac{4n\pi}{11}+\sin\frac{6n\pi}{11}+\sin\frac{8n\pi}{11}-\sin\frac{12n\pi}{11}.$$

Using the infinite product representations of theta functions and some simple calculation, we can find that

$$\prod_{r=1}^{5}\theta_1\left(\frac{r\pi}{11}\Big|\tau\right)=\sqrt{11\prod_{n=1}^{\infty}\frac{(1-q^n)^{11}}{(1-q^{11n})}\times\frac{\vartheta_1'(11\tau)}{\vartheta_1'(\tau)}},$$

$$\prod_{r=1}^{5}\theta_j\left(\frac{r\pi}{11}\Big|\tau\right)=\sqrt{\prod_{n=1}^{\infty}\frac{(1-q^n)^{11}}{(1-q^{11n})}\times\frac{\vartheta_j(11\tau)}{\vartheta_j(\tau)}}\quad\text{for}\quad j=2,3,4.$$

Substituting the above into (3.10) and simplifying, we immediately find that

$$C + 4\sum_{n=1}^{\infty} A(n)\frac{q^n}{1-q^n}$$

$$= -\sqrt{\frac{\vartheta_1'(\tau)^3}{11\vartheta_1'(11\tau)}}\left(\sqrt{\frac{\vartheta_2(11\tau)}{\vartheta_2^3(\tau)}} + \sqrt{\frac{\vartheta_3(11\tau)}{\vartheta_3^3(\tau)}} - \sqrt{\frac{\vartheta_4(11\tau)}{\vartheta_4^3(\tau)}}\right). \tag{3.11}$$

Now we begin to determine C and $A(n)$ from the above and we do not need compute them from other resource. Letting $q \to 0$, we can find that $C = \sqrt{11}$. Equating the coefficients of q gives $2A(1) = \sqrt{11}$, which implies $2A(n) = \chi_{11}(n)\sqrt{11}$. Substituting these values back to (3.11) completes the proof of the proposition. □

4. Theta functions and the Weierstrass ℘-function

Ramanujan's Eisenstein series $L(\tau)$ is defined as

$$L(\tau) = 1 - 24\sum_{n=1}^{\infty}\frac{nq^n}{1-q^n}.$$

If $\Omega = \{m\pi + n\pi\tau : (m,n) \in \mathbb{Z}^2\}$ is a lattice, then the Weierstrass ℘-elliptic function attached to this lattice is defined by (see, for example [2, p. 10])

$$\wp(z|\tau) = \frac{1}{z^2} + \sum_{\substack{\omega\in\Omega\\ \omega\neq 0}} \frac{1}{(z-\omega)^2} - \frac{1}{\omega^2},$$

which has primitive periods π and $\pi\tau$. Also, it has one pole at $z = 0$ of order 2 and no other poles in the fundamental period parallelogram. It is easily shown that θ_1 and ℘ satisfy the relation [20, p. 460]

$$\wp(z|\tau) = -\frac{1}{3}L(\tau) - (\log\theta_1)''(z|\tau). \tag{4.1}$$

In many cases, it is convenient to use the Jacobi theta functions to study the Weierstrass ℘-elliptic function with this relation.

In this section we provide the following theta function representation of the Weierstrass elliptic function, which has many applications:

Theorem 4.1. *Let ℘ be the Weierstrass elliptic function. Then we have*

$$\wp(u_1|\tau) + \wp(u_2|\tau) + \wp(u_3|\tau)$$

$$= \frac{\vartheta_1'(\tau)^2}{\vartheta_2^2(\tau)}\prod_{r=1}^{3}\frac{\vartheta_2^2(u_r|\tau)}{\vartheta_1^2(u_r|\tau)} - \frac{\vartheta_1'(\tau)^2}{\vartheta_3^2(\tau)}\prod_{k=1}^{3}\frac{\vartheta_3^2(u_r|\tau)}{\vartheta_1^2(u_r|\tau)} + \frac{\vartheta_1'(\tau)^2}{\vartheta_4^2(\tau)}\prod_{r=1}^{3}\frac{\vartheta_4^2(u_r|\tau)}{\vartheta_1^2(u_r|\tau)}. \tag{4.2}$$

Proof. We use Corollary 1.2 to prove this Theorem. For any three complex numbers u_1, u_2 and u_3, in Corollary 1.2, we can take

$$f(z) = \prod_{r=1}^{3} \theta_1(z + u_r|\tau)\theta_1(z - u_r|\tau).$$

With the aid of (2.1), we readily find, by direct computation, that

$$f(0) = -\prod_{r=1}^{3} \theta_1^2(u_r|\tau), \quad f\left(\frac{\pi + \pi\tau}{2}\right) = q^{-3/4}\prod_{r=1}^{3} \theta_3^2(u_r|\tau),$$

$$f\left(\frac{\pi}{2}\right) = \prod_{r=1}^{3} \theta_2^2(u_r|\tau), \quad f\left(\frac{\pi\tau}{2}\right) = -q^{-3/4}\prod_{r=1}^{3} \theta_4^2(u_r|\tau).$$

Since $(\log\theta_1)'(z|\tau)$ is an odd function of z and $(\log\theta_1)''(z|\tau)$ is an even function of z, we conclude that

$$(\log f)'(0) = \sum_{r=1}^{6}(\log\theta_1)'(u_r|\tau) = 0 \quad \text{and} \quad (\log f)''(0) = 2\sum_{r=1}^{3}(\log\theta_1)''(u_r|\tau).$$

If these equations are substituted into (1.3), we immediately find that

$$L(\tau) + (\log\theta_1)''(u_1|\tau) + (\log\theta_1)''(u_2|\tau) + (\log\theta_1)''(u_3|\tau)$$
$$= -\frac{\vartheta_1'(\tau)^2}{\vartheta_2^2(\tau)}\prod_{r=1}^{3}\frac{\vartheta_2^2(u_r|\tau)}{\vartheta_1^2(u_r|\tau)} + \frac{\vartheta_1'(\tau)^2}{\vartheta_3^2(\tau)}\prod_{r=1}^{3}\frac{\vartheta_3^2(u_r|\tau)}{\vartheta_1^2(u_r|\tau)} - \frac{\vartheta_1'(\tau)^2}{\vartheta_4^2(\tau)}\prod_{r=1}^{3}\frac{\vartheta_4^2(u_r|\tau)}{\vartheta_1^2(u_r|\tau)}$$

which is equivalent to (4.2) since $\wp(u_r|\tau) = -\frac{1}{3}L(\tau) - (\log\theta_1)''(u_r|\tau)$. □

When $u_1 = u_2 = u_3 = z$, Theorem 4.1 immediately reduces to the following corollary:

Corollary 4.1.

$$\wp(z|\tau) = \frac{\vartheta_1'(\tau)^2}{3\theta_1^6(z|\tau)}\left(\frac{\theta_2^6(z|\tau)}{\vartheta_2^2(\tau)} - \frac{\theta_3^6(z|\tau)}{\vartheta_3^2(\tau)} + \frac{\theta_4^6(z|\tau)}{\vartheta_4^2(\tau)}\right).$$

Remark 4.1. The referee kindly pointed out that Theorem 4.1 is equivalent to the following proposition:

Proposition 4.1. *If e_1, e_2, and e_3 be the Weierstrassian invariants, then we have*

$$\begin{aligned}(e_2 - e_3)&(\wp(x|\tau) - e_1)(\wp(y|\tau) - e_1)(\wp(z|\tau) - e_1)\\ &+ (e_3 - e_1)(\wp(x|\tau) - e_2)(\wp(y|\tau) - e_2)(\wp(z|\tau) - e_2)\\ &+ (e_1 - e_2)(\wp(x|\tau) - e_3)(\wp(y|\tau) - e_3)(\wp(z|\tau) - e_3)\\ &= -(e_1 - e_2)(e_2 - e_3)(e_3 - e_1)(\wp(x|\tau) + \wp(y|\tau) + \wp(z|\tau)).\end{aligned}$$

Proof. By direct computation, we find that if f is any function and e_1, e_2, and e_3 are any complex numbers, then

$$\begin{aligned}
&(e_2-e_3)(f(x)-e_1)(f(y)-e_1)(f(z)-e_1)\\
&\quad+(e_3-e_1)(f(x)-e_2)(f(y)-e_2)(f(z)-e_2)\\
&\quad+(e_1-e_2)(f(x)-e_3)(f(y)-e_3)(f(z)-e_3)\\
&=(e_1-e_2)(e_2-e_3)(e_3-e_1)(e_1+e_2+e_3-f(x)-f(y)-f(z)).
\end{aligned}$$

If we now put $f=\wp$ and let e_1, e_2 and e_3 be the Weierstrassian invariants so that $e_1+e_2+e_3=0$, then the result simplifies to Proposition 4.1. □

The left hand side of the identity in Proposition 4.1 can be expressed in terms of sigma functions by using [20, p. 451, Example 4]. The sigma functions can be converted to theta functions using Sections 21.43 and 21.9 of [20]. Thus, Theorem 4.1 is really equivalent to Proposition 4.1.

Corollary 4.1 can be used to derive the following Eisenstein series identity:

Proposition 4.2.

$$\frac{1}{2}(3L(3\tau)-L(\tau))=\frac{\vartheta_1'(3\tau)^5}{\vartheta_1'(\tau)^3}\left(\frac{\vartheta_2^3(\tau)}{\vartheta_2^5(3\tau)}-\frac{\vartheta_3^3(\tau)}{\vartheta_3^5(3\tau)}+\frac{\vartheta_4^3(\tau)}{\vartheta_4^5(3\tau)}\right).$$

Proof. Using the infinite product representation of θ_1, we can deduce that

$$\theta_1(z+\pi\tau|3\tau)\theta_1(z-\pi\tau|3\tau)=\prod_{n=1}^{\infty}\frac{(1-q^{3n})^3}{(1-q^n)}\frac{\theta_1(z|\tau)}{\theta_1(z|3\tau)}.$$

Differentiating the above equation with respect to z, two times, and then putting $z=0$, we find that

$$(\log\theta_1)''(\pi\tau|3\tau)=\frac{1}{6}(L(3\tau)-L(\tau)).$$

Thus we have

$$\wp(\pi\tau|3\tau)=-\frac{1}{3}L(3\tau)-(\log\theta_1)''(\pi\tau|3\tau)=\frac{1}{6}\left(L(\tau)-3L(3\tau)\right).$$

If we replace τ with 3τ in Corollary 4.1, set $z=\pi\tau$, and use the above equation in the resulting equation, we find that

$$L(\tau)-3L(3\tau)=\frac{2\vartheta_1'(3\tau)^2}{\theta_1^6(\pi\tau|3\tau)}\left(\frac{\theta_2^6(\pi\tau|3\tau)}{\vartheta_2^2(3\tau)}-\frac{\theta_3^6(\pi\tau|\tau)}{\vartheta_3^2(3\tau)}+\frac{\theta_4^6(\pi\tau|3\tau)}{\vartheta_4^2(3\tau)}\right).\tag{4.3}$$

Using the infinite product representations of theta functions, we deduce that

$$\theta_1^2(\pi\tau|3\tau)=-\prod_{n=1}^{\infty}\frac{(1-q^{3n})^3}{(1-q^n)}\times\frac{\vartheta_1'(\tau)}{\vartheta_1'(3\tau)},$$

$$\theta_j^2(\pi\tau|3\tau)=\prod_{n=1}^{\infty}\frac{(1-q^{3n})^3}{(1-q^n)}\times\frac{\vartheta_j(\tau)}{\vartheta_j(3\tau)}\quad\text{for}\quad j=2,3,4.$$

Substituting the above two equations into (4.3) and simplifying, we complete the proof of Proposition 4.2. □

In the same way, by taking $u_1 = \frac{\pi}{7}, u_2 = \frac{2\pi}{7}$, and $u_3 = \frac{3\pi}{7}$ in Theorem 4.1, we can get the following proposition:

Proposition 4.3.

$$\frac{1}{6}(7L(7\tau) - L(\tau)) = \frac{\vartheta_1'(\tau)^3}{49\vartheta_1'(7\tau)}\left(\frac{\vartheta_2(7\tau)}{\vartheta_2^3(\tau)} - \frac{\vartheta_3(7\tau)}{\vartheta_3^3(\tau)} + \frac{\vartheta_4(7\tau)}{\vartheta_4^3(\tau)}\right).$$

One may prove the Jacobi imaginary transformation formulas of theta functions employing the Poisson summation formula [4, pp. 7–11].

Lemma 4.1. *If z is complex,* $\mathrm{Im} z > 0$, *and* $\sqrt{-\tau i} = +1$ *for* $\tau = i$, *then we have*

$$\theta_1\left(\frac{z}{\tau}\middle| -\frac{1}{\tau}\right) = -i\sqrt{-i\tau}\exp(iz^2/(\pi\tau))\theta_1(z|\tau),$$

$$\theta_2\left(\frac{z}{\tau}\middle| -\frac{1}{\tau}\right) = \sqrt{-i\tau}\exp(iz^2/(\pi\tau))\theta_4(z|\tau),$$

$$\theta_4\left(\frac{z}{\tau}\middle| -\frac{1}{\tau}\right) = \sqrt{-i\tau}\exp(iz^2/(\pi\tau))\theta_2(z|\tau),$$

$$\theta_3\left(\frac{z}{\tau}\middle| -\frac{1}{\tau}\right) = \sqrt{-i\tau}\exp(iz^2/(\pi\tau))\theta_3(z|\tau). \tag{4.4}$$

In particular, by setting $z = 0$ in these formulas we find that

$$\vartheta_1'\left(-\frac{1}{\tau}\right) = -i\tau\sqrt{-i\tau}\vartheta_1'(\tau),$$

$$\vartheta_2\left(-\frac{1}{\tau}\right) = \sqrt{-i\tau}\vartheta_4(\tau),$$

$$\vartheta_4\left(-\frac{1}{\tau}\right) = \sqrt{-i\tau}\vartheta_2(\tau),$$

$$\vartheta_3\left(-\frac{1}{\tau}\right) = \sqrt{-i\tau}\vartheta_3(\tau). \tag{4.5}$$

Lemma 4.2.

$$L(-1/\tau) = -6i\tau/\pi + \tau^2 L(\tau).$$

Proof. Taking the logarithmic derivative of the first equation in (4.4) with respect to z, we obtain

$$(\log\theta_1)'\left(\frac{z}{\tau}\middle| -\frac{1}{\tau}\right) = \frac{2iz}{\pi} + \tau(\log\theta_1)'(z|\tau).$$

Equating the coefficients of z, we complete the proof of the lemma (see also [2, p. 24]). □

Using Proposition 4.2 we can derive the following proposition:

Proposition 4.4.
$$L(\sqrt{3}i) = \frac{\sqrt{3}}{\pi} + \frac{\sqrt{3}}{4}\vartheta_3(\sqrt{3}i)^4.$$

Proof. Putting $\tau = -1/(\sqrt{3}i)$ in Proposition 4.2, we obtain

$$\begin{aligned}
&\frac{1}{2}(3L(\sqrt{3}i) - L(-1/(\sqrt{3}i))) \\
&\quad = \frac{\vartheta_1'(\sqrt{3}i)^5}{\vartheta_1'\left(-\frac{1}{\sqrt{3}i}\right)^3}\left(\frac{\vartheta_2^3\left(-\frac{1}{\sqrt{3}i}\right)}{\vartheta_2^5(\sqrt{3}i)} - \frac{\vartheta_3^3\left(-\frac{1}{\sqrt{3}i}\right)}{\vartheta_3^5(\sqrt{3}i)} + \frac{\vartheta_4^3\left(-\frac{1}{\sqrt{3}i}\right)}{\vartheta_4^5(\sqrt{3}i)}\right). \qquad (4.6)
\end{aligned}$$

Using the imaginary transformation formulas in (4.5), we deduce that

$$\vartheta_1'\left(-\frac{1}{\sqrt{3}i}\right) = \sqrt[4]{27}\vartheta_1'(\sqrt{3}i), \quad \vartheta_2\left(-\frac{1}{\sqrt{3}i}\right) = \sqrt[4]{3}\vartheta_4(\sqrt{3}i),$$
$$\vartheta_3\left(-\frac{1}{\sqrt{3}i}\right) = \sqrt[4]{3}\vartheta_3(\sqrt{3}i), \quad \vartheta_4\left(-\frac{1}{\sqrt{3}i}\right) = \sqrt[4]{3}\vartheta_2(\sqrt{3}i).$$

Substituting these equations into the right hand side of (4.6), we find that

$$\frac{\vartheta_1'(\sqrt{3}i)^2}{\sqrt{27}}\left(\frac{\vartheta_4^3(\sqrt{3}i)}{\vartheta_2^5(\sqrt{3}i)} - \frac{1}{\vartheta_3^2(\sqrt{3}i)} + \frac{\vartheta_2^3(\sqrt{3}i)}{\vartheta_4^5(\sqrt{3}i)}\right),$$

which reduces to the following after using the Jacobi quartic identity:

$$\frac{\vartheta_1'(\sqrt{3}i)^2}{\sqrt{27}}\left(\frac{\vartheta_3^8(\sqrt{3}i) - 2\vartheta_2^4(\sqrt{3}i)\vartheta_4^4(\sqrt{3}i)}{\vartheta_2^5(\sqrt{3}i)\vartheta_4^5(\sqrt{3}i)} - \frac{1}{\vartheta_3^2(\sqrt{3}i)}\right).$$

If we use this to replace the right hand side of (4.6), then we have

$$\begin{aligned}
&\frac{1}{2}(3E_2(\sqrt{3}i) - E_2(-1/(\sqrt{3}i))) \\
&\quad = \frac{\vartheta_1'(\sqrt{3}i)^2}{\sqrt{27}}\left(\frac{\vartheta_3^8(\sqrt{3}i) - 2\vartheta_2^4(\sqrt{3}i)\vartheta_4^4(\sqrt{3}i)}{\vartheta_2^5(\sqrt{3}i)\vartheta_4^5(\sqrt{3}i)} - \frac{1}{\vartheta_3^2(\sqrt{3}i)}\right). \qquad (4.7)
\end{aligned}$$

To simplify the right hand side of the above equation, we set $\tau = -1/(\sqrt{3}i)$ in the Legendre identity

$$\vartheta_2(\tau)\vartheta_2(3\tau) + \vartheta_4(\tau)\vartheta_4(3\tau) = \vartheta_3(\tau)\vartheta_3(3\tau),$$

and find that

$$\vartheta_2\left(-\frac{1}{\sqrt{3}i}\right)\vartheta_2(\sqrt{3}i) + \vartheta_4\left(-\frac{1}{\sqrt{3}i}\right)\vartheta_4(\sqrt{3}i) = \vartheta_3\left(-\frac{1}{\sqrt{3}i}\right)\vartheta_3(\sqrt{3}i).$$

Combining this equation with (4.5), we deduce that

$$2\vartheta_2(\sqrt{3}i)\vartheta_4(\sqrt{3}i) = \vartheta_3(\sqrt{3}i)^2. \tag{4.8}$$

Substituting this equation into (4.7) and simplifying, we conclude that

$$3L(\sqrt{3}i) - L(-1/(\sqrt{3}i)) = \frac{3\sqrt{3}}{2}\vartheta_3^4(\sqrt{3}i).$$

We take $\tau = \sqrt{3}i$ in Lemma 4.2 to find that

$$L(-1/(\sqrt{3}i)) = \frac{6\sqrt{3}}{\pi} - 3L(\sqrt{3}i).$$

Eliminating $L(-1/(\sqrt{3}i))$ from the above two equations, we complete the proof of Proposition 4.4. □

Proposition 4.5.

$$L(\sqrt{7}i) = \frac{3}{\sqrt{7}\pi} + \frac{27}{16\sqrt{7}}\vartheta_3^4(\sqrt{7}i).$$

Proof. We use Proposition 4.3 to prove Proposition 4.5. Let $\tau = -1/(\sqrt{7}i)$ in Proposition 4.3, we find that

$$\begin{aligned}&\frac{1}{6}(7E_2(\sqrt{7}i) - E_2(-1/(\sqrt{7}i))\\ &= \frac{\vartheta_1'(\sqrt{7}i)^3}{\vartheta_1'\left(-\frac{1}{\sqrt{7}i}\right)}\left(\frac{\vartheta_2\left(-\frac{1}{\sqrt{7}i}\right)}{\vartheta_2^3(7\tau)} - \frac{\vartheta_3\left(-\frac{1}{\sqrt{7}i}\right)}{\vartheta_3^3(7\tau)} + \frac{\vartheta_4\left(-\frac{1}{\sqrt{7}i}\right)}{\vartheta_4^3(7\tau)}\right).\end{aligned} \tag{4.9}$$

Using the imaginary transformation formulas, we can deduce that

$$\vartheta_1'\left(-\frac{1}{\sqrt{7}i}\right) = \sqrt[4]{343}\vartheta_1'(\sqrt{7}i), \quad \vartheta_2\left(-\frac{1}{\sqrt{7}i}\right) = \sqrt[4]{7}\vartheta_4(\sqrt{7}i),$$

$$\vartheta_3\left(-\frac{1}{\sqrt{7}i}\right) = \sqrt[4]{7}\vartheta_3(\sqrt{7}i), \quad \vartheta_4\left(-\frac{1}{\sqrt{7}i}\right) = \sqrt[4]{7}\vartheta_2(\sqrt{7}i).$$

Substituting these equations into the right hand side of (4.9), appealing to Jacobi's quartic identity, we are led to

$$\begin{aligned}&\frac{\vartheta_1'(\sqrt{7}i)^2}{\sqrt{7}}\left(\frac{\vartheta_4(\sqrt{7}i)}{\vartheta_2^3(\sqrt{7}i)} - \frac{1}{\vartheta_3^2(\sqrt{7}i)} + \frac{\vartheta_2(\sqrt{7}i)}{\vartheta_4^3(\sqrt{7}i)}\right)\\ &= \frac{\vartheta_1'(\sqrt{7}i)^2}{\sqrt{7}}\left(\frac{\vartheta_3^4(\sqrt{7}i)}{\vartheta_2^3(\sqrt{7}i)\vartheta_4^3(\sqrt{7}i)} - \frac{1}{\vartheta_3^2(\sqrt{7}i)}\right).\end{aligned} \tag{4.10}$$

To simplify this we set $\tau = -1/(\sqrt{7}i)$ in the Guetslafff identity, and find that

$$\sqrt{\vartheta_2(\sqrt{7}i)\vartheta_2(-1/(\sqrt{7}i))} + \sqrt{\vartheta_4(\sqrt{7}i)\vartheta_4(-1/(\sqrt{7}i))} = \sqrt{\vartheta_3(\sqrt{7}i)\vartheta_3(-1/(\sqrt{7}i))}.$$

Using the last three identities of (4.5) in this equation, we deduce that

$$\vartheta_3(\sqrt{7}i) = 2\sqrt{\vartheta_2(\sqrt{7}i)\vartheta_4(\sqrt{7}i)}. \tag{4.11}$$

With the help of this identity, we find (4.10) can be simplified to $\frac{9\sqrt{7}}{16}\vartheta_3^4(\sqrt{7}i)$. Thus we have

$$\frac{1}{6}(7E_2(\sqrt{7}i) - E_2(-1/(\sqrt{7}i)) = \frac{9\sqrt{7}}{16}\vartheta_3^4(\sqrt{7}i).$$

We take $\tau = \sqrt{7}i$ in Lemma 4.2, and immediately find that

$$E_2(-1/(\sqrt{7}i)) = \frac{6\sqrt{7}}{\pi} - 7E_2(\sqrt{7}i).$$

Eliminating $E_2(-1/(\sqrt{7}i))$ from the above two equations, we complete the proof of Proposition 4.5. □

Now we use Propositions 4.4 and 4.5 to give simple derivation of the following two Ramanujan's series for $1/\pi$:

Theorem 4.2 (Ramanujan).

$$\frac{4}{\pi} = \sum_{n=0}^{\infty} \frac{(6n+1)\left(\frac{1}{2}\right)_n^3}{(n!)^3}\left(\frac{1}{4}\right)^n, \tag{4.12}$$

$$\frac{16}{\pi} = \sum_{n=0}^{\infty} \frac{(42n+5)\left(\frac{1}{2}\right)_n^3}{(n!)^3}\left(\frac{1}{64}\right)^n. \tag{4.13}$$

In his famous paper [18], Ramanujan recorded a total of 17 series for $1/\pi$, the above are the first two of these series. The Borweins [7] provided rigorous proofs of Ramanujan for the first time and obtained many new series for $1/\pi$. Since then many new results of the Ramanujan type series for $1/\pi$ have arisen, see for example, [3,8,11,14,10].

Proof. For simplicity, in the sequent part of this section we denote

$$\alpha := \alpha(\tau) = \vartheta_2^4(\tau)/\vartheta_3^4(\tau).$$

The functions α, ϑ_3, and L satisfy the Ramanujan identity [6, p. 149, Eq. (9. 10)]

$$L(\tau) = (1-3\alpha)\vartheta_3^2 + 12\alpha(1-\alpha)\vartheta_3\frac{d\vartheta_3}{d\alpha}. \tag{4.14}$$

Clausen's identity which involves the hypergeometric function ${}_2F_1(a,b;c;t)$ and the generalized hypergeometric function ${}_3F_2(a,b,c;d,e;t)$ is given by (see for example, [1, p. 116])

$$\left[{}_2F_1\left(\begin{matrix}a, b\\ c\end{matrix}; t\right)\right]^2 = {}_3F_2\left(\begin{matrix}2a, a+b, 2b\\ a+b+\frac{1}{2}, 2a+2b\end{matrix}; t\right). \tag{4.15}$$

Recall the Jacobi identity [19], [20, p. 499]

$$\vartheta_3^2 = {}_2F_1\left(\frac{1}{2}, \frac{1}{2}; 1; \alpha\right). \tag{4.16}$$

Putting $a = b = \frac{1}{2}$ and $c = 1$ and replacing t by α in (4.15), combining the resulting equation with (4.16), we find that

$$\vartheta_3^4 = {}_3F_2\left(\frac{1}{2}, \frac{1}{2}, \frac{1}{2}; 1, 1; 4\alpha(1-\alpha)\right). \tag{4.17}$$

If we differentiate the above equation with respect to α, we deduce that

$$2\vartheta_3 \frac{d\vartheta_3}{d\alpha} = 4(1-2\alpha){}_3F_2'\left(\frac{1}{2}, \frac{1}{2}, \frac{1}{2}; 1, 1; 4\alpha(1-\alpha)\right). \tag{4.18}$$

Substituting the above two equations into (4.14) and simplifying, we find that

$$L(\tau) = (1-2\alpha)\sum_{n=0}^{\infty} \frac{(3n+1)\left(\frac{1}{2}\right)_n^3}{(n!)^3}(4\alpha(1-\alpha))^n, \quad 0 \le \alpha \le \frac{1}{2}. \tag{4.19}$$

Setting $\tau = \sqrt{3}i$ in the above, in the light of (4.8), we find α becomes $(2-\sqrt{3})/4$ and $4\alpha(1-\alpha)$ gives $1/4$. Thus we have

$$L(\sqrt{3}i) = \frac{\sqrt{3}}{2}\sum_{n=0}^{\infty} \frac{(3n+1)\left(\frac{1}{2}\right)_n^3}{(n!)^3}\left(\frac{1}{4}\right)^n,$$

and when $\tau = \sqrt{3}i$, (4.17) reduces to

$$\vartheta_3^4(\sqrt{3}i) = \sum_{n=0}^{\infty} \frac{\left(\frac{1}{2}\right)_n^3}{(n!)^3}\left(\frac{1}{4}\right)^n.$$

Substituting the above two equations into Proposition 4.4 and simplifying, we arrive at the Ramanujan formula (4.12).

Now we begin to derive (4.13). From (4.11) we can easily find that

$$1 - 2\alpha(\sqrt{7}i) = \frac{3\sqrt{7}}{8}, \quad \alpha(\sqrt{7}i)(1-\alpha(\sqrt{7}i)) = \frac{1}{256}. \tag{4.20}$$

Putting $\tau = \sqrt{7}i$ in (4.19), appealing to the above, we conclude that

$$L(\sqrt{7}i) = \frac{3\sqrt{7}}{8}\sum_{n=0}^{\infty} \frac{(3n+1)\left(\frac{1}{2}\right)_n^3}{(n!)^3}\left(\frac{1}{64}\right)^n, \tag{4.21}$$

and when $\tau = \sqrt{7}i$, (4.17) reduces to

$$\vartheta_3^4(\sqrt{7}i) = \sum_{n=0}^{\infty} \frac{\left(\frac{1}{2}\right)_n^3}{(n!)^3} \left(\frac{1}{64}\right)^n.$$

Substituting the above two equations into Proposition 4.5 and simplifying, we arrive at the Ramanujan formula (4.13) to complete the proof of Theorem 4.2. □

5. Conclusion

We conclude this paper with the following theorem involving a general entire function and theta functions, which may have many applications to the theory of elliptic theta functions:

Theorem 5.1. *Suppose $z_1, z_2, \ldots, z_r$ are r complex numbers such that $z_1 + z_2 + \cdots + z_r = 0$ and $z_j - z_k \not\equiv 0 \pmod{\Omega}$ for $j \neq k$; and further suppose $f(z)$ is an entire function of z satisfying the functional equations*

$$f(z) = (-1)^r f(z+\pi) = (-1)^r q^{(r+4)/2} e^{2(r+4)iz} f(z+\pi\tau).$$

Then $f(z)$ and the Jacobi theta functions satisfy the identity

$$\sum_{k=1}^{r} \frac{2f(z_k)}{\theta_1(2z_k|\tau)\prod_{j\neq k}\theta_1(z_k - z_j|\tau)}$$

$$= \frac{(-1)^{r-1}f(0)}{\prod_{j=1}^{r}\theta_1(z_k|\tau)} + \frac{f\left(\frac{\pi}{2}\right)}{\prod_{k=1}^{r}\theta_2(z_k|\tau)} - \frac{q^{\frac{r+4}{8}} f\left(\frac{\pi+\pi\tau}{2}\right)}{\prod_{k=1}^{r}\theta_3(z_k|\tau)} + \frac{i^{3r} q^{\frac{4+r}{8}} f\left(\frac{\pi\tau}{2}\right)}{\prod_{k=1}^{r}\theta_4(z_k|\tau)}.$$

It is obvious that the case $r = 2$ of the above theorem is Theorem 1.1. The proof of Theorem 5.1 is similar to that of Theorem 1.1, so we omit it.

Acknowledgements

The author would like to thank the referee and the editors for many useful comments and suggestions.

References

[1] G. E. Andrews and R. Askey and R. Roy, Special functions, Encyclopedia of Mathematics and its Applications, 71. Cambridge University Press, Cambridge (1999).

[2] T. M. Apostol, Modular functions and Dirichlet series in number theory, Springer-Verlag, New York (1990).

[3] N. D. Baruah and B. C. Berndt, Ramanujan's series for $1/\pi$ arising from his cubic and quartic theories of elliptic functions, *J. Math. Anal. Appl.*, **341** (2008) 357–371.

[4] R. Bellman, A brief introduction to theta functions, Holt Rinehart and Winston, New York (1961).

[5] B. C. Berndt, Ramanujan's Notebooks, Part III, Springer-Verlag, New York (1991).

[6] B. C. Berndt, Ramanujan's Notebooks, Part V, Springer-Verlag, New York (1998).

[7] J. M. Borwein and P. B. Borwein, Pi and the AGM, Wiley, New York (1987).

[8] H. H. Chan, S. H. Chan and Z. Liu, Domb's numbers and Ramanujan-Sato type series for $1/\pi$, *Adv. Math.*, **186** (2004) 396–410.

[9] K. Chandrasekharan, Elliptic Functions, Springer-Verlag, Berlin Heideberg (1985).

[10] K. S. Chua, Extremal modular lattices, McKay Thompson series, quadratic iterations, and new series for $1/\pi$, *Experiment. Math.*, **14** (2005) 343–357.

[11] D. V. Chudnovsky and G. V. Chudnovsky, Approximation and complex multiplication according to Ramanujan, in: G. E. Andrews, R. A. Askey, B. C. Berndt, K. G. Ramanathan, R. A. Rankin (Eds.), Ramanujan Revisited, Academic Press, Boston (1988) pp. 375–472.

[12] S. Cooper, The quintuple product identity, *Int. J. Number Theory*, **2** (2006) 115–161.

[13] C. Guetslaff, Aequatio modularis pro transformatione functionum ellipticarum septimi ordinis, *J. Reine Angew. Math.*, **12** (1834) 173–177.

[14] J. Guillera, A new method to obtain series for $1/\pi$ and $1/\pi^2$, *Experiment. Math.*, **15** (2006) 83–89.

[15] A. M. Legendre, Traité des Fonctions Elliptiques, Huzard-Courcier, Paris (1828).

[16] Z.-G. Liu, Residue theorem and theta function identities, *Ramanujan J.*, **5** (2001) 129–151.

[17] Z.-G. Liu, An extension of the quintuple product identity and its applications, *Pacific J. Math.*, **246** (2010) 345–390.

[18] S. Ramanujan, Modular functions and approximations to π, *Quart. J. Math. (Oxford)*, **45** (1914) 350–372.

[19] L.-C. Shen, On a definite integral of a hypergeometric function, *Proc. Amer. Math. Soc.*, **120** (1994) 1131–1136.

[20] E. T. Whittaker and G. N. Watson, A course in modern analysis, 4th ed., Cambridge University Press, Cambridge (1990).

Ramanujan Rediscovered No. 14, 2010, pp. 185–199.

Some New Schläfli-Type Modular Equations in the Quartic Theory

M. S. Mahadeva Naika* and S. Chandan Kumar

Department of Mathematics, Bangalore University, Central College Campus, Bangalore 560 001, India
e-mail: msmnaika@rediffmail.com; chandan.s17@gmail.com

Abstract. In this paper, we establish several new Schläfli-type modular equations in the theory of signature 4. We establish modular relations between the quartic class invariants h_n with h_{k^2n} and H_n with H_{k^2n}. We also establish several new explicit evaluations of the quartic class invariants and quartic singular moduli.

Keywords. Modular equation, theta-function, quartic class invariant, quartic singular modulus.

2000 Subject Classification: 11B65, 33D10; Secondary: 11F27

Dedicated to Prof. K. Venkatachaliengar on the occasion of his 100th Birthday

1. Introduction

The ordinary hypergeometric series ${}_2F_1(a, b; c; x)$ is defined by

$$ {}_2F_1(a, b; c; x) = \sum_{n=0}^{\infty} \frac{(a)_n (b)_n x^n}{(c)_n n!}, \qquad |x| < 1, $$

where

$$ (a)_0 = 1, (a)_n = a(a+1)(a+2)\ldots(a+n-1), \qquad \text{for } n \geq 1. $$

Let

$$ Z(r) := Z(r; x) := {}_2F_1\left(\frac{1}{r}, \frac{r-1}{r}; 1; x\right) $$

and

$$ q_r := q_r(x) := \exp\left(-\pi \csc\left(\frac{\pi}{r}\right) \frac{{}_2F_1\left(\frac{1}{r}, \frac{r-1}{r}; 1; 1-x\right)}{{}_2F_1\left(\frac{1}{r}, \frac{r-1}{r}; 1; x\right)}\right), $$

where $r = 2, 3, 4, 6$ and $0 < x < 1$.

---

*Research supported by DST grant SR/S4/MS:509/07, Govt. of India.

Let n denote a fixed natural number, and assume that

$$n\frac{{}_2F_1\left(\frac{1}{4},\frac{3}{4};1;1-\alpha\right)}{{}_2F_1\left(\frac{1}{4},\frac{3}{4};1;\alpha\right)}=\frac{{}_2F_1\left(\frac{1}{4},\frac{3}{4};1;1-\beta\right)}{{}_2F_1\left(\frac{1}{4},\frac{3}{4};1;\beta\right)}. \tag{1.1}$$

Then a quartic modular equation of degree n in the theory of elliptic functions of signature 4 is a relation between α and β induced by (1.1). We say β is of degree n over α, and $m(4) = \frac{Z(4;\alpha)}{Z(4;\beta)}$ is called the multiplier. We also use the notations $Z_1 := Z_1(4;\alpha) := Z(4;\alpha)$ and $Z_n := Z_n(4;\beta) := Z(4;\beta)$ to indicate β has degree n over α. When the context is clear, we omit the argument 4 in q_4, $Z(4)$ and $m(4)$. We define quartic class invariants h_n and H_n which are analogous to Ramanujan-Weber class invariants in the classical theory as follows:

$$h_n = \{\alpha(1-\alpha)^{-1}\}^{-\frac{1}{24}} \tag{1.2}$$

and

$$H_n = \{\alpha(1-\alpha)\}^{-\frac{1}{24}}, \tag{1.3}$$

where $q = e^{-\pi\sqrt{n}}$, n a positive rational number.

On pages 257–262 of his second notebook [23], Ramanujan gives an outline of theories of elliptic functions to alternate bases corresponding to the classical theory by way of statements of some results. K. Venkatachaliengar [24] examined some of these results. On page 261 of his second notebook [23], Ramanujan recorded several modular equations in the theory of signature 4. Proofs of all these identities can be found in [8] and [9, pp.122–123]. In [10], B. C. Berndt, H. H. Chan and W. C. Liaw have derived quartic versions of duplication and dimidiation formulas and also proved many theorems of Ramanujan.

In [17], Mahadeva Naika has established several cubic modular equations of degree 2 akin to Ramanujan and as an application he established some new P–Q eta-function identities. In [1], C. Adiga, Taekyun Kim and Mahadeva Naika have also established several cubic modular equations and some P–Q eta-function identities. In [16], Mahadeva Naika has established some Russell-type cubic mixed modular equations and some P–Q eta-function identities.

Recently, [14] T. Kim, M. S. Mahadeva Naika, S. Chandan Kumar, Lee-Chae Jang, Young-Hee Kim and Byungje Lee have established several new Schläfli-type modular equations in the Theory of Signature 3 and also established several explicit evaluations of cubic class invariants and cubic singular moduli. For more details about Schläfli-type modular equations and class invariants one can see [2]–[5], [7], [11]–[15], [18]–[20], [22] and [25].

2. Preliminary results

In this section, we collect some identities which are useful in proving our main results.

Theorem 2.1. *We have*

$$q^{1/24}f(-q) = \sqrt{Z}2^{-1/4}x^{1/24}(1-x)^{1/12}, \tag{2.1}$$

where $f(-q) := \prod_{n=1}^{\infty}(1-q^n),\ |q| < 1.$

$$q^{1/12}f(-q^2) = \sqrt{Z}2^{-1/4}x^{1/12}(1-x)^{1/24}. \tag{2.2}$$

If $\alpha\beta = \pi^2$, *then*

$$\sqrt[4]{\alpha}e^{-\alpha/12}f(-e^{-2\alpha}) = \sqrt[4]{\beta}e^{-\beta/12}f(-e^{-2\beta}). \tag{2.3}$$

If β *is of degree 3 over* α, *then*

$$\{\alpha\beta\}^{1/2} + \{(1-\alpha)(1-\beta)\}^{1/2} + 4\{\alpha\beta(1-\alpha)(1-\beta)\}^{1/4} = 1. \tag{2.4}$$

If β *is of degree 5 over* α, *then*

$$\{\alpha\beta\}^{1/2} + \{(1-\alpha)(1-\beta)\}^{1/2} + 8\{\alpha\beta(1-\alpha)(1-\beta)\}^{1/6}[\{\alpha\beta\}^{1/6} + \{(1-\alpha)(1-\beta)\}^{1/6}] = 1. \tag{2.5}$$

If β *is of degree 7 over* α, *then*

$$\{\alpha\beta\}^{1/2} + \{(1-\alpha)(1-\beta)\}^{1/2} + 20\{\alpha\beta(1-\alpha)(1-\beta)\}^{1/4} + 8\sqrt{2}\{\alpha\beta(1-\alpha)(1-\beta)\}^{1/8}[\{\alpha\beta\}^{1/4} + \{(1-\alpha)(1-\beta)\}^{1/4}] = 1. \tag{2.6}$$

If β *is of degree 11 over* α, *then*

$$\{\alpha\beta\}^{1/2} + \{(1-\alpha)(1-\beta)\}^{1/2} + 68\{\alpha\beta(1-\alpha)(1-\beta)\}^{1/4} + 16\{\alpha\beta(1-\alpha)(1-\beta)\}^{1/12}[\{\alpha\beta\}^{1/3} + \{(1-\alpha)(1-\beta)\}^{1/3}] + 48\{\alpha\beta(1-\alpha)(1-\beta)\}^{1/6}[\{\alpha\beta\}^{1/6} + \{(1-\alpha)(1-\beta)\}^{1/6}] = 1. \tag{2.7}$$

For a proof of the equation (2.1), see [8, Theorem 9.9, p.148], for a proof of the equation (2.2), see [8, Theorem 9.10, p. 148], for a proof of the equation (2.3), see [7, Entry 27 (iii), p. 43], for a proof of the equation (2.4), see [8, Theorem 10.1, p. 153], for a proof of the equation (2.5), see [8, Theorem 10.2, p. 154], for a proof of the equation (2.6), see [8, Theorem 10.3, p. 155], for a proof of the equation (2.7), see [8, Theorem 10.4, p. 155].

3. Schläfli-type modular equations

In this section, we establish several new Schläfli-type modular equations using Ramanujan's modular equations. Set

$$P = \{\alpha\beta(1-\alpha)(1-\beta)\}^{\frac{1}{24}} \tag{3.1}$$

and

$$Q = \left\{ \frac{\beta(1-\beta)}{\alpha(1-\alpha)} \right\}^{\frac{1}{24}}, \tag{3.2}$$

where β is of degree n over α.

From the equations (3.1) and (3.2), we find that

$$PQ = \{\beta(1-\beta)\}^{\frac{1}{12}} \tag{3.3}$$

and

$$\frac{P}{Q} = \{\alpha(1-\alpha)\}^{\frac{1}{12}}. \tag{3.4}$$

From the equations (3.3) and (3.4), we deduce that

$$\alpha = \frac{1+A}{2}, \quad A = \pm\sqrt{1 - \frac{4P^{12}}{Q^{12}}} \tag{3.5}$$

and

$$\beta = \frac{1+B}{2}, \quad B = \pm\sqrt{1 - 4P^{12}Q^{12}}. \tag{3.6}$$

Theorem 3.1. *If P and Q are defined as in the equation* (3.1) *and* (3.2), *then*

$$\frac{1}{P^3}\left(Q^{12} + \frac{1}{Q^{12}}\right) + 8\left(24P^9 - \frac{1}{P^9}\right) + 2\left(\frac{39}{P^3} - 112P^3\right) = 0, \tag{3.7}$$

where β is of degree 3 over α.

Proof. Employing the equations (3.1) and (3.2) in the equation (2.4), we find that

$$\{\alpha\beta\}^{\frac{1}{2}} + \frac{P^{12}}{\{\alpha\beta\}^{\frac{1}{2}}} = 1 - 4P^6. \tag{3.8}$$

Squaring both sides of the equation (3.8), we find that

$$\{\alpha\beta\}^2 - \{(1-4P^6)^2 - 2P^{12}\}\alpha\beta + P^{24} = 0. \tag{3.9}$$

Employing the equations (3.1) with $n = 3$, (3.5) and (3.6) in the equation (3.9), after eliminating A and B, we obtain the equation (3.7). □

Theorem 3.2. *If P and Q are defined as in the equations* (3.1)–(3.2), *then*

$$\left(Q^{18} + \frac{1}{Q^{18}}\right) + 3775\left(Q^6 + \frac{1}{Q^6}\right) + 103680P^{12}\left(Q^6 + \frac{1}{Q^6}\right)$$
$$+ 130\left(Q^{12} + \frac{1}{Q^{12}}\right) + 64\left(1944P^{12} - \frac{1}{P^{12}}\right) + 11900 = 0, \tag{3.10}$$

where β is of degree 5 over α.

Proof. Employing the equations (3.1) and (3.2) with $n = 5$ in the equation (2.5), we find that

$$\{\alpha\beta\}^{\frac{1}{2}} + \frac{P^{12}}{\{\alpha\beta\}^{\frac{1}{2}}} + 8P^4\left(\{\alpha\beta\}^{\frac{1}{6}} + \frac{P^4}{\{\alpha\beta\}^{\frac{1}{6}}}\right) = 1. \tag{3.11}$$

Employing the equations (3.3), (3.4), (3.5) and (3.6), the above equation (3.11) reduces to

$$\begin{aligned}
&(P^{12} + 3775P^{12}Q^{12} + 103680P^{24}Q^{12} + 3775P^{12}Q^{24} + 103680P^{24}Q^{24}\\
&+ Q^{36}P^{12} - 124416P^{24}Q^{18} + 64Q^{18} - 130Q^6P^{12} - 11900P^{12}Q^{18}\\
&- 130Q^{30}P^{12})(P^{12} + 3775P^{12}Q^{12} + 103680P^{24}Q^{12} + 3775P^{12}Q^{24}\\
&+ 103680P^{24}Q^{24} + Q^{36}P^{12} + 124416P^{24}Q^{18} - 64Q^{18} + 130Q^6P^{12}\\
&+ 11900P^{12}Q^{18} + 130Q^{30}P^{12}) = 0.
\end{aligned} \tag{3.12}$$

By examining the behaviour of first factor near $q = 0$, it can be seen that there is a neighbourhood about the origin, where the first factor is not zero. Then the second factor is zero in this neighbourhood. By the Identity Theorem second factor vanishes identically. On simplification, we obtain the equation (3.10). □

Theorem 3.3. *If P and Q are defined as in the equation* (3.1)–(3.2), *then*

$$\begin{aligned}
&28\left(\frac{1}{Q^{12}} + Q^{12}\right)\left(1982880P^9 + 230832P^3 + \frac{7329}{P^3} + \frac{52}{P^9}\right) + \frac{1}{P^3}\left(Q^{24} + \frac{1}{Q^{24}}\right)\\
&= 188116992P^{21} + 170201088P^{15} + \frac{512}{P^{21}} - 131072256P^9 - 42166656P^3\\
&+ \frac{3152954}{P^3} + \frac{531552}{P^9} - \frac{35392}{P^{15}}.
\end{aligned} \tag{3.13}$$

where β is of degree 7 over α.

Proof. Employing the equations (3.1) and (3.2) with $n = 7$ in the equation (2.6), we find that

$$\{\alpha\beta\}^{\frac{1}{2}} + \frac{P^{12}}{\{\alpha\beta\}^{\frac{1}{2}}} + 20P^6 + 8\sqrt{2}P^3\left(\{\alpha\beta\}^{\frac{1}{4}} + \frac{P^6}{\{\alpha\beta\}^{\frac{1}{4}}}\right) = 1. \tag{3.14}$$

Employing the equations (3.3), (3.4), (3.5) and (3.6), the above equation (3.14) reduces to

$$\begin{aligned}
&(-P^{18} - 1456P^{12}Q^{12} - 42166656P^{24}Q^{24} - 6463296P^{24}Q^{12}\\
&- 205212P^{18}Q^{12} - 55520640P^{30}Q^{12} + 188116992P^{42}Q^{24} + 512Q^{24}
\end{aligned}$$

$$+170201088P^{36}Q^{24}-131072256P^{30}Q^{24}-6463296P^{24}Q^{36}$$
$$-205212P^{18}Q^{36}-55520640P^{30}Q^{36}-1456P^{12}Q^{36}+531552P^{12}Q^{24}$$
$$+3152954P^{18}Q^{24}-P^{18}Q^{48}-35392Q^{24}P^{6})^2(-P^{18}+1616P^{12}Q^{12}$$
$$-512Q^{24}-234588544P^{24}Q^{24}+33050816P^{24}Q^{12}-430492P^{18}Q^{12}$$
$$-780266880P^{30}Q^{12}+1886810112P^{42}Q^{24}-858451968P^{36}Q^{24}$$
$$-954040576P^{30}Q^{24}+33050816P^{24}Q^{36}-430492P^{18}Q^{36}$$
$$+1616P^{12}Q^{36}-780266880P^{30}Q^{36}-3038112P^{12}Q^{24}$$
$$+53574714P^{18}Q^{24}-P^{18}Q^{48}+67008Q^{24}P^{6})^2=0. \tag{3.15}$$

By examining the behaviour of second factor near $q = 0$, it can be seen that there is a neighbourhood about the origin, where the second factor is not zero. Then the first factor is zero in this neighbourhood. By the Identity Theorem first factor vanishes identically. On simplification, we obtain the equation (3.13). □

Theorem 3.4. *If P and Q are defined as in the equations* (3.1)–(3.2), *then*

$$\frac{1}{P^3}\left(Q^{36}+\frac{1}{Q^{36}}\right)+\left(Q^{24}+\frac{1}{Q^{24}}\right)$$
$$\times\left(1899390211799616P^9-2981005147296P^3+\frac{1378118378}{P^3}-\frac{164864}{P^9}\right)$$
$$+1301797103468544P^{33}-4660847651192832P^{27}$$
$$+4568332198256640P^{21}-4743319404552192P^{15}+564317587338624P^9$$
$$-2493407003253696P^3+\frac{1213770519844748}{P^3}-\frac{99666540942224}{P^9}$$
$$+\frac{2184867437184}{P^{15}}-\frac{17649625600}{P^{21}}+\frac{52289536}{P^{27}}-\frac{32768}{P^{33}}$$
$$=\left(Q^{12}+\frac{1}{Q^{12}}\right)\left(3253301270433792P^{21}-\frac{164864}{P^{21}}\right.$$
$$+2521615039739136P^9-278109156534912P^3+\frac{188214244472497}{P^3}$$
$$\left.-\frac{263374174048}{P^9}+\frac{286658368}{P^{15}}-5333256318947328P^{15}\right). \tag{3.16}$$

where β is of degree 11 over α.

Proof. Since the proof of the above Theorem 3.4 is similar to the proof of Theorem 3.3, we omit the details. □

4. Modular relations between quartic class invariants h_n and h_{k^2n}

In this section, we establish several modular relations between the quartic class invariants h_n and h_{k^2n}.

Theorem 4.1. *We have*

$$h_{2n}h_{2/n} = 1. \tag{4.1}$$

Proof. From the equations (2.1), (2.2) and (2.3), we obtain the equation (4.1). □

Theorem 4.2. *We have*

$$H_n^{12} = h_n^{12} + \frac{1}{h_n^{12}}. \tag{4.2}$$

Proof. From the equations (1.2) and (1.3), we obtain the equation (4.2). □

Theorem 4.3. *We have*

$$H_{2n} = H_{2/n}. \tag{4.3}$$

Proof. Employing the equation (4.1) in the equation (4.2), we obtain the equation (4.3). □

Theorem 4.4. *We have*

$$\alpha_{2n} + \alpha_{2/n} = 1. \tag{4.4}$$

Proof. From the equations (1.2), (1.3), (4.1) and (4.3), we obtain the equation (4.4). □

Theorem 4.5. *If* $X = h_n^3 h_{9n}^3$ *and* $Y = \frac{h_n^6}{h_{9n}^6}$*, then*

$$\frac{1}{Y} - Y = 2\sqrt{2}\left(X + \frac{1}{X}\right). \tag{4.5}$$

Proof. The equation (2.4) can be written as

$$1 + \left\{\frac{(1-\alpha)(1-\beta)}{\alpha\beta}\right\}^{\frac{1}{2}} + 4\left\{\frac{(1-\alpha)(1-\beta)}{\alpha\beta}\right\}^{\frac{1}{4}} = \frac{1}{\sqrt{\alpha\beta}}. \tag{4.6}$$

Employing the equation (1.2) in the equation (4.6), we find that

$$18 + 8X^2 + \frac{8}{X^2} = Y^2 + \frac{1}{Y^2}. \tag{4.7}$$

On simplification, we obtain the equation (4.5). □

Corollary 4.1. *We have*

$$h_6 = (\sqrt{2}+1)^{\frac{1}{6}} \tag{4.8}$$

and

$$h_{\frac{2}{3}} = (\sqrt{2}-1)^{\frac{1}{6}}. \tag{4.9}$$

Proof. Putting $n = \frac{2}{3}$ in the equation (4.5), we find that

$$X = 1. \tag{4.10}$$

Employing the equation (4.10) in the equation (4.5), we find that

$$\frac{1}{Y} - Y = 4\sqrt{2}. \tag{4.11}$$

Solving the above equation (4.11), we obtain the equations (4.8) and (4.9). □

Corollary 4.2. *We have*

$$h_{18} = (\sqrt{3} + \sqrt{2})^{1/3} \tag{4.12}$$

and

$$h_{2/9} = (\sqrt{3} - \sqrt{2})^{1/3}. \tag{4.13}$$

Proof. Putting $n = 2$ in the equation (4.5), we find that

$$h_{18}^3 - \frac{1}{h_{18}^3} = 2\sqrt{2}. \tag{4.14}$$

On simplification, we obtain the equation (4.12). □

Theorem 4.6. *If $X = h_n^2 h_{25n}^2$ and $Y = \frac{h_n^3}{h_{25n}^3}$, then*

$$\frac{1}{Y} - Y = 2\left(X + \frac{1}{X}\right). \tag{4.15}$$

Proof. The equation (2.5) can be written as

$$1 + A^{\frac{1}{2}} + 8A^{\frac{1}{6}}(1 + A^{\frac{1}{6}}) = \frac{1}{\sqrt{\alpha\beta}}, \tag{4.16}$$

where

$$A = \frac{(1-\alpha)(1-\beta)}{\alpha\beta}.$$

Employing the equation (1.2) in the equation (4.16), we find that

$$\left(Y + 2X + \frac{2}{X} - \frac{1}{Y}\right)\left(Y - 2X - \frac{2}{X} - \frac{1}{Y}\right)$$
$$\times \left(Y^2 + 4X^{\frac{4}{3}} + 10 + \frac{4}{X^{\frac{4}{3}}} + \frac{1}{Y^2}\right) = 0. \tag{4.17}$$

Since $Y^2 + 4X^{\frac{4}{3}} + 10 + \frac{4}{X^{\frac{4}{3}}} + \frac{1}{Y^2} \neq 0$. By examining the behaviour of second factor near $q = 0$, it can be seen that there is a neighbourhood about the origin, where the second factor is not zero, whereas the first factor is zero in this neighbourhood. By the Identity Theorem first factor vanishes identically. Hence

$$Y + 2X + \frac{2}{X} - \frac{1}{Y} = 0.$$

This completes the proof. □

Corollary 4.3. *We have*

$$h_{10} = \left(\frac{\sqrt{5}+1}{2}\right)^{\frac{1}{2}} \tag{4.18}$$

and

$$h_{\frac{2}{5}} = \left(\frac{\sqrt{5}-1}{2}\right)^{\frac{1}{2}}. \tag{4.19}$$

Proof. Putting $n = \frac{2}{5}$ in the equation (4.15), we find that

$$X = 1. \tag{4.20}$$

Employing the equation (4.20) in the equation (4.15), we find that

$$\frac{1}{Y} - Y = 4. \tag{4.21}$$

On solving the above equation (4.21), we obtain the equations (4.18) and (4.19). □

Theorem 4.7. *If* $X = h_n^3 h_{49n}^3$ *and* $Y = \frac{h_n^4}{h_{49n}^4}$, *then*

$$\frac{1}{Y} + Y = 2\sqrt{2}\left(X + \frac{1}{X}\right) + 7. \tag{4.22}$$

Proof. The equation (2.6) can be written as

$$1 + A^{\frac{1}{2}} + 20A^{\frac{1}{4}} + 8\sqrt{2}A^{\frac{1}{8}}(1 + A^{\frac{1}{4}}) = \frac{1}{\sqrt{\alpha\beta}}, \tag{4.23}$$

where

$$A = \frac{(1-\alpha)(1-\beta)}{\alpha\beta}.$$

Employing the equation (1.2) in the equation (4.23), we find that

$$\begin{aligned}(h_n^{16} &+ 8h_n^2 h_{49n}^2 + 64h_n^8 h_{49n}^8 + 2\sqrt{2}h_n h_{49n}^9 + 2\sqrt{2}h_{49n}^{15} h_n^7 + 7h_{49n}^4 h_n^{12} + h_{49n}^{16} \\ &+ 2\sqrt{2}h_n^9 h_{49n} + 28h_{49n}^{11} h_n^{11}\sqrt{2} + 2h_n^{15}\sqrt{2}h_{49n}^7 + 28\sqrt{2}h_n^5 h_{49n}^5 + 7h_n^4 h_{49n}^{12} \\ &+ 8h_{49n}^{14} h_n^{14})(-h_n^8 - h_{49n}^8 + 2\sqrt{2}h_n h_{49n} + 7h_n^4 h_{49n}^4 + 2\sqrt{2}h_{49n}^7 h_n^7) = 0. \end{aligned} \tag{4.24}$$

By examining the behaviour of first factor near $q = 0$, it can be seen that there is a neighbourhood about the origin, where the first factor is not zero, whereas the second factor is zero in this neighbourhood. By the Identity Theorem second factor vanishes identically. Hence

$$-h_n^8 - h_{49n}^8 + 2\sqrt{2}h_n h_{49n} + 7h_n^4 h_{49n}^4 + 2\sqrt{2}h_{49n}^7 h_n^7 = 0.$$

On simplification, we obtain the equation (4.22). □

Corollary 4.4. *We have*

$$h_{14} = \sqrt{\frac{1+\sqrt{2}+\sqrt{2\sqrt{2}-1}}{2}} \tag{4.25}$$

and

$$h_{\frac{2}{7}} = \sqrt{\frac{1+\sqrt{2}-\sqrt{2\sqrt{2}-1}}{2}}. \tag{4.26}$$

Proof. Putting $n = \frac{2}{7}$ in the equation (4.22), we find that

$$\frac{1}{Y} + Y = 4\sqrt{2} + 7. \tag{4.27}$$

On solving the above equation (4.27), we obtain the equations (4.25) and (4.26). □

Theorem 4.8. *If* $X = h_n h_{121n}$ *and* $Y = \frac{h_n^6}{h_{121n}^6}$, *then*

$$\frac{1}{Y} - Y = 2\sqrt{2}\left(\frac{2}{X^5} + \frac{11}{X^3} + \frac{22}{X} + 22X + 11X^3 + 2X^5\right). \tag{4.28}$$

Proof. The equation (2.7) can be written as

$$1 + A^{\frac{1}{2}} + 68A^{\frac{1}{4}} + 16A^{\frac{1}{4}}(1 + A^{\frac{1}{4}}) + 48A^{\frac{1}{6}}(1 + A^{\frac{1}{6}}) = \frac{1}{\sqrt{\alpha\beta}}, \tag{4.29}$$

where

$$A = \frac{(1-\alpha)(1-\beta)}{\alpha\beta}.$$

Employing the equation (1.2) in the equation (4.29), we find that

$$\begin{aligned} &1672\left(\frac{1}{h_n^6 h_{121n}^6} + h_n^6 h_{121n}^6\right) + 32\left(\frac{1}{h_n^{10} h_{121n}^{10}} + h_n^{10} h_{121n}^{10}\right) \\ &+ 352\left(\frac{1}{h_n^8 h_{121n}^8} + h_n^8 h_{121n}^8\right) + 4576\left(\frac{1}{h_n^4 h_{121n}^4} + h_n^4 h_{121n}^4\right) \\ &+ 8096\left(\frac{1}{h_n^2 h_{121n}^2} + h_n^2 h_{121n}^2\right) - \frac{h_{121n}^{12}}{h_n^{12}} - \frac{h_n^{12}}{h_{121n}^{12}} + 9746 = 0. \end{aligned} \tag{4.30}$$

On simplification, we obtain the equation (4.28). □

Corollary 4.5. *We have*

$$h_{22} = \sqrt{1+\sqrt{2}} \tag{4.31}$$

and

$$h_{\frac{2}{11}} = \sqrt{-1+\sqrt{2}}. \tag{4.32}$$

Proof. *Putting* $n = \frac{2}{11}$ *in the equation* (4.15), *we find that*

$$\frac{1}{Y} - Y = 140\sqrt{2}. \tag{4.33}$$

On solving the above quadratic equation (4.33), *we obtain the equations* (4.31) *and* (4.32). □

5. Modular relations between quartic class invariants H_n and h_n

In this section, we establish several modular relations between the quartic class invariants H_n and h_n. We also establish several explicit evaluations of H_n and h_n.

Theorem 5.1. *If* $P = h_n^6 h_{9n}^6$ *and* $Q = H_n^6 H_{9n}^6$, *then*

$$\left(P + \frac{1}{P}\right) + 4 = Q. \tag{5.1}$$

Proof. From the equations (1.2) and (1.3), we find that

$$\alpha^{\frac{1}{12}} = \frac{1}{h_n H_n} \quad \text{and} \quad \beta^{\frac{1}{12}} = \frac{1}{h_{9n} H_{9n}}. \tag{5.2}$$

and

$$(1-\alpha)^{\frac{1}{12}} = \frac{h_n}{H_n} \quad \text{and} \quad (1-\beta)^{\frac{1}{12}} = \frac{h_{9n}}{H_{9n}}. \tag{5.3}$$

Employing the equations (5.2) and (5.3) in the equation (2.4), we obtain (5.1). □

Theorem 5.2. *If* $P = h_n^2 h_{25n}^2$ *and* $Q = H_n^6 H_{25n}^6$, *then*

$$\left(P^3 + \frac{1}{P^3}\right) + 8\left(P + \frac{1}{P}\right) = Q. \tag{5.4}$$

Theorem 5.3. *If* $P = h_n^3 h_{49n}^3$ *and* $Q = H_n^6 H_{49n}^6$, *then*

$$\left(P^2 + \frac{1}{P^2}\right) + 8\sqrt{2}\left(P + \frac{1}{P}\right) + 20 = Q. \tag{5.5}$$

Theorem 5.4. *If* $P = h_n^2 h_{121n}^2$ *and* $Q = H_n^6 H_{121n}^6$, *then*

$$\left(P^3 + \frac{1}{P^3}\right) + 16\left(P^2 + \frac{1}{P^2}\right) + 48\left(P + \frac{1}{P}\right) + 68 = Q. \tag{5.6}$$

Since the proofs of Theorems 5.2–5.4 are similar to the proof of Theorem 5.1, we omit the details.

Theorem 5.5. *We have*

$$H_2 = \sqrt[12]{2}, \tag{5.7}$$

$$H_6 = \sqrt[12]{6}, \tag{5.8}$$

$$H_{10} = \sqrt[12]{18}, \tag{5.9}$$

$$H_{14} = \sqrt[12]{2(11+8\sqrt{2})}, \tag{5.10}$$

$$H_{18} = \sqrt[12]{98}, \tag{5.11}$$

$$H_{22} = \sqrt[12]{198}, \tag{5.12}$$

$$H_{30} = \sqrt[12]{6(5+4\sqrt{2})^2}, \tag{5.13}$$

$$H_{\frac{10}{3}} = \sqrt[12]{6(-5+4\sqrt{2})^2}, \tag{5.14}$$

$$H_{42} = \sqrt[12]{6(7\sqrt{3}+8\sqrt{2})^2}, \tag{5.15}$$

$$H_{\frac{6}{7}} = \sqrt[12]{6(7\sqrt{3}-8\sqrt{2})^2}, \tag{5.16}$$

$$H_{70} = \sqrt[12]{126(8\sqrt{2}+5\sqrt{5})^2}, \tag{5.17}$$

$$H_{\frac{10}{7}} = \sqrt[12]{126(8\sqrt{2}-5\sqrt{5})^2}, \tag{5.18}$$

and

$$H_{78} = \sqrt[12]{6(75+52\sqrt{2})^2}. \tag{5.19}$$

Proof of (5.7). Putting $n = 1$ in the equation (4.1), we find that

$$h_2 = 1. \tag{5.20}$$

Employing the equation (5.20) in the equation (4.2) with $n = 2$, we obtain (5.7).

Since the proofs of the equations (5.8)–(5.19) are similar to the proof of the equation (5.7), we omit the details.

Remark. In their paper [6], Baruah and Berndt found the evaluations in (5.8)–(5.12).

6. Explicit evaluations of the quartic singular moduli

In this section, we establish several explicit evaluations of the quartic singular moduli using the values of h_n and H_n.

Theorem 6.1. *We have*

$$\alpha_2 = \frac{1}{2}, \tag{6.1}$$

$$\alpha_{\frac{2}{3}} = \frac{(\sqrt{2}+1)^2}{6}, \tag{6.2}$$

$$\alpha_6 = \frac{(\sqrt{2}-1)^2}{6}, \tag{6.3}$$

$$\alpha_{\frac{2}{9}} = \frac{(\sqrt{3}+\sqrt{2})^4}{98}, \tag{6.4}$$

$$\alpha_{18} = \frac{(\sqrt{3}-\sqrt{2})^4}{98}, \tag{6.5}$$

$$\alpha_{\frac{2}{5}} = \frac{1}{18}\left(\frac{\sqrt{5}+1}{2}\right)^6, \tag{6.6}$$

$$\alpha_{10} = \frac{1}{18}\left(\frac{\sqrt{5}-1}{2}\right)^6, \tag{6.7}$$

$$\alpha_{\frac{2}{7}} = \frac{(8\sqrt{2}-11)}{14}\left(\frac{1+\sqrt{2}+\sqrt{2\sqrt{2}-1}}{2}\right)^6, \tag{6.8}$$

$$\alpha_{14} = \frac{(8\sqrt{2}-11)}{14}\left(\frac{1+\sqrt{2}-\sqrt{2\sqrt{2}-1}}{2}\right)^6, \tag{6.9}$$

$$\alpha_{\frac{2}{11}} = \frac{(\sqrt{2}-1)^6}{198}, \tag{6.10}$$

$$\alpha_{22} = \frac{(\sqrt{2}+1)^6}{198}, \tag{6.11}$$

$$\alpha_{30} = \frac{(\sqrt{5}-2)^2(\sqrt{10}-3)^2(4\sqrt{2}-5)^2}{294}, \tag{6.12}$$

$$\alpha_{\frac{2}{15}} = \frac{(\sqrt{5}+2)^2(\sqrt{10}+3)^2(4\sqrt{2}-5)^2}{294}, \tag{6.13}$$

$$\alpha_{\frac{10}{3}} = \frac{(\sqrt{5}+2)^2(\sqrt{10}-3)^2(4\sqrt{2}+5)^2}{294}, \tag{6.14}$$

$$\alpha_{\frac{6}{5}} = \frac{(\sqrt{5}-2)^2(\sqrt{10}+3)^2(4\sqrt{2}+5)^2}{294}, \tag{6.15}$$

$$\alpha_{42} = \frac{(2\sqrt{2}-7)^2(7\sqrt{3}-8\sqrt{2})^2}{2166}\left(\frac{\sqrt{7}-\sqrt{3}}{2}\right)^6, \tag{6.16}$$

$$\alpha_{\frac{6}{7}} = \frac{(2\sqrt{2}-7)^2(7\sqrt{3}+8\sqrt{2})^2}{2166}\left(\frac{\sqrt{7}-\sqrt{3}}{2}\right)^6, \tag{6.17}$$

$$\alpha_{70} = \frac{(\sqrt{2}-1)^6(8\sqrt{2}-5\sqrt{5})^2}{353934}\left(\frac{3-\sqrt{5}}{2}\right)^6, \tag{6.18}$$

$$\alpha_{\frac{10}{7}} = \frac{(\sqrt{2}-1)^6(8\sqrt{2}+5\sqrt{5})^2}{353934}\left(\frac{3-\sqrt{5}}{2}\right)^6, \tag{6.19}$$

and

$$\alpha_{78} = \frac{(\sqrt{26}-5)^2(75-52\sqrt{2})^2}{282534}\left(\frac{\sqrt{13}-3}{2}\right)^6. \tag{6.20}$$

Proof of (6.1). From the equations (1.2) and (1.3), we find that

$$\alpha_n = \frac{1}{h_n^{12}H_n^{12}}. \tag{6.21}$$

Putting $n = 2$ in the equation (6.21) and then using the equations (5.7) and (5.20), we obtain the equation (6.1).

Since the proofs of the equations (6.2)–(6.20) are similar to the proof of the equation (6.1), we omit the details.

Remark. Baruah and Berndt [6] found the evaluations in (6.3), (6.5), (6.7), (6.9) and (6.11).

Acknowledgement

The authors are thankful to the referee for very detailed and helpful comments.

References

[1] C. Adiga, Taekyun Kim and M. S. Mahadeva Naika, Modular equations in the theory of signature 3 and P–Q identities, *Adv. Stud. Contemp. Math.*, **7(1)** (2003) 33–40.

[2] C. Adiga, M. S. Mahadeva Naika and K. Shivashankara, On some P–Q eta-function identities of Ramanujan, *Indian J. Math.*, **44(2)** (2002) 253–267.

[3] K. Sushan Bairy, Some theorems on the explicit evaluations of singular moduli, *General Math.*, **17(1)** (2009) 71–87.

[4] N. D. Baruah, A few theta-function identities and some of Ramanujan's modular equations, *Ramanujan J.*, **4(3)** (2000) 239–250.

[5] N. D. Baruah, On some of Ramanujan's Schläfli-type "mixed" modular equations, *J. Number Theory*, **100(2)** (2003) 270–294.

[6] N. D. Baruah and B. C. Berndt, Ramanujan's series for $1/\pi$ arising from his cubic and quartic theories of elliptic functions, *J. Math. Anal. Appl.*, **341** (2008) 357–371.

[7] B. C. Berndt, Ramanujan's Notebooks, Part III, Springer-Verlag, New York (1991).

[8] B. C. Berndt, Ramanujan's Notebooks, Part V, Springer-Verlag, New York (1998).

[9] B. C. Berndt, S. Bhargava and F. G. Garvan, Ramanujan's theories of elliptic functions to alternative bases, *Trans. Amer. Math. Soc.*, **347** (1995) 4163–4244.

[10] B. C. Berndt, H. H. Chan and W. C. Liaw, On Ramanujan's quartic theory of elliptic functions, *J. Number Theory*, **88** (2001) 129–156.

[11] S. Bhargava, C. Adiga and M. S. Mahadeva Naika, A new class of modular equations akin to Ramanujan's P–Q eta-function identities and some evaluations there from, *Adv. Stud. Contemp. Math.*, **5(1)** (2002) 37–48.

[12] S. Bhargava, C. Adiga and M. S. Mahadeva Naika, A new class of modular equations in Ramanujan's alternative theory of elliptic function of signature 4 and some new P–Q eta-function identities, *Indian J. Math.*, **45(1)** (2003) 23–39.

[13] William Hart, Schläfli modular equations for generalised Weber functions, *Ramanujan J.*, **15** (2008) 435–468.

[14] T. Kim, M. S. Mahadeva Naika, S. Chandan Kumar, Lee-Chae Jang, Young-Hee Kim and Byungje Lee, On some new Schläfli-type cubic modular equations, *Adv. Stud. Contemp. Math.*, **20(1)** (2010) 63–80.

[15] M. S. Mahadeva Naika, P–Q eta-function identities and computation of Ramanujan-Weber class invariants, *J. Indian Math. Soc.*, **70(1–4)** (2003) 121–134.

[16] M. S. Mahadeva Naika, On some cubic Russell-type mixed modular equations, *Adv. Stud. Contemp. Math.*, **10(2)** (2005) 219–231.

[17] M. S. Mahadeva Naika, A note on cubic modular equations of degree 2, *Tamsui Oxf. J. Math. Sci.*, **22(1)** (2006) 1–8.

[18] M. S. Mahadeva Naika, Some new explicit values for Ramanujan class invariants, *Adv. Stud. Contemp. Math.*, **20(4)** (2010), to appear.

[19] M. S. Mahadeva Naika and K. Sushan Bairy, On some new explicit evaluations of class invariants, *Vietnam J. Math.*, **36(1)** (2008) 103–124.

[20] M. S. Mahadeva Naika and K. Sushan Bairy, On some new Schläfli-type mixed modular equations (Communicated).

[21] M. S. Mahadeva Naika, M. C. Maheshkumar and K. Sushan Bairy, Certain quotient of eta-function identities, *Adv. Stud. Contemp. Math.*, **16(1)** (2008) 121–136.

[22] K. G. Ramanathan, Ramanujan's modular equations, *Acta. Arith.*, **53** (1990) 403–420.

[23] S. Ramanujan, Notebooks (2 volumes), Tata Institute of Fundamental Research, Bombay (1957).

[24] K. Venkatachaliengar, Development of elliptic functions according to Ramanujan, Technical Report 2, Madurai Kamaraj University, Madurai (1988).

[25] E. T. Whittaker and G. N. Watson, A course of modern analysis, 4th ed., Cambridge University Press, Cambridge (1966).

Made in United States
Orlando, FL
04 November 2026

90209485R00136